Algorithmic Differentiation
of Pragma-Defined Parallel Regions

Michael Förster

Algorithmic Differentiation of Pragma-Defined Parallel Regions

Differentiating Computer Programs
Containing OpenMP

 Springer Vieweg

Michael Förster
RWTH Aachen University
Aachen, Germany

D 82, Dissertation RWTH Aachen University, Aachen, Germany, 2014

ISBN 978-3-658-07596-5   ISBN 978-3-658-07597-2 (eBook)
DOI 10.1007/978-3-658-07597-2

The Deutsche Nationalbibliothek lists this publication in the Deutsche Nationalbibliografie;
detailed bibliographic data are available in the Internet at http://dnb.d-nb.de.

Library of Congress Control Number: 2014951338

Springer Vieweg
© Springer Fachmedien Wiesbaden 2014

Printed on acid-free paper

Springer Vieweg is a brand of Springer DE.
Springer DE is part of Springer Science+Business Media.
www.springer-vieweg.de

Abstract

The goal of this dissertation is to develop a source code transformation that exploits the knowledge that a given input code is parallelizable in a way that it generates derivative code efficiently executable on a supercomputer environment.

There is barely a domain where optimization does not play a role. Not only in science and engineering, also in economics and industry it is important to find optimal solutions for a given problem. The size of these optimization problems often requires large-scale numerical techniques that are capable of running on a supercomputer architecture.

For continuous optimization problems the calculation of derivative values of a given function is crucial. If these functions are given as a computer code implementation Q then techniques known as algorithmic differentiation (AD) alias automatic differentiation can be used to obtain an implementation Q' that is capable of computing the derivative of a given output of Q with respect to a certain input. This thesis focuses on algorithmic differentiation by source transformation. The implementation Q is transformed into Q' such that Q' contains assignments for computing the derivative values.

On the one side, the size of optimization problems is rising. On the other side, the number of cores per central processing unit (CPU) in modern computers is growing. A typical supercomputer node has up to 32 cores or even more in case that multiple physical nodes form a compound. In order to allow Q to compute its output values efficiently, the implementation of Q should exploit the underlying multicore computer architecture. An easy approach of using parallel programming is to declare a certain code region inside of Q as parallelizable. This declaration is done by setting a certain kind of pragma in front of the corresponding code region. The pragma is a compiler directive and in our case this special directive informs the compiler that the corresponding code region should be executed concurrently. This code region is denoted as a parallel region P and the parallel instances which execute P are called threads.

There are two fundamental modes in AD, the forward and the reverse mode. We present source transformation rules for a simplified programming language, called *SPL*. In addition, we show that these rules provide derivative code either in forward or in reverse mode. One crucial goal of this work is that the knowledge that the original code contains a parallel region P leads to a parallel region P'

in the derivative code. This allows a concurrent computation of the derivative values. We exhibit a proof to ensure that the parallel execution of P' is correct. In case that the user of AD wants to achieve higher derivative code the possibility of reapplying the source transformation is important. Therefore, we exhibit that the source transformation is closed in the sense that the output code language is the same as the input language.

The reverse mode of AD builds the so called adjoint code. The term 'reverse' indicates that the adjoint code requires a data flow reversal of the execution of P. Suppose that P consists of code where a memory location is read by multiple threads. The data flow reversal of P leads to the situation that the corresponding derivative component of this memory location is a target of multiple store operations from different threads during the execution of the adjoint code. These store operations must be synchronized. Otherwise, the adjoint code would have a race condition at runtime. Conservatively, one could assume that all memory locations in P are read by multiple threads, which leads to the result that the adjoint source transformation generates a lot of synchronization constructs to ensure a correct parallel execution. In the worst case the synchronization overhead leads to a concurrent runtime of the derivative code P' that is bigger than the sequential runtime. In order to avoid as much synchronization as possible, we develop a static program analysis that collects information about P at compile time whether or not a memory location is exclusively read by a thread. Is a memory location read exclusively, the adjoint source transformation does not need to emit a synchronization method for the corresponding derivative computation. This can make a major difference.

We demonstrate how the context-free grammar for the language SPL can be extended in order to recognize pragmas defined in the OpenMP standard. Beside the extension of the grammar we present source transformation rules for these OpenMP constructs. With the source transformation rules for constructs such as the barrier, the critical, or the worksharing loop construct, this work provides rules for generating derivative code for most of the occurring OpenMP parallel regions.

The approach of this work has been implemented in a tool, called simple parallel language compiler (SPLc). We give evidence that our approach is applicable through the implementation of two optimization problems. On the one hand, we use a first derivative code provided by SPLc to solve a nonlinear least-squares problem. On the other hand, a nonlinear constrained optimization problem has been solved with the second derivative code provided by SPLc as well.

Acknowledgments

First and foremost, I would like to thank my adviser Prof. Dr. Uwe Naumann for his thoughtful guidance and constant encouragement. He has been inspiring since my days as a computer science student and I am grateful that I did get the chance to make my PhD at his institute.

Special thanks goes to my co-supervisor Prof. Dr. Christian Bischof for his support and constructive suggestions. I also would like to thank Prof. Dr. Thomas Noll for the fruitful discussions about static program analysis.

Further thanks goes to all current and former employees of the Lehr- und Forschungsgebiet Computer Science 12 at the RWTH Aachen University. We had a lot of fun and many discussions mainly about algorithmic differentiation but besides about all sorts of things. I would especially like to thank Johannes Lotz, Lukas Razik, Michel Schanen, Arindam Sen, and Markus Towara for reading early drafts of this manuscript and for finding most of the contained errors and unintelligible explanations.

Many thanks to the IT center of the RWTH Aachen University. The HPC group provided a lot of useful material about parallel programming and in particular about OpenMP. At this point, a special thank goes to Sascha Bücken for providing us access to the SUN cluster that had served its time, but has never lost its capabilities for being a good environment for parallel programming.

I really appreciate all my family and I am grateful for all their support during all these years. I wish my father could read this but he passed away much too early. Anyway, thanks for all the fun during our successful motocross years. The ups and downs during my sports career have been a major influence on my further life.

Last, but not least, I would like to thank my wife Mirjam. Without her understanding and support during the past few years, I may not have finished this thesis. Many thanks for all the pleasure that we had together with our two little daughters Emmy and Matilda.

Contents

1 Motivation and Introduction

1.1 Numerical Optimization in the Multicore Era

Todays problems in science and engineering are often solved with the help of parallel computing since the sequential computation would take weeks, months or even years until presenting results. Another reason for parallel computation is that the problem size is too big to process the data with a single computer. An example is a simulation of physical phenomena. Such problems can often be divided into subproblems which can be processed independently and in parallel. Another reason for a growing computation complexity is the natural trend to develop simulations which are more precise as their predecessors. But more precise results is almost always related to a bigger computation complexity.

As a result of this growing need for performance, the computer hardware engineers had to find ways to increase computer performance. Throughout the single-processor era the strategy was to increase the number of transistors per die. This is better known as "Moore's law" which says that the density of transistors on integrated circuits is doubling every 18 months. Until the mid 2000s, this higher amount of transistors per die allowed to increase the performance from one processor generation to the next. However, the transistor size became so small that physical boundaries such as power consumption and heat generation made it infeasible to further increase the transistor density on die. Instead, modern processor generations increase the number of cores on each die. Processors with multiple cores on one chip are called *multicore* processors. Typical desktop processors nowadays consists of two or four cores. In the area of high performance computing (HPC), there are often processors with even more cores per chip. These multicore machines can be interconnected with the help of a network which results in a cluster system. A single computer in this cluster is called node and can be seen as a single processing unit. However, each node is able to process tasks in parallel due to its multicore processor.

Another trend in HPC is to combine CPU architectures with special purpose FPGAs or Graphics Processors (GPUs). Similar to the approach to put many multicore machines into one cluster, one can equip each multicore machine with powerful FPGAs or GPUs and increase the number of overall cores in the cluster even further. The above mentioned method of organizing computer architectures leads to systems with an huge peak performance assuming that a program could make use of all the provided cores at the same time. The peak performance satisfies most of the numerical simulations but to write a code that utilizes parallelism as much as possible is all but an easy task.

There is a long history of approaches where software engineers were equipped with programming features that allowed code to run in parallel on a certain computer architecture. However, what remains important is the knowledge about the data dependencies of the code. In this context the memory organization of the parallel computer is important.

In parallel programming, there are two important classifications of memory organization, the *shared memory* machines and the *distributed memory* machines [66]. In the distributed memory model, each parallel processing unit has its own local memory. The parallel tasks are distributed among cluster nodes. The communication between the parallel instances is done by exchanging messages. The de facto standard application programming interface (API) for distributed memory parallel programming is the *message passing interface* (MPI)[54].

In shared memory parallel programming the multiprocessing is performed on a multicore machine. The parallel instances communicate through *shared variables* accessible to all instances. One of the early standards for shared memory parallel programming was the POSIX thread standard (Pthreads) [37]. From now on, we will only consider parallel programming for shared memory machines. Therefore, we speak about threads when the simultaneous execution of a code region is meant opposing to processes which is the commonly used term when speaking in the context of distributed memory.

The Pthreads API is still in use nowadays but being a low-level API means that writing Pthreads programs is often error-prone. Most of the engineers who write numerical kernels are not interested in learning a low level API. Instead, they want to run certain parts of their code in parallel with the minimal required effort. Therefore, an important innovation was the approach where the developer provides information about the code with the help of compiler directives. These compiler directives are given in form of pragmas and serve as information about the code where they are defined. For example, an #include is a preprocessor directive that instructs the preprocessor to include the content of a given file at this point in the code. A pragma such as #pragma parallel may inform the compiler about a parallel code region as long as the compiler knows this kind of pragma. If the compiler does not know this pragma it ignores this pragma. The pragmas are not meant as extensions of the programming languages but they only serve as information for the compiler.

Automatic parallelization is a good example for the use of pragmas. Rice's theorem [68] states that all nontrivial semantic questions cannot be checked automatically. Therefore, the compiler often has to be conservative in its decisions. This can mean that the compiler cannot decide whether or not a certain loop is parallelizable or not. This decision can be made much easier by the developer of the code. The pragma approach allows the developer to extend the loop with a pragma to declare the loop as parallelizable. An example for such a pragma is the one supported by the Intel Compiler, as shown in the Example 1.

Example 1. (Compiler Specific Auto Parallelization):
The Intel Compiler provides a pragma for auto-parallelization. With this pragma, one can declare a loop as explicitly parallelizable.

```
1 #include <math.h>
2
3 void f(int n, double *x, double *y)
4 {
5    int i,j,k;
6    #pragma parallel
7    for(i=0;i<n;i++)
8      for(j=0;j<n;j++)
9        for(k=0;k<n;k++)
10          y[i]=y[i]+x[i]*sin(x[j])*cos(x[k]);
11 }
```

Since the loop is declared as parallelizable, the loop body is parallelized in a way that the set of iterations are distributed among a group of threads. Therefore, we regard the loop body from line 8 to line 10 as a parallel region. The iterations of the outermost loop do not have any data dependencies among each other, or in other words, this loop is parallelizable. $\triangle$

The pragma in Example 1 is compiler specific. It would not be very beneficial if all compiler developers defined their own pragmas. Therefore, the OpenMP standard [63] was established at the end of the 1990s to provide a simpler way for the software engineer for writing parallel programs than with the Pthreads standard. In this work, we focus on the OpenMP standard version 3.1 which was published in July 2011. OpenMP 3.1 is well supported by most compilers[1]. Therefore, we

---

[1]See for example http://openmp.org/wp/openmp-compilers/

consider OpenMP as a prime choice for shared memory parallel programming. However, in case that the reader is interested in more details about shared memory parallel programming with Pthreads then the textbooks [13] and [9] serve as a good introduction into this topic.

With OpenMP a loop can be declared as parallelizable as follows:

```
#pragma omp parallel for
for( i←0; i<n; i←i+1) { /* some code */ }
```

The pragma declares the counting loop as parallelizable and the OpenMP enabled compiler distributes the iterations among the group of threads executing the program. To get a respective Pthreads implementation, one would have to write the code completely different. The loop body would be put into an external function and the distribution among the threads has to be implemented by the developer. The only thing that differentiate the above code from its sequential version is the pragma. The developer has to be aware whether or not a loop is parallelizable. With OpenMP, the way how the iterations are distributed among the threads can be determined by the backend compiler since it generates the associated code.

Some successful approaches where computationally intensive numerical kernels have been parallelized with the help of OpenMP can be found in [2], [41] or [40]. In addition to this, the coming subsections introduces some numerical problems where OpenMP can be used to allow simultaneous computation on a multicore machine.

A similar development as from Pthreads to OpenMP can be observed in the domain of high performance computing on GPUs. The high performance computing on GPUs was started around the year 2002. This was first more or less a workaround since the typical graphic related operations were exploited for doing linear algebra computations such as the dot product. In order to allow a general purpose programming on their GPUs, NVIDIA publishes the CUDA SDK [61, 69] in 2007. In 2008, the Khronos group publishes the OpenCL framework [44, 25] that was meant to write programs for heterogeneous platforms as, for example, a system with a CPU and a FPGA. In November 2011, a new API standard was published called OpenACC [62]. OpenACC makes extensive use of pragmas to abstract the programming on a lower level with CUDA and OpenCL. If the OpenACC standard will have the same success as the OpenMP standard is uncertain at this time, in particular, because the recent OpenMP 4.0 standard introduces also a couple of new pragmas to support accelerators and vectorization.

A typical application in numerical simulations is, in addition to just simulate a certain problem, finding input parameters minimizing or maximizing a certain output of the simulation. Since the underlying problems are typically of nonlinear nature, the domain of finding solutions for these kinds of optimization problems is called *nonlinear optimization*. In nonlinear optimization derivative values play an important role. As pointed out in [60, ch. 8], these derivative values can either be approximated by *finite differences* or they can be computed with *algorithmic differentiation* (AD) [29, 57]. We will introduce these two methods in Section 1.2. AD has the advantage that it provides the derivative values with an accuracy up to machine precision. The manual differentiation of a given code is error prone and often infeasible for some reason, for example, because of the code's size or because each change to the original code implies a change in the derivative code and therefore the code's maintenance is difficult.

The two main methods of AD are source transformation and overloading. The overloading method is a runtime solution in the sense that a basic variable type is substituted by a customized type that propagates the derivative information along the dataflow. The source transformation on the other side is a compile time solution since a function implementation is transformed into another implementation that can be used to compute projections of the Jacobian of this function. Since the pragma-based compiler directives are only available at compile time, the overloading approach cannot take these pragmas into account, and valuable information about the code is neglected. The source transformation method has the downside that the input code must be read by a parser that fully understands the code's programming language. Hence, the parser must support a certain subset of the programming language to cover most of the codes written as a numerical kernel.

Currently, AD source transformation tools do not have a satisfying OpenMP support. Hence, the knowledge about the inherent parallelism is lost. The fact that the derivative codes only uses a sequential execution can be an huge disadvantage because in nonlinear optimization almost all the algorithms consist of an iterative search method where the derivative values are computed in each iteration again. Especially considering the increasing number of cores in recent computer architectures, the exploitation of the parallelism is crucial. Hence, we present an approach for an AD source transformation for pure parallel regions. This means that a certain code region is declared as parallel by a pragma but inside the parallel region there exists no further pragmas. This approach allows to examine the source transformation requirements for a common parallel region without any API dependent features. Subsequently, we focus more on OpenMP and examine the source transformation of parallel regions containing various OpenMP constructs.

To motivate this approach with two practical examples, we present two typical problems in numerical optimization. The first is a nonlinear least-squares problem which makes use of the first derivative code of an implementation that contains an OpenMP parallel region. Subsequently, we introduce algorithmic differentiation together with an application from the domain of biotechnology where our derivative code compiler dcc plays an important role. In this area, the second derivative code is crucial since the usage of the second derivative values instead of approximating these values by a quasi-Newton method often leads to a more robust behavior and a faster convergence of the optimization algorithm. Since the examples from the biotechnology are beyond the scope of this work, we simplify this example to a common nonlinear optimization problem with constraints. The associated first- and second derivative codes are used to calculate a solution for this constrained optimization problem using the open source software Ipopt[2] [82].

1.1.1 A Nonlinear Least-Squares Problem

The least-squares problem is a mathematical problem that commonly occurs in practice. The underlying optimization problem appears whenever observed measurement data should be described by a mathematical model function, but the exact parameters of the model function are unknown.

We consider a simple spring-mass system over a period of time. This kind of problem is a common example in mechanic textbooks. We do not go into details and the reader may find a detailed description in [20, 19, 83, 67]. As mentioned in [20], the vibration of this system can mathematically modeled by the differential equation

$$\frac{\partial^2 u}{\partial^2 t} + \frac{b}{m}\frac{\partial u}{\partial t} + \frac{D}{m}u = 0.$$

m is the mass, D is the spring constant and b is the damping constant. Solutions of this equation have the form

$$u(t) = x_1 \exp(-x_2 t)\sin(x_3 t + x_4),\qquad(1.1)$$

where the meaning of the parameters x_1, x_2, x_3, x_4 are, as mentioned above, not of interest here. We consider these parameters as unknown, and the goal is to find the parameter values which minimize the discrepancy between the model function and the observed data.

Let us first consider, how a solution can be obtained for this single equation. Afterwards, we extend the problem by having not only one single equation of the form (1.1), but n equations of this form. The objective function of our least-squares problem is $\phi : \mathbb{R}^4 \to \mathbb{R}$ and is defined as:

$$\phi(\mathbf{x}) := \frac{1}{2}\|f(\mathbf{x})\|_2^2,$$

---

[2]https://projects.coin-or.org/Ipopt

where $f : \mathbb{R}^4 \to \mathbb{R}^m$, and $f_i(\mathbf{x}) := y(t_i; \mathbf{x}) - b_i$, $i = 1, \ldots, m$ are called the *residuals*. The input space dimension is given by the number of unknown parameters and the output space dimension m is the number of measurement values given by the observed data. The model function, here denoted by y, represents the function shown in (1.1). y has the argument t_i that represents the point in time where the measurement was taken. In addition, y has four unknown parameters $\mathbf{x} = (x_1, x_2, x_3, x_4)$. The observed measurement value at time t_i is referred to as b_i.

The goal is to minimize the discrepancy between the model function and the residuals. Usually, measurements have errors, and therefore the residuals seldom are zero, but they should be as small as possible. This brings us to the unconstrained optimization problem, where we have to find $\mathbf{x}^*$, such that

$$\phi(\mathbf{x}^*) = \min_{\mathbf{x} \in \mathbb{R}^4} \phi(\mathbf{x}),$$

with

$$\phi(\mathbf{x}) = \frac{1}{2}\|f(\mathbf{x})\|_2^2 = \frac{1}{2} f(\mathbf{x})^T f(\mathbf{x}) = \frac{1}{2} \sum_{i=1}^{m} (y(t_i; \mathbf{x}) - b_i)^2.$$

At this point we enter the domain of numerical optimization where most of the algorithms require the knowledge of derivatives. Therefore, we have to introduce some closely connected notions.

Theorem 1 (Chain Rule of Differential Calculus). *Suppose that $g_1 : \mathbb{R}^n \to \mathbb{R}^k$ with $\mathbf{z} = g_1(\mathbf{x})$ is differentiable at $\mathbf{x}$ and $g_2 : \mathbb{R}^k \to \mathbb{R}^m$ with $\mathbf{y} = g_2(\mathbf{z})$ is differentiable at $\mathbf{z}$. Then $f(\mathbf{x}) = g_2(g_1(\mathbf{x}))$ is differentiable at $\mathbf{x}$ with*

$$\frac{\partial f}{\partial \mathbf{x}} = \frac{\partial g_2}{\partial \mathbf{z}} \cdot \frac{\partial g_1}{\partial \mathbf{x}} = \frac{\partial \mathbf{y}}{\partial \mathbf{z}} \cdot \frac{\partial \mathbf{z}}{\partial \mathbf{x}}$$

Proof. See, for example, [3]. □

Definition 2. The derivative of $f(\mathbf{x})$, $f : \mathbb{R}^n \to \mathbb{R}$, $\mathbf{x} = (x_i)_{i=1,\ldots,n}$ at point $\mathbf{x}_0$ with respect to x_j is denoted as

$$f_{x_j}(\mathbf{x}_0) \equiv \frac{\partial f}{\partial x_j}(\mathbf{x}_0).$$

The *gradient* of f at point $\mathbf{x}_0$ is defined by

$$\nabla f(\mathbf{x}_0) \equiv \begin{pmatrix} f_{x_1}(\mathbf{x}_0) \\ \vdots \\ f_{x_n}(\mathbf{x}_0) \end{pmatrix} \in \mathbb{R}^n.$$

Considering the function $F : \mathbb{R}^n \to \mathbb{R}^m$,

$$\nabla F(\mathbf{x}_0) \equiv \begin{pmatrix} \nabla F_1(\mathbf{x}_0)^T \\ \vdots \\ \nabla F_m(\mathbf{x}_0)^T \end{pmatrix}$$

is called the *Jacobian* of F at point $\mathbf{x}_0$. The *Hessian* of f at the point $\mathbf{x}_0$ is defined as

$$\nabla^2 f(\mathbf{x}_0) := (f_{x_i x_j}(\mathbf{x}_0)) = \begin{pmatrix} f_{x_1 x_1}(\mathbf{x}_0) & f_{x_1 x_2}(\mathbf{x}_0) & \cdots & f_{x_1 x_n}(\mathbf{x}_0) \\ \vdots & \vdots & \ddots & \vdots \\ f_{x_n x_1}(\mathbf{x}_0) & f_{x_n x_2}(\mathbf{x}_0) & \cdots & f_{x_n x_n}(\mathbf{x}_0) \end{pmatrix}$$

$\square$

The gradient of ϕ concerning our example is given by

$$\nabla \phi(\mathbf{x}) = (\nabla f(\mathbf{x}))^T f(\mathbf{x}), \tag{1.2}$$

where $\nabla f(\mathbf{x})$ is the Jacobian matrix of f evaluated at point $\mathbf{x}$. The Hessian matrix of ϕ is

$$\nabla^2 \phi(\mathbf{x}) = (\nabla f(\mathbf{x}))^T \nabla f(\mathbf{x}) + \sum_{i=1}^m \nabla^2 f_i(\mathbf{x}) f_i(\mathbf{x}) \tag{1.3}$$

The following algorithm indicates how a solution $\mathbf{x}^*$ can be computed with the help of the Gauss-Newton, or the Levenberg-Marquardt method [50, 53, 51, 60]. As pointed out in Algorithm 1, the Gauss-Newton method only approximates the second derivative of ϕ by omitting the second derivative information of f in (1.3). Matrix A represents the second-order approximation of ϕ in Algorithm 1.

The Levenberg-Marquardt method, also known as the damped least-squares (DLS) method, is similar to the Gauss-Newton method, but it ensures that matrix A has full rank $4n$. This method adds the identity matrix multiplied by a factor μ^2 to the matrix A. This full-rank property is not necessarily given when using the Gauss-Newton method. The values of μ are adjusted during the optimization process as indicated in Algorithm 1. The actual value or adjustment method for μ is not part of this work and it only influences the convergence properties of Algorithm 1.

Algorithm 1 Calculate $\mathbf{x}^*$ per Gauss-Newton or Levenberg-Marquardt method

---

Require: $\mathbf{x}^0 \in \mathbb{R}^4, \varepsilon \in \mathbb{R}, \mu > 0$
Ensure: $\mathbf{x}^* = \min\limits_{\mathbf{x} \in \mathbb{R}^4} \ \frac{1}{2}\|f(\mathbf{x})\|_2^2$

$k \leftarrow 0$
$r \leftarrow 2\varepsilon$
while $r > \varepsilon$ **do**
  $\mathbf{b} \leftarrow (\nabla f(\mathbf{x}^k))^T f(\mathbf{x}^k)$
  $A \leftarrow (\nabla f(\mathbf{x}^k))^T \nabla f(\mathbf{x}^k)$
  if Levenberg-Marquardt method **then**
    $\mu \leftarrow$ adjust_mu(μ)
    $A \leftarrow A + \mu^2 I$
  end if
  Solve linear system $A\mathbf{s}^k = -\mathbf{b}$
  $\mathbf{x}^{k+1} \leftarrow \mathbf{s}^k + \mathbf{x}^k$
  $r \leftarrow \|b\|_2$
  $k \leftarrow k+1$
end while

Next, we extend the above system from one equation to n equations. This means instead of considering only a single spring-mass system, we approximate parameters for n independent spring-mass systems. All the n equations are of the form (1.1). As in the single equation case, the individual four unknown parameters will be approximated with the help of m measurement values. The residual function maps $4n$ parameter values to mn measurement values. The objective function $\Phi : \mathbb{R}^{4n} \to \mathbb{R}$ becomes

$$\Phi(\mathbf{x}) = \frac{1}{2}\|F(x_{1,1}, x_{1,2}, x_{1,3}, x_{1,4}, x_{2,1}, \ldots, x_{n,4})\|_2^2 \qquad (1.4)$$

$$= \frac{1}{2}\sum_{j=1}^{n}\sum_{i=1}^{m}(x_{j,1}e^{-x_{j,2}t_i}\sin(x_{j,3}t_i + x_{j,4}) - b_{j,i})^2 \quad ,$$

with the residual function

$$F : \mathbb{R}^{4n} \to \mathbb{R}^{nm} \qquad (1.5)$$

and

$$F_{j,i}(\mathbf{x}) := y(t_i; x_{j,1}, x_{j,2}, x_{j,3}, x_{j,4}) - b_i, \quad j = 1, \ldots, n,$$
$$i = 1, \ldots, m \quad .$$

The optimization problem is now

$$\Phi(\mathbf{x}^*) = \min_{\mathbf{x} \in \mathbb{R}^{4n}} \Phi(\mathbf{x}) \quad , \tag{1.6}$$

and the gradient of Φ is

$$\nabla\Phi(\mathbf{x}) = (\nabla F(\mathbf{x}))^T F(\mathbf{x}) \quad .$$

The Hessian of Φ is given by

$$\nabla^2\Phi(\mathbf{x}) = \nabla F(\mathbf{x})^T \nabla F(\mathbf{x}) + \sum_{j=1}^{n} \sum_{i=1}^{m} \nabla^2 F_{j,i}(\mathbf{x}) \nabla F_{j,i}(\mathbf{x}) \tag{1.7}$$

The goal is to find a solution $\mathbf{x}^*$ that satisfies $\nabla\Phi(\mathbf{x}^*) = 0$, and while $\nabla^2\Phi(\mathbf{x}^*) \in \mathbb{R}^{4n \times 4n}$ is symmetric positive definite. As in the single equation case, we approximate the Hessian by omitting the second-order information.

$$\nabla^2\Phi(\mathbf{x}) \approx \nabla F(\mathbf{x})^T \nabla F(\mathbf{x}) \tag{1.8}$$

A possible implementation of function F in (1.5) may be as shown in Listing 1.1. The outer loop in line 11 iterates through all n equations. After collecting the four values of $\mathbf{x}$ for the current equation (lines 16 to 17), the inner loop computes all the residuals for the j-th equation (line 18). Since any two sets of measurement data from two distinct equations are independent of each other, we can parallelize the outer loop with a OpenMP pragma. The pragma in line 10 declares the counting loop as parallelizable which leads to the execution of the loop iterations by a group of threads.

```
1  void F(const int n, const int m,
2          double* t, double *x, double* b, double* y)
3  // n: number of equations
4  // m: number of measurement values per equation
5  // t: vector with m*n timestamps
6  // x: current x vector with 4*n values
7  // b: measurement values taken at timestamps in t
8  // y: output vector containing m*n values
9  {
10 #pragma omp parallel for
11    for(int j=0;j<n;j++)
12    {
13      int i=0, xbase, mbase;
14      double yy, x0, x1, x2, x3, t_i, b_i;
```

```
15      xbase=j*4;
16      x0=x[xbase+0]; x1=x[xbase+1];
17      x2=x[xbase+2]; x3=x[xbase+3];
18      while(i<m) {
19        mbase=j*m;
20        t_i=t[mbase+i]; b_i=b[mbase+i];
21        yy=x0*exp(0.-x1*t_i)*sin(x2*t_i+x3);
22        y[mbase+i]=yy-b_i;
23        i=i+1;
24      }
25    }
26 }
```

Listing 1.1: Residual function F.

For solving the optimization problem (1.6), we also need the values of the Jacobian ∇F at point $\mathbf{x}$. One could use finite differences or AD. Finite differences would need $4(n+1)$ evaluations of the code shown in Listing 1.1. One major benefit of AD is that truncation is avoided. This can make a big difference in the area of ODEs, in particular when we consider stiff ODE systems [18]. Therefore, we show how the least-squares problem can be solved by using AD.

The next section presents the basics of AD. Afterwards, we are able to reformulate Algorithm 1 in a way that uses AD for calculating $\mathbf{b}$ and A. We will show how we obtain the first derivative code of Listing 1.1 from dcc assuming that we omit the OpenMP pragma in line 10 because dcc does not support OpenMP at the moment. In addition we will present an application of dcc in biotechnology.

1.2 Algorithmic Differentiation

In practice there are three main methods for calculating derivative values [60, ch. 8]. Suppose a function $F : \mathbb{R}^n \to \mathbb{R}^m$ is given such that it is too complicated to calculate the derivatives by hand. One way to calculate the derivative values is to use a finite differences approximation. We can approximate the j-th partial derivative $\partial F_j / \partial \mathbf{x}_i$ at a given point $\mathbf{x}$ by *forward-difference* approximation, defined as

$$\frac{\partial F_j}{\partial \mathbf{x}_i}(\mathbf{x}) \approx \frac{F_j(\mathbf{x}+h\mathbf{e}_i)-F_j(\mathbf{x})}{h} \quad , i = 1,2,\ldots,n \tag{1.9}$$

where $j \in \{1,2,\ldots,m\}$, h is a small scalar, and $\mathbf{e}_i$ is the i-th unit vector. We can approximate the i-th column of the Jacobian by applying (1.9) for $j = 1,2,\ldots,m$.

With

$$\frac{\partial F}{\partial \mathbf{x}_i}(\mathbf{x}) \approx \frac{F(\mathbf{x} + h\mathbf{e}_i) - F(\mathbf{x})}{h}$$

at hand, the whole Jacobian matrix can be calculated by applying this formula for $i = 1, 2, \ldots, n$.

Another method for computing the derivative values of F is *symbolic differentiation*. This method can be made by hand or one uses an algebra system like Mathematica [81] or Maple [16]. Since this method uses a much bigger amount of computer resources than AD and it is therefore of limited use for calculating higher derivatives, we do not discuss this method here.

We focus on calculating the derivative values with the third main method, namely algorithmic differentiation (AD) [29, 57]. AD exploits the fact that the chain rule of differential calculus holds and takes the view that a given implementation of F as computer code can be seen as a composition of elementary arithmetic operations. As an example, let us consider the assignment

```
y=sin(x2*t_i+x3);
```

as an example. This assignment occurred in similar form in Listing 1.1 line 21. The computation of the right-hand side can be displayed by a *directed acyclic graph* (DAG) as illustrated in Figure 1.1a.

We associate each node of the DAG with an auxiliary variable v, as shown in Figure 1.1b. The DAG in Figure 1.1b is called linearized, because each edge is labeled with the local partial derivative of the edge's target node with respect to its predecessor. For example, node v_5 represents the value $\sin(v_4)$, and the partial derivative of v_5 with respect to its predecessor v_4 is $\cos(v_4)$. Therefore, the edge from v_4 to v_5 is labeled with $\cos(v_4)$.

Figure 1.1b induces a semantically equivalent representation of the assignment y=sin(x2*t_i+x3) as *single assignment code* (SAC). Each node v in Figure 1.1b can be written as an assignment with the auxiliary variable v on the left-hand side. The predecessors of node v in the linearized graph represent the right-hand side in form of an expression. The following SAC and the assignment y=sin(x2*t_i+x3) are semantically equivalent.

```
v0 = x2;
v1 = t_i;
v2 = x3;
v3 = v0*v1;
v4 = v3+v2;
v5 = sin(v4);
```

```
y = v5;
```

Listing 1.2: The assignment y=sin(x2*t_i+x3); can be represented semantically equivalent as single assignment code (SAC).

AD has two modes, the *forward mode* and the *reverse mode*. An application of the forward mode to a given implementation P of a mathematical function $F : \mathbb{R}^n \to \mathbb{R}^m$ builds a *tangent-linear model* of P that is defined in Definition 3. The reverse mode provides the *adjoint model* of P which is defined in Definition 6.

Definition 3 ([57], p. 39). The Jacobian $\nabla F = \nabla F(\mathbf{x})$ induces a linear mapping $\nabla F : \mathbb{R}^n \to \mathbb{R}^m$ defined by

$$\mathbf{x}^{(1)} \to \nabla F \cdot \mathbf{x}^{(1)}.$$

The function $F^{(1)} : \mathbb{R}^{2 \cdot n} \to \mathbb{R}^m$, defined as

$$\mathbf{y}^{(1)} = F^{(1)}(\mathbf{x}, \mathbf{x}^{(1)}) \equiv \nabla F(\mathbf{x}) \cdot \mathbf{x}^{(1)}, \tag{1.10}$$

is referred to as the *tangent-linear model* of F. $\square$

The tangent-linear model $F^{(1)}(\mathbf{x}, \mathbf{x}^{(1)})$ augments each assignment of a given SAC with an additional assignment where the partial derivative of the augmented assignment is computed. The computational complexity of computing the first

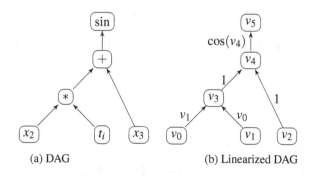

(a) DAG (b) Linearized DAG

Figure 1.1: The expression $\sin(x_2 \cdot t_i + x_3)$ can be represented by a DAG (Figure 1.1a). The linearized DAG of this expression is shown in Figure 1.1b.

derivative with the tangent-linear model is $\mathcal{O}(n) \cdot Cost(F)$ where $Cost(F)$ denoted the cost for evaluating the implementation of F. As an example, we apply the forward mode locally to the SAC shown in Listing 1.2. Each assignment can be seen as a scalar function $g : \mathbb{R}^2 \to \mathbb{R}$. According to (1.10), the tangent-linear model of g is $g^{(1)} : \mathbb{R}^4 \to \mathbb{R}$, where $g^{(1)}(\mathbf{x}, \mathbf{x}^{(1)}) \equiv \nabla g \cdot \mathbf{x}^{(1)}$. The components of $\mathbf{x}^{(1)}$ are represented in the code by the prefix 't1_'. For example, the variable x2 has an associated derivative component t1_x2. Listing 1.3 shows the tangent-linear model of Listing 1.2.

```
t1_v0 = t1_x2;
v0 = x2;
t1_v1 = t1_t_i;
v1 = t_i;
t1_v2 = t1_x3;
v2 = x3;
t1_v3 = v1*t1_v0+v0*t1_v1;
v3 = v0*v1;
t1_v4 = t1_v3+t1_v2;
v4 = v3+v2;
t1_v5 = cos(v4)*t1_v4;
v5 = sin(v4);
t1_y = t1_v5;
y = v5;
```

Listing 1.3: Tangent-linear model of Listing 1.2.

Definition 4 ([57], p. 53). The adjoint of the linear operator $\nabla F : \mathbb{R}^n \to \mathbb{R}^m$ is defined as $(\nabla F)^* : \mathbb{R}^m \to \mathbb{R}^n$ where

$$\langle (\nabla F)^* \cdot \mathbf{y}_{(1)}, \mathbf{x}^{(1)} \rangle_{\mathbb{R}^n} = \langle \mathbf{y}_{(1)}, \nabla F \cdot \mathbf{x}^{(1)} \rangle_{\mathbb{R}^m}, \tag{1.11}$$

and where $\langle ., . \rangle_{\mathbb{R}^n}$ and $\langle ., . \rangle_{\mathbb{R}^m}$ denote the scalar products in $\mathbb{R}^n$ and $\mathbb{R}^m$, respectively. $\qquad\square$

Theorem 5 ([57], p. 53). $(\nabla F)^* = (\nabla F)^T$.

Definition 6 ([57], p. 54). The Jacobian $\nabla F = \nabla F(\mathbf{x})$ induces a linear mapping $\mathbb{R}^m \to \mathbb{R}^n$ defined by

$$\mathbf{y}_{(1)} \to \nabla F^T \cdot \mathbf{y}_{(1)}.$$

The function $F_{(1)} : \mathbb{R}^{n+m} \to \mathbb{R}^n$ defined as

$$\mathbf{x}_{(1)} = F_{(1)}(\mathbf{x}, \mathbf{y}_{(1)}) \equiv \nabla F(\mathbf{x})^T \cdot \mathbf{y}_{(1)} \qquad (1.12)$$

is referred to as the *adjoint model* of F.   □

The adjoint model $F_{(1)}(\mathbf{x}, \mathbf{y}_{(1)})$ in AD yields an adjoint code with the important property that its computational complexity $\mathcal{O}(m) \cdot Cost(F)$ grows with the number of dependent variables m in contrast to the tangent-linear model whose computational complexity grows with the number of independent variables n. Mathematical models in physics, chemistry or economic science often have thousands of inputs and only some values or even only one value as output. Hence, the adjoint code is frequently the only possibility to compute the derivative values of mathematical models in feasible time. A downside of the adjoint code is that the adjoint values are computed during a complete dataflow reversal of the original function evaluation. This dataflow reversal is often connected to an high usage of memory resources.

We apply the reverse mode to our example SAC from Listing 1.2. An assignment from the SAC is again represented as $g : \mathbb{R}^2 \to \mathbb{R}$. The adjoint model of g is $g_{(1)} : \mathbb{R}^3 \to \mathbb{R}^2$ with $g_{(1)}(\mathbf{x}, \mathbf{y}_{(1)}) \equiv (\nabla g)^T \cdot \mathbf{y}_{(1)}$.

```
1  v0 = x2;
2  v1 = t_i;
3  v2 = x3;
4  v3 = v0*v1;
5  v4 = v3+v2;
6  v5 = sin(v4);
7  y = v5;
8  a1_v5 = a1_y;
9  a1_y = 0;
10 a1_v4 = cos(v4)*a1_v5;
11 a1_v3 = a1_v4;
12 a1_v2 = a1_v4;
13 a1_v0 = v1*a1_v3;
14 a1_v1 = v0*a1_v3;
15 a1_x3 += a1_v2;
16 a1_t_i += a1_v1;
17 a1_x2 += a1_v0;
```

Listing 1.4: Adjoint model of Listing 1.2.

When we apply (1.12) to each assignment in Listing 1.2, we obtain the code shown in Listing 1.4, which contains two distinct sections. The *forward section* comprises the lines 1 to 7 and this section is for this example equivalent with the original code from Listing 1.2. The adjoint computation is done during the *reverse section* (line 8 to line 17). The adjoint components are preceded by the prefix 'a1_'. These listings already indicate that differentiating a given code is error-prone, particularly in the adjoint model case. But the source transformation can be undertaken by a tool and there are already several tools providing AD by source transformation, for example, OpenAD[3], Tapenade[4], and dcc[5]. Unfortunately, none of these tools provide at this time an AD source transformation support for OpenMP. It is ongoing work to enable dcc to support OpenMP [23].

With this information about AD at hand we are able to present how AD can be utilized to solve the previously introduced least-squares problem from Section 1.1.1.

Example 2. Utilizing AD to solve a least-squares problem (Continuation of Section 1.1.1):
We saw in Algorithm 1 that we need the product of the transposed Jacobian of the residual function times a vector from the output space to obtain **b**.

$$\mathbf{b} \leftarrow (\nabla F(\mathbf{x}^k))^T F(\mathbf{x}^k) \quad . \tag{1.13}$$

However, we do not compute the whole Jacobian $\nabla F(\mathbf{x}) \in R^{nm \times 4n}$, since the adjoint model (1.12) provides exactly what we need for acquiring **b**. Therefore, we rewrite (1.13) in order to use the adjoint model.

$$\mathbf{y} \leftarrow F(\mathbf{x})$$
$$\mathbf{b} \leftarrow F_{(1)}(\mathbf{x}, \mathbf{y})$$

To solve the linear system of equations occurring in Algorithm 1, we additionally need the matrix $A \in \mathbb{R}^{4n \times 4n}$, which is the product of the Jacobian of the residual function with itself:

$$A \leftarrow (\nabla F(\mathbf{x}^k))^T \nabla F(\mathbf{x}^k) \quad .$$

We use a combination of the tangent-linear model with the adjoint model to com-

---

[3]http://www.mcs.anl.gov/OpenAD/
[4]http://tapenade.inria.fr:8080/tapenade/index.jsp
[5]http://www.stce.rwth-aachen.de/software/dcc.html

pute the matrix A column by column:

$$\textbf{for } i \in \{1,\ldots,4n\} :$$
$$\mathbf{y}^{(1)} \leftarrow F^{(1)}(\mathbf{x},\mathbf{e}_i)$$
$$A_i \leftarrow F_{(1)}(\mathbf{x},\mathbf{y}^{(1)}) \quad .$$

First, we compute the i-th column of the Jacobian $\nabla F(\mathbf{x})$ by calling the tangent-linear model $F^{(1)}$ with $\mathbf{x}$ and the i-th euclidean basis vector $\mathbf{e}_i$ as arguments. Afterwards, we use this result to obtain the i-th line of A by calling the adjoint model $F_{(1)}$. Algorithm 2 shows the adjusted version of Algorithm 1. $\triangle$

Algorithm 2 Adjusted pendant to Algorithm 1, where we use AD to obtain $\mathbf{b}$ and A.

Require: $\mathbf{x}^0 \in \mathbb{R}^4, \varepsilon \in \mathbb{R}, \mu > 0$
Ensure: $\mathbf{x}^* = \min\limits_{\mathbf{x} \in \mathbb{R}^4} \ \frac{1}{2}\|f(\mathbf{x})\|_2^2$

  $k \leftarrow 0$
  $r \leftarrow 2\varepsilon$
  while $r > \varepsilon$ **do**
    $\mathbf{y} \leftarrow F(\mathbf{x}^k)$
    $\mathbf{b} \leftarrow F_{(1)}(\mathbf{x}^k,\mathbf{y})$
    for all $i = 1,\ldots,4n$ **do**
      $\mathbf{y}^{(1)} \leftarrow F^{(1)}(\mathbf{x}^k,\mathbf{e}_i)$
      $A_i \leftarrow F_{(1)}(\mathbf{x}^k,\mathbf{y}^{(1)})$
    end for
    if Levenberg-Marquardt method **then**
      $\mu \leftarrow$ adjust_mu(μ)
      $A \leftarrow A + \mu^2 I$
    end if
    Solve linear system $A\mathbf{s}^k = -\mathbf{b}$
    $\mathbf{x}^{k+1} \leftarrow \mathbf{s}^k + \mathbf{x}^k$
    $r \leftarrow \|b\|_2$
    $k \leftarrow k+1$
  end while

1.2.1 Second Derivative Code

In numerical optimization many algorithms use an approximation of the second derivative values. This approximation can be achieved by a quasi-Newton method. The downside of this approach is that the convergence during the iterative search for the minimum suffers. Hence, algorithms that use real second derivative values often tend to be more robust. Therefore, we consider how the second derivative code is obtained by source transformation.

An implementation of derivative code that computes the second derivative values of F is achieved by various combinations of applying the forward mode or the reverse mode. More specifically, the source transformation is first applied to the given implementation of F what results in the first derivative code. Afterwards, the source transformation tool is reapplied to its own output code which results in the tangent-linear model or the adjoint model of the first derivative code. This allows different combinations depending on what model is applied to the original function F, and which model is applied to the first derivative code. The easiest combination is the *forward-over-forward mode* where the tangent-linear model is applied twice. Its computational complexity is $\mathcal{O}(n^2) \cdot Cost(F)$.

As notation for higher derivatives we use the notation introduced in [57] as indicated in the following definitions.

Definition 7 ([57], p. 92). Let $A \in \mathbb{R}^{m \times n}$ a matrix alias a 2-tensor. A *first-order tangent-linear projection* of A in direction $v \in \mathbb{R}^n$ is defined as the usual matrix vector product $A \cdot v$. This can be expressed by using the inner product notation.

$$\mathbf{b} = \langle A, \mathbf{v} \rangle \in \mathbb{R}^m .$$

A *first-order adjoint projection*

$$\mathbf{c} = \langle \mathbf{w}, A \rangle \in \mathbb{R}^n$$

of A in direction $\mathbf{w} \in \mathbb{R}^m$ is defined as

$$\mathbf{c} = \mathbf{w}^T \cdot A$$

where $\mathbf{w}^T$ is the transpose of vector $\mathbf{w}$. $\qquad\qquad\square$

Definition 8 ([57], p. 95). Consider a symmetric 3-tensor $A \in \mathbb{R}^{m \times n \times n}$. A *second-order tangent-linear projection* of A in directions $\mathbf{u}, \mathbf{v} \in \mathbb{R}^n$ is defined as

$$\langle A, \mathbf{u}, \mathbf{v} \rangle \equiv \langle \langle A, \mathbf{u} \rangle \mathbf{v} \rangle$$

where $\langle A, \mathbf{u} \rangle$ is the projection of A in direction $\mathbf{u}$ and $\langle \langle A, \mathbf{u} \rangle \mathbf{v} \rangle$ is the first-order tangent-linear projection of $\langle A, \mathbf{u} \rangle$ in direction $\mathbf{v} \in \mathbb{R}^n$.

A *second-order adjoint projection* of A in directions $\mathbf{w} \in \mathbb{R}^m$ and $\mathbf{v} \in \mathbb{R}^n$ is defined as

$$\langle \mathbf{v}, \mathbf{w}, A \rangle \equiv \langle \mathbf{v}, \langle \mathbf{w}, A \rangle \rangle \in \mathbb{R}^n$$

where $\langle \mathbf{v}, \langle \mathbf{w}, A \rangle \rangle$ is a first-order adjoint projection of $\langle \mathbf{w}, A \rangle$ in direction $\mathbf{v}$ and $\langle \mathbf{w}, A \rangle$ is the projection of A in direction $\mathbf{w} \in \mathbb{R}^m$. $\square$

Lemma 9. *Let* $A \in \mathbb{R}^{m \times n \times n}$ *be a symmetric 3-tensor,* $\mathbf{u}, \mathbf{v} \in \mathbb{R}^n$, *and* $\mathbf{w} \in \mathbb{R}^m$. *Then,*

$$\langle A, \mathbf{u}, \mathbf{v} \rangle = \langle A, \mathbf{v}, \mathbf{u} \rangle$$

$$\langle \mathbf{v}, \mathbf{w}, A \rangle = \langle \mathbf{w}, A, \mathbf{v} \rangle$$

$$\langle \mathbf{w}, A, \mathbf{v} \rangle = \langle \langle \mathbf{w}, A \rangle, \mathbf{v} \rangle = \langle \mathbf{w}, \langle A, \mathbf{v} \rangle \rangle$$

Proof. [57], p. 95ff $\square$

Definition 10 ([57], p. 111). The Hessian $\nabla^2 F = \nabla^2 F(\mathbf{x}) \in \mathbb{R}^{m \times n \times n}$ of a multivariate vector function $\mathbf{y} = F(\mathbf{x})$, $F : \mathbb{R}^n \to \mathbb{R}^m$, induces a bilinear mapping $\mathbb{R}^n \times \mathbb{R}^n \to \mathbb{R}^m$ defined by

$$(\mathbf{u}, \mathbf{v}) \mapsto \langle \nabla^2 F, \mathbf{u}, \mathbf{v} \rangle.$$

The function $F^{(1,2)} : \mathbb{R}^{3 \cdot n} \to \mathbb{R}^m$, that is defined as

$$F^{(1,2)}(\mathbf{x}, \mathbf{u}, \mathbf{v}) \equiv \langle \nabla^2 F(\mathbf{x}), \mathbf{u}, \mathbf{v} \rangle, \tag{1.14}$$

is referred to as the *second-order tangent-linear model* of F. $\square$

The remaining three combinations to the generation of second derivative code involve at least one application of the adjoint model. Depending on the order in which we apply the forward mode or the reverse mode, we obtain the forward-over-reverse, reverse-over-forward, or the reverse-over-reverse mode. The corresponding second-order adjoint models are denoted by $F_{(1)}^{(2)}$, $F_{(2)}^{(1)}$, and $F_{(1,2)}$. They all have the same computational complexity of $O(n \cdot m) \cdot cost(F)$ for the accumulation of the whole Hessian.

Definition 11 ([57], p. 118). The Hessian $\nabla^2 F = \nabla^2 F(\mathbf{x}) \in \mathbb{R}^{m \times n \times n}$ of a multivariate vector function $\mathbf{y} = F(\mathbf{x})$, $F : \mathbb{R}^n \to \mathbb{R}^m$, induces a bilinear mapping $\mathbb{R}^n \times \mathbb{R}^m \to \mathbb{R}^n$ defined by

$$(\mathbf{v}, \mathbf{w}) \mapsto \langle \mathbf{w}, \nabla^2 F, \mathbf{v} \rangle .$$

The function $F'' : \mathbb{R}^{2 \cdot n + m} \to \mathbb{R}^n$ that is defined as

$$F''(\mathbf{x}, \mathbf{v}, \mathbf{w}) \equiv \langle \mathbf{w}, \nabla^2 F(\mathbf{x}), \mathbf{v} \rangle \tag{1.15}$$

is referred to as the *second-order adjoint model* of F. $\square$

Without going into further details, we give the theorems from [57] that this source transformation is actually correct. Further details about higher derivatives can be found in this textbook.

Theorem 12 ([57], p. 111). *The application of forward mode AD to the tangent-linear model yields the second-order tangent-linear model.*

Theorem 13 ([57], p. 119). *The application of forward mode AD to the adjoint model yields an implementation of the second-order adjoint model.*

Theorem 14 ([57], p. 120). *The application of reverse mode AD to the tangent-linear model yields an implementation of the second-order adjoint model.*

Theorem 15 ([57], p. 123). *The application of reverse mode AD to the adjoint model yields an implementation of the second-order adjoint model.*

In case that the reader is interested in more information about AD we recommend the textbooks [29, 57]. The next section introduces the software tool dcc that is able to generate first- and higher derivative codes by source transformation. The results from this dissertation will be integrated into this tool.

1.2.2 dcc - A Derivative Code Compiler

In case that the reader is interested in writing an AD source transformation tool from scratch then [57] describes a good way for achieving a prototype that is able to parse a simple language, called *SL*. This prototype served as a base for our AD source transformation tool dcc[6], which takes *SL* as its input language and generates also *SL* as output code. *SL* covers a subset of the C programming language. Pointers, for example, are left out for keeping aliasing effects easy for the AD source transformation.

The fact that dcc uses *SL* as its input language and as its output language as well is an important fact because the reapplication of dcc to its own output allows the creation of higher derivative code. It is ongoing work to extend the language *SL* such that it covers the arising codes in several applications [32, 70, 33]. For each extension of *SL*, we have to keep in mind that this extension influences the possible input of the source transformation as well as the possible output. For example, if the source transformation generates code that contains a certain statement *s* but the context-free grammar that describes the input language, does not contain a production rule for this statement *s*, we would loose the so called *closure property*. This term will be defined in detail in the next chapter. At this point, we only need to know that the loss of the closure property leads to loss of the ability to generate higher derivative codes for an arbitrary code input code *P*.

We explain the application of dcc to the Listing 1.1 that shows the implementation of the objective function for our least-squares problem. Let us illustrate how we apply dcc to obtain the tangent-linear model $F^{(1)}$ and the adjoint model $F_{(1)}$ of F. First of all, we have to adjust Listing 1.1 since the dcc input language does not allow OpenMP pragmas at the moment. The signature of F in Listing 1.1 is

```
void F(const int n, const int m,
       double* t, double *x, double* b, double* y)
```

Assuming that file F.c contains the adjusted implementation of F, we apply the forward mode by calling dcc as follows:

```
dcc F.c -t
```

The result is a file t1_F.c which contains the implementation of $F^{(1)}$. Analogously, the adjoint model $F_{(1)}$ is obtained by the command line

```
dcc F.c -a
```

---

[6]http://www.stce.rwth-aachen.de/software/dcc.html

which results in a generated file named a1_F.c. When speaking about the implementations of the tangent-linear or the adjoint model we often speak about the tangent-linear and the adjoint code, respectively. We only list their signatures since we already showed brief examples of tangent-linear and adjoint code. The implementation of $F^{(1)}$ can be called by using the following signature

```
void t1_F(const int n, const int m,
          double* t, double* t1_t, double* x, double* t1_x,
          double* b, double* t1_b, double* y, double* t1_y)
```

whereby

```
void a1_F(int bmode_1, const int n, const int m,
          double* t, double* a1_t, double* x, double* a1_x,
          double* b, double* a1_b, double* y, double* a1_y)
```

implements $F_{(1)}$. The first parameter of a1_F, namely bmode_1, is not of importance at this point. The remaining list of parameters in both signatures reveal that all the floating-point parameters of F are augmented by an additional parameter. This additional parameter serves as derivative association. The tangent-linear associates $x^{(1)}$ are named t1_x in the tangent-linear code, whereby the adjoint associates $x_{(1)}$ are referred to as a1_x in the adjoint code.

As mentioned in Section 1.2.1, to obtain an implementation of the second derivative code, we have different possibilities. Depending on what AD mode we apply to F and depending on what mode we apply afterwards to achieve the second derivative code, we obtain different combinations for the second-order implementation. As an example we generate a second-order adjoint code for F in reverse-over-forward mode. This code is referred to as $F^{(1)}_{(2)}$ and it is achieved by applying dcc two times:

```
dcc F.c −t
dcc t1_F.c −a −d 2
```

First, we create the tangent-linear code $F^{(1)}$ in t1_F.c. Subsequently, we apply dcc to its own output by providing the option -a that defines that the reverse mode should be applied. For technical reasons, we have to inform the compiler that the outcome is a second derivative code (-d 2). These two calls provides us the implementation $F^{(1)}_{(2)}$ whose signature looks as follows:

```
void a2_t1_F(int bmode_2, const int n, const int m,
          double* t, double* a2_t, double* t1_t, double* a2_t1_t,
          double* x, double* a2_x, double* t1_x, double* a2_t1_x,
          double* b, double* a2_b, double* t1_b, double* a2_t1_b,
          double* y, double* a2_y, double* t1_y, double* a2_t1_y)
```

Without going into details, the reader recognizes again that each floating-point parameter is augmented by another derivative component. $x^{(1)}_{(2)}$ (alias a2_t1_x) is, for example, the corresponding component associated to $x^{(1)}$ (alias t1_x).

A Relevant Example from Biotechnology

To explain why it is crucial to have AD source transformation rules for pragma-based parallel regions, we present the project *Jülich Aachen Dynamic Optimization Environment* (JADE) [33], where the dcc plays an important role. This project is a scientific cooperation that brings together the domains of biotechnology, process engineering, and computer science. The work flow of JADE is shown in Figure 1.2. The first step is to model a natural process, for example, a model for the bacterium Escherichia coli. A graphical user interface software, called OMIX[7] [21] supports this development phase. The graphical description of the model is transformed into an equation-based code language predefined by the software Modelica[8] [24]. An interface to access model properties and procedures for calculating solutions are given by several utility functions written in C++.

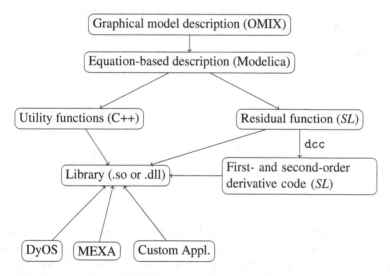

Figure 1.2: The work flow of modeling, simulating, and optimizing a natural process with JADE.

---

[7]The OMIX website is www.omix-visualization.com.

[8]Official Modelica website: www.modelica.org.

A solution for the corresponding model can be computed by transforming the Modelica model into a residual function where the input values must fulfill certain constrains. This residual function is written in *SL* since the dcc must be able to parse this code. The reason that dcc plays an important role at this point is because one crucial goal of JADE is to find optimal parameters for a given model. The residual function is typically nonlinear which means that we have to solve a nonlinear constrained optimization problem. Therefore, we need the first- and second derivative codes of the residual function. These derivative codes are the contribution of dcc.

All these codes are compiled into a library which can be accessed through a well defined interface, called equation set object (ESO) [7]. We emphasize two applications that have successfully used such a library. The first application is DyOS[9] which is a software tool for the solution of dynamic optimization problems. The second software is MEXA[10] which was developed for the efficient model-based experimental analysis of kinetic phenomena.

To give the reader an impression of the process, let us consider a model for the bacterium Escherichia coli. This model consist of 1488 differential algebraic equations (DAE). The model description in the Modelica format has a size of 4100 lines of code. This Modelica model is transformed into a residual function that consist of 19.000 lines of *SL* code:

```
1  void res(double* yy, double* der_x, double* x,
2            double* p, int &n_x, int &n_p)
3  {
4      ...
18838   #pragma ad equation
18839   {
18840       // scalar equation 1488
18841       yy[i_E] = var_Xyl5P_y5_0-var_Xyl5P_11111;
18842       i_E = i_E+1;
18843   }
18844  }
```

The important fact of this code is that the function is structured in a way that it contains 1488 equations where the start and the end of the equation are determined by a pragma. In case that all equations do not have any data dependencies, they can be computed simultaneously. A possible OpenMP implementation of this is that we transform, for example,

---

[9]The DyOS website is http://www.avt.rwth-aachen.de/AVT/index.php?id=484&L=1.

[10]The MEXA website is http://www.avt.rwth-aachen.de/AVT/index.php?id=891&L=1.

```
#pragma ad equation
{ /* equation 1 */ }
#pragma ad equation
{ /* equation 2 */ }
```

into an OpenMP sections construct inside a parallel region:

```
#pragma omp parallel sections
{
    #pragma omp section
    { /* equation 1 */ }
    #pragma omp section
    { /* equation 2 */ }
}
```

This work will define transformation rules for OpenMP codes as the above and therefore the parallelism is not lost in the derivative code creation. Once the AD source transformation for OpenMP pragmas is well defined, one can think about an approach where we transform the ad equation structure into code that uses heterogeneous parallel programming. For example, a solution would be that we use distributed memory parallel programming on an outer level and shared memory parallel programming on an inner level. The number of equations is, for example, divided into halves and each halve is processed by one MPI process running on different cluster nodes. Inside of the two MPI processes, we continue to exploit the parallelism of the equations by using shared memory parallel programming with OpenMP or OpenACC. As a result, this heterogeneous approach allows to use the maximum number of cores that a certain cluster configuration has available, and the number of equations is not limited by the machine's hardware constraints because the equations are distributed among cluster nodes with the help of MPI. Hybrid approaches with MPI/OpenMP usually show good scaling properties as presented for example in [49].

Mathematical models in process engineering become more and more complex and with an increasing complexity the number of equations usually grows as well. With an increasing number of equations there is no way around the approach of letting the equations be calculated simultaneously. To emphasize this, we consider the sizes of the derivative codes from the above listing. Please note that we created the derivative codes of res by ignoring all ad equation pragmas. The function res has 19.000 lines of code. The application of dcc to the function res in tangent-linear mode results in 73.000 lines of code whereby the application of the adjoint model gives us 78.000 lines of code. The second derivative code is obtained by reapplying dcc to the first derivative code. The application of the tangent-linear model to the first-order tangent-linear code (forward-over-forward) results in 361.000 lines of code. The adjoint model applied to the tangent-linear code (reverse-over-forward) yields 554.000 lines of code. Considering these code sizes and the fact that the Escherichia coli model with over 1000 equations is a relative small model in biotechnology, one can imagine that from a certain number of equations on a numerical optimization process is only feasible by using parallel computation in both, the residual function and in the corresponding derivative code.

Despite the fact that we will consider the transformation of OpenMP parallel regions in particular, the base approach of transforming a parallel region of a shared memory model can be applied to any pragma-based approach. Besides OpenMP, OpenACC is a relative new example for a pragma-based API.

The next section will introduce a nonlinear constrained optimization problem where we do not only need the first-order but also the second derivative code.

1.2.3 A Nonlinear Constrained Optimization Problem

This kind of problem is about minimizing a function subject to certain constraints on the variables. A collection of examples for these problems is the Hock-Schittkowski test suite [36]. We consider the problem number 71 from this collection. This problem is defined as

$$
\begin{aligned}
\min_{\mathbf{x} \in \mathbb{R}^4} \quad & x_1 x_4 (x_1 + x_2 + x_3) + x_3 \\
\text{subject to} \quad & x_1 \cdot x_2 \cdot x_3 \cdot x_4 \geq 25 \\
& x_1^2 + x_2^2 + x_3^2 + x_4^2 = 40 \\
& 1 \leq x_1, x_2, x_3, x_4 \leq 5 \quad .
\end{aligned}
\tag{1.16}
$$

The calculation of a solution for this constrained optimization problem can be done by using the *Ipopt* software[11] [82]. This software makes use of a subclass of the *interior point method* known as *primal-dual method* [64, 60]. We define

$$
x^0 = (1, 5, 5, 1)
$$

as starting point and the optimal solution is known to be

$$
x^* = (1.00000000, 4.74299963, 3.82114998, 1.37940829).
$$

---

[11]https://projects.coin-or.org/Ipopt

For this small example, the gradient of the objective function $f(\mathbf{x})$ and the Jacobian of the constraints $g(\mathbf{x})$ can be deduced by symbolic differentiation. Nevertheless, as we saw in the previous section, it is often the case that the constrained optimization problem consists of many independent equations and all these equations must be differentiated. The symbolic differentiation approach is therefore often not applicable due to the size of the implementation of the objective function. Hence, we apply AD to obtain the first- and second derivative values of the objective function f and the corresponding constraints g. The Lagrangian function of (1.16) is defined as $f(\mathbf{x}) + g(\mathbf{x})^T \lambda$ and the Hessian of the Lagrangian function is $\nabla^2 f(\mathbf{x}) + \sum_{i=1}^{m} \lambda_i \nabla^2 g_i(\mathbf{x})$, where m is the number of constraints.

As we did for the least-squares problem, we assume that we have given a constrained optimization problem which consists of N independent problems. In practice, the independent problems often vary in the number of variables and in the computational complexity. This would involve the runtime results in an unbalanced way because some threads process their work while the rest of the threads have already finished their work. Since we are only interested in scalability results of this work's approach, we give all the threads a similar amount of work to process. This is achieved by defining the N subproblems as given in (1.16). The overall optimization problem can be described as

$$\min_{\mathbf{x} \in \mathbb{R}^{4N}} \quad \sum_{i=1}^{N} x_{i,1} x_{i,4} \left(x_{i,1} + x_{i,2} + x_{i,3} \right) + x_{i,3}$$

$$\text{subject to} \quad x_{i,1} \cdot x_{i,2} \cdot x_{i,3} \cdot x_{i,4} \geq 25$$

$$x_{i,1}^2 + x_{i,2}^2 + x_{i,3}^2 + x_{i,4}^2 = 40 \tag{1.17}$$

$$1 \leq x_{i,1}, x_{i,2}, x_{i,3}, x_{i,4} \leq 5 \quad .$$

A possible implementation of the objective function is to use a OpenMP loop construct such as we did in the least-squares example. Instead, we just use one pragma in Listing 1.5 to define a parallel region and inside this region we distribute the work to each thread by an explicit data decomposition. This does not need to be clear in all details since the next section provides an explanation for this. At this point it is sufficient to know that all equations are partitioned into chunks defined by a lower bound lb and an upper bound ub. Afterwards, a while-loop iterates over the elements of each chunk and the corresponding thread processes these elements. Since each thread computes only its local sum the overall value of the objective function is computed by adding all these local sums after the parallel region.

```
#pragma omp parallel
{
    int i,j,tid,nt,c,lb,ub;
```

```
double y←0.;

tid ← omp_get_thread_num();
nt  ← omp_get_num_threads();
c ← n/nt;
lb ← tid*c;
ub ← (tid+1)*c−1;
i←lb;
while( i ≤ ub ) {
    j ← i*4;
    y +←  (x_j*x_{j+3}*(x_j+x_{j+1}+x_{j+2})+x_{j+2});
    i← i+1;
}
thread_result_{tid} ← y;
}
for( i←0; i<omp_get_max_threads(); i←i+1) {
    obj +←thread_result_i;
}
```

Listing 1.5: Portion of a possible implementation of the objective function f. We assume that there are n independent equations that should be processed in parallel.

The Ipopt API expects the implementation of the objective function in the C function eval_f and Listing 1.5 would therefore be part of this code. The user of Ipopt needs to provide an implementation which computes the first derivative values of f and this code belongs into eval_grad_f. Analogously, the values for the constraints are computed in eval_g, while the Jacobian values of the constraints are computed in eval_jac_g. The values of the Hessian of the Lagrangian function are computed in eval_h. All these eval_* functions must be provided by the user of Ipopt since it does not use any AD features. Hence, we need an AD source transformation tool at hand capable of transforming OpenMP parallel regions such as given in Listing 1.5. Once we have these transformations, we simply insert them into the predefined framework of Ipopt.

This concludes the motivation part of this introduction. We saw some applications where AD is very useful but as soon as pragma-based parallel regions are used it gets impossible to use AD since the required transformation rules are missing at this time. This dissertation fills this gap and since we mainly focus on the OpenMP standard 3.1, we now introduce the parts of the standard that are treated in this work.

1.3 OpenMP Standard 3.1

OpenMP 3.1 was released in 2011 and consists of minor changes to OpenMP 3.0 which was published in 2008. In contrast to this, OpenMP 4.0 was brought out in 2013 and it contains many new features including the support of heterogeneous systems and more possibilities to exploit compute and data locality. It is not clear at this time which architecture will become accepted in the future. Probably, it will be an architecture having accelerator devices of some sort as, for example, GPUs (NVIDIA, AMD), manycore (Intel Xeon Phi), or FPGAs (Xilinx). However, in this work we focus on OpenMP 3.1 since it is supported quite well and its features are sufficient for most of the shared memory architectures used nowadays.

This section gives an overview about the OpenMP 3.1 standard [63] where we only consider these constructs that later serve as possible input for our source transformation tool. A common introduction into parallel programming with OpenMP can be found in the textbooks [15] and [14]. For a precise description of the OpenMP constructs, we will cite the original document[12].

For the reader who is already familiar with OpenMP, we still recommend to go through the examples because they will later be used as test cases for getting runtime results of the performed source transformation. Another reason is that the typical code pattern of an explicit data decomposition is shown in one example and the static program analysis that we will introduce in Chapter 3 is based on the idea that this code pattern is often used in shared memory parallel programming, explicitly or implicitly.

Throughout this work, we will assume that the used computer architecture is a shared memory multicore machine. Typical home computers are nowadays equipped with two or four computing cores. OpenMP can also be used on these computers but the general use is the domain of high performance computing (HPC). These machines are at this time equipped with up to 64 (SUN T5120) or 128 (Bull BCS) cores and have a maximum of one terabyte of memory.

The memory model of OpenMP consist of two types of memory references, shared and private memory references. The shared memory references are accessible from the whole group of threads which execute the current parallel region. The private references are stored in a memory location that is associated to exactly one thread.

OpenMP provides multiple runtime library routines. The most important ones are:

1. int omp_get_max_threads(): This routine returns the maximum number

---

[12]This can be downloaded from http://www.openmp.org/mp-documents/OpenMP3.1.pdf.

of threads that could be used to form a new team using a parallel construct.

2. int omp_get_num_threads(): The return value is the number of threads attending the execution of the current parallel region.

3. omp_set_num_threads(int): This routine defines the number of threads that execute the consecutive parallel regions.

4. int omp_get_thread_num(): Each thread has an unique thread identifier number which can be obtained by this routine.

The most important construct is the parallel construct which starts the parallel execution. The parallel construct is associated with a sequence of statements that is executed by a team of threads. Each thread in this team has an unique identifier number starting with zero. The thread with the identifier zero is called master or initial thread. Unless otherwise specified, we assume the size of the group of threads to be p.

parallel Construct

In order to be a close as possible to the description in the standard, we cite the description in the standard instead of repeating in other words. The first citation describes the parallel construct.

OpenMP 3.1 Citation 16. parallel Construct (p. 33)
"This fundamental construct starts parallel execution.

```
#pragma omp parallel [clauses]
    structured-block
```

where *clause* is one of the following:

1. if(*scalar-expression*)
2. num_threads(*scalar-expression*)
3. default(shared|none)
4. private(*list*)
5. firstprivate(*list*)
6. shared(*list*)
7. copyin(*list*)
8. reduction(*operator: list*)

[...]

When a thread encounters a parallel construct, a team of threads is created to execute the parallel region [...] . The thread that encountered the parallel construct becomes the master thread of the new team, with a thread number of zero for the duration of the new parallel region. All threads in the new team, including the master thread, execute the region. Once the team is created, the number of threads in the team remains constant for the duration of that parallel region.

Within a parallel region, thread numbers uniquely identify each thread. Thread numbers are consecutive whole numbers ranging from zero for the master thread up to one less than the number of threads in the team. A thread may obtain its own thread number by a call to the omp_get_thread_num library routine.

A set of implicit tasks, equal in number to the number of threads in the team, is generated by the encountering thread. The structured block of the parallel construct determines the code that will be executed in each implicit task. Each task is assigned to a different thread in the team and becomes tied. The task region of the task being executed by the encountering thread is suspended and each thread in the team executes its implicit task. Each thread can execute a path of statements that is different from that of the other threads.

The implementation may cause any thread to suspend execution of its implicit task at a task scheduling point, and switch to execute any explicit task generated by any of the threads in the team, before eventually resuming execution of the implicit task [...] .

There is an implied barrier at the end of a parallel region. After the end of a parallel region, only the master thread of the team resumes execution of the enclosing task region.

If a thread in a team executing a parallel region encounters another parallel directive, it creates a new team, according to the rules in [...], and it becomes the master of that new team.

If execution of a thread terminates while inside a parallel region, execution of all threads in all teams terminates. The order of termination of threads is unspecified. All work done by a team prior to any barrier that the team has passed in the program is guaranteed to be complete. The amount of work done by each thread after the last barrier that it passed and before it terminates is unspecified."   □

To describe a given code on an abstract level, we use S as an abbreviation for a sequence of statements. This allows us to show the structure of the given code. For example, in the code

S_0
#pragma omp parallel
{
  S_1
}
S_2

are three structured blocks S_0, S_1, S_2. These blocks are sequences of statements, whereby the specific kind of statement is not of importance. S_1 is declared as to be evaluated in parallel by a team of threads. Consider Listing 1.3 and assume that S_0 and S_2 contain only one statement. The parallel region is represented by $S_1 = (s_1; \ldots; s_q)$ comprising of q statements. The evaluation of Listing 1.3 is illustrated in Figure 1.3. The q statements are executed by a team of p threads. The following notation is used for expressing that the i-th statement in a sequence of statements is executed by thread t:

$$s_i^t$$

For example s_1^0 is the first statement in a sequence and it is evaluated by thread 0. s_1^3 denotes an instance of s_1 that is executed by thread 3. All the p instances of the first statement that are executed by the team of threads are denoted by

$$s_1^0 \quad s_1^1 \quad \ldots \quad s_1^{p-1} \quad .$$

These instances are shown on the same horizontal level in Figure 1.3. This expresses that they are theoretically executed simultaneously.

The topmost node in Figure 1.3 is only executed by the master thread, which is indicated with the superscript 0. Then, following the arrows in the figure from top to bottom, a team of p threads is created and each thread processes the statements of sequence S_1, namely $(s_1; \ldots; s_q)$. After each thread has encountered the end of the parallel region by finishing statement s_q, the master thread continues the sequential execution with the statement s_{q+1}. An implicit barrier at the end of the parallel region ensures that before statement s_{q+1} is processed, each thread has finished its execution of the parallel region. This is called a *join* operation.

Example 3. (Memory model of OpenMP)
This example displays the use of private and shared variables inside a parallel region. We define two floating-point variables x and y outside the parallel region. In OpenMP, these two variables are per default shared variables inside the scope of the parallel region. Therefore, each thread can read and write to the memory location of x and y. The parallel region is executed by two threads as defined by omp_set_num_threads(2). Each thread prints its thread identifier number and

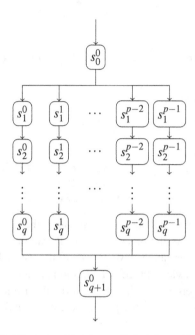

Figure 1.3: The execution of a parallel region with p threads.

the addresses of x and y on the screen through a function call to fprintf. After the parallel region the master thread prints its memory addresses of x and y through another call to fprintf.

```
fprintf(stderr, "Inside/Outside␣of␣␣␣␣␣\n");
fprintf(stderr,
"parallel␣region␣␣␣␣␣␣␣␣␣␣␣␣&x␣␣␣␣␣␣␣␣␣␣␣␣␣␣␣␣␣␣&y\n");
    fprintf(stderr,
"————————————————————————————————————————————————\n"
    );
float x;
float y;
omp_set_num_threads(2);
#pragma omp parallel
{
    float x;
    fprintf(stderr,
"Inside␣(thread␣%d)␣␣%15p␣␣%15p\n", omp_get_thread_num()
    , &x, &y);
```

```
}
fprintf(stderr,
 "Outside␣(thread␣%d)␣%15p␣␣%15p\n", omp_get_thread_num()
  , &x, &y);
```

The output looks as follows:

Inside/Outside of parallel region	&x	&y
Inside (thread 1)	0x2b6c3d9ffe2c	0x7ffff35a6238
Inside (thread 0)	0x7ffff35a620c	0x7ffff35a6238
Outside (thread 0)	0x7ffff35a623c	0x7ffff35a6238

We have two lines showing the addresses from inside the parallel region, one is for the thread with the ID zero, the other one is for thread one. The third line shows the addresses, which the variables have outside of the parallel region.

Variable x is defined twice. Once outside the parallel region, once inside the parallel region. The memory model of OpenMP determines that all references defined inside the parallel region are private references and only visible for each individual thread.

The column with the header line labeled with '&x' displays that the code has three instances of x. One for each thread and one that is accessed only from outside of the parallel region. The latter one is not accessible from inside the parallel region since all references to x are linked to the private instance of x.

The variable y is defined only once outside the parallel region. Without any additional clauses defining the status of y, this means that y is shared and accessible for the whole group of threads which execute the parallel region. This means that we only have one instance of y. The output of this code displays this in the column with the header '&y' by showing three times the same address in opposite to the column of &x.

Please note, that the memory location where the private variables are placed is compiler related and is not determined in the OpenMP standard. For example, for those who are familiar with the memory model of Linux, the instances of x of thread 0 are both placed on the stack whereby the instance of thread 1 is located inside the data segment of the executing process. This may be different for distinct compilers and the developer cannot rely on this.   △

Example 4. (Explicit data decomposition)
The following code declares a parallel region. A group of threads is created by the runtime system and all statements from line 3 to 19 are evaluated by all threads. Given the two n-dimensional arrays x and y, the code implements

$$y_i = 2 \cdot x_i \cdot x_i \text{ , where } i = 0, \ldots, n-1.$$

The code from line 3 to line 15 defines a lower bound (lb) and an upper bound (ub). These boundaries are unique for each thread and define a range of the input data that is processed by a certain thread. The lines 12 and 15 are for adjusting the chunk size and the upper bound for the case that the data cannot be partitioned into equal parts. This is the case when the data size n is not divisible by the number of threads p.

```
1  #pragma omp parallel
2  {
3    int i←0;
4    int tid←0;
5    int lb←0;
6    int ub←0;
7    int chunk_size←0;
8    tid←omp_get_thread_num();
9    p←omp_get_num_threads();
10   chunk_size←n/p;
11   i←chunk_size*p;
12   if(i ≠ n) { chunk_size←chunk_size+1; }
13   lb←tid*chunk_size;
14   ub←(tid+1)*chunk_size-1;
15   if( ub ≥ n ) { ub←n-1; }
16   for(i←lb; i ≤ ub; i++) {
17     y_i←2*x_i*x_i;
18     x_i←0.;
19   }
20 }
```

Listing 1.6: Data decomposition for a typical SPMD problem.

The method shown in Listing 1.6 is called *data decomposition* and allows an execution of the program where each processor fetches its own instructions and operates on its own data. The underlying model is called the *Single Program Multiple Data* (SPMD) model which is very often used in parallel programming [66]. Each thread gets its own chunk of data but the whole group of threads execute the same set of instructions only with different offsets for accessing the data. However,

OpenMP could not call itself an API for abstracting low-level thread programming if it did not provide mechanisms to do this data decomposition implicitly. The loop construct that is introduced in the section about work sharing constructs is a typical example where implicit data decomposition is used.   △

The next example illustrates that knowledge about data dependencies is still necessary when using OpenMP. One can easily write wrong code when the data dependencies are ignored by the software engineer. Another reason can be that the developer just forgets to share the work as shown in Example 5. The compiler will not prevent the compilation with an error when the code is not parallelizable, it will just create code according to the pragmas that the software developer has provided. In the best case, the compiler gives a warning but this presupposes that the compiler can expect what the developer probably wanted to do, which is often not the case.

Example 5. (Wrong OpenMP code)
Although at first glance the code below looks fine, the result that elements in y are zero may be surprising. The developer probably wanted to share the work among the threads but instead each thread evaluates the exact same code.

```
1 #pragma omp parallel
2 {
3     int i←0;
4     for( i←0; i<n; i++) {
5         yi←2*xi*xi;
6         xi←0.;
7     }
8 }
```

All threads simultaneously write to the memory locations $y+\text{sizeof}(y_0)*i$, where $i \in \{0,1,\ldots,n-1\}$. In the previous example the range $0,1,\ldots,n-1$ was partitioned by an explicit data decomposition and therefore each thread in the team was responsible for only one partition. Here, each thread processes all n iterations of the loop. In iteration i, each thread sets first the value of y_i and then it sets the component x_i to zero. The value of y_i depends on the value of x_i and therefore it depends on the fact whether or not the assignment that sets x_i to zero has been already executed by another thread. The result in y_i is decided by a race between read and store operations from different threads. Therefore, this situation at runtime is called a *race condition*. The reason for this race condition is the *critical reference* x_i that is read and stored by multiple threads.   △

Worksharing Constructs

The developer of Example 5 probably wanted to share the store operations among the threads. This is being done by defining a worksharing directive as explained in this section.

OpenMP 3.1 Citation 17. (p. 38)
"A worksharing construct distributes the execution of the associated region among the members of the team that encounters it. Threads execute portions of the region in the context of the implicit tasks each one is executing. If the team consists of only one thread then the worksharing region is not executed in parallel.

A worksharing region has no barrier on entry; however, an implied barrier exists at the end of the worksharing region, unless a nowait clause is specified. If a nowait clause is present, an implementation may omit the barrier at the end of the worksharing region. In this case, threads that finish early may proceed straight to the instructions following the worksharing region without waiting for the other members of the team to finish the worksharing region, and without performing a flush operation. [...]
Each worksharing region must be encountered by all threads in a team or by none at all. The sequence of worksharing regions and barrier regions encountered must be the same for every thread in a team." □

The next three paragraphs will introduce the worksharing constructs of OpenMP. The first is applicable to a counting loop, the sections construct is for defining different code sections which share no data dependencies to one another. Finally, the single construct is for defining a sequence of statements to be executed by only one thread, the remaining threads jump over this sequence of statements.

Loop construct

OpenMP 3.1 Citation 18. Loop Construct, p. 39
"The loop construct specifies that the iterations of one or more associated loops will be executed in parallel by threads in the team in the context of their implicit tasks. The iterations are distributed across threads that already exist in the team executing the parallel region to which the loop region binds.
The syntax of the loop construct is as follows:

```
#pragma omp for [clause[[,] clause] [...]] new-line
    for-loops
```

where *clause* is one of the following:

1. private(*list*)
2. firstprivate(*list*)
3. lastprivate(*list*)
4. reduction(*operator: list*)
5. schedule(*kind[, chunk_size]*)
6. collapse(*n*)
7. ordered
8. nowait

The for directive places restrictions on the structure of all associated *for-loops*.
[...]
The canonical form allows the iteration count of all associated loops to be computed before executing the outermost loop. [...]
The loop construct is associated with a loop nest consisting of one or more loops that follow the directive.

There is an implicit barrier at the end of a loop construct unless a nowait clause is specified. [...]
A worksharing loop has logical iterations numbered 0,1,[...],N-1 where N is the number of loop iterations, and the logical numbering denotes the sequence in which the iterations would be executed if the associated loop(s) were executed by a single thread."   □

Example 6. (Implicit data decomposition)
This example illustrates the usage of the loop construct of OpenMP. The code implements the same computation as the code from Example 4, or in other words the code from Example 4 and the current code are semantically equivalent. The difference is that in the current implementation the data decomposition is done implicitly by code that is generated by the OpenMP enabled compiler. This means that the data decomposition is handled by the OpenMP runtime system.

```
1 #pragma omp parallel
2 {
3   #pragma omp for
4   for ( i←0; i<n ; i++) {
5     yi←2*xi*xi ;
6     xi←0 . ;
7   }
8 }
```

Obviously, the implicit data decomposition makes the code much easier to read, which was one big objective of developing the OpenMP standard.   △

The next worksharing construct that we introduce is the sections construct.

sections Construct

This construct is more suitable in cases where the threads should execute code where the individual control flow of the codes differs from thread to thread. The easiest example is that two procedure calls can be executed simultaneously since the two subroutines do not have any data dependencies. In this case each section contains one subroutine call.

OpenMP 3.1 Citation 19. p. 48

"The sections construct is a noniterative worksharing construct that contains a set of structured blocks that are to be distributed among and executed by the threads in a team. Each structured block is executed once by one of the threads in the team in the context of its implicit task.

The syntax of the sections construct is as follows:

```
#pragma omp sections [clause[[,] clause] [...]] new-line
{
[ #pragma omp section  new-line]
    structured-block
[ #pragma omp section  new-line
    structured-block ]
\mbox{[$\ldots$]}
}
```

where *clause* is one of the following:

1. private(*list*)
2. firstprivate(*list*)
3. lastprivate(*list*)
4. reduction(*operator: list*)
5. nowait

[...] Each structured block in the sections construct is preceded by a section directive except possibly the first block, for which a preceding section directive is optional. [...] There is an implicit barrier at the end of a sections construct unless a nowait clause is specified."   □

Example 7. (Parallel code sections)
Suppose we have three vectors $\mathbf{x}, \mathbf{y}, \mathbf{z} \in \mathbb{R}^n$. A componentwise computation of

$$\mathbf{y}_i \leftarrow \mathbf{x}_i \cdot \mathbf{x}_i$$
$$\mathbf{z}_i \leftarrow \sin(\mathbf{x}_i) \cdot \mathbf{x}_i \quad , \text{ where } i \in \{1, \ldots, n\}$$

can be implemented with two sections.

```
 1 #pragma omp parallel
 2 {
 3   #pragma omp sections
 4   {
 5     #pragma omp section
 6     {
 7       for(int i←0;i<n;i++)
 8         yᵢ←2*xᵢ*xᵢ;
 9     }
10     #pragma omp section
11     {
12       for(int i←0;i<n;i++)
13         zᵢ←sin(xᵢ);
14     }
15   }
16 }
```

The computation of the two vectors is executed simultaneously.   △

single Construct

In case that we define a structured block that is only executed by one thread and not the whole group of threads, then the single construct can be used. This can, for example, be useful for changing shared scalar data.

OpenMP 3.1 Citation 20. p. 50

"The single construct specifies that the associated structured block is executed by only one of the threads in the team (not necessarily the master thread), in the context of its implicit task. The other threads in the team, which do not execute the block, wait at an implicit barrier at the end of the single construct unless a nowait clause is specified. The syntax of the single construct is as follows:

```
#pragma omp single [clause[[,] clause] [...]] new-line
    structured-block
```

where *clause* is one of the following:

1. private(*list*)
2. firstprivate(*list*)
3. copyprivate(*list*)
4. nowait

[...]
The method of choosing a thread to execute the structured block is implementation defined. There is an implicit barrier at the end of the single construct unless a nowait clause is specified." □

There are another two worksharing constructs which are a combination of the parallel construct and a worksharing construct. The reason for defining a separate construct is that it often occurs that, for example, a loop is parallelizable. Without the combined version the developer would have to use the parallel construct first, and the loop construct would have to be placed inside of the associated structured code block. To make this more compact and readable, OpenMP knows the combined worksharing constructs.

Combined Worksharing Constructs

OpenMP 3.1 Citation 21. p. 55
"Combined parallel worksharing constructs are shortcuts for specifying a worksharing construct nested immediately inside a parallel construct. The semantics of these directives are identical to that of explicitly specifying a parallel construct containing one worksharing construct and no other statements." □

Parallel Loop Construct

OpenMP 3.1 Citation 22. Loop Construct, p. 56
"The parallel loop construct is a shortcut for specifying a parallel construct containing one or more associated loops and no other statements.
  The syntax of the parallel loop construct is as follows:

```
#pragma omp parallel for [clause[[,] clause] [...]] new-line
    for-loop
```

where *clause* can be any of the clauses accepted by the parallel or for directives, except the nowait clause, with identical meanings and restrictions." □

Example 8. In this example we use the combined parallel loop construct from OpenMP.

```
1 #pragma omp parallel for
2 for ( i←0 ; i<n ; i++) {
3     yᵢ←2*xᵢ*xᵢ;
4     xᵢ←0.;
5 }
```

It is semantically equivalent to Example 6 and Example 4.   △

Parallel sections Construct

OpenMP 3.1 Citation 23. Loop Construct, p. 57

"The parallel sections construct is a shortcut for specifying a parallel construct containing one sections construct and no other statements. The syntax of the parallel sections construct is as follows:

```
#pragma omp parallel sections [clause[[,] clause] [...]] new-line
{
[ #pragma omp section new-line]
    structured-block
[ #pragma omp section new-line
    structured-block ]
\mbox{[$\ldots$]}
}
```

where *clause* can be any of the clauses accepted by the parallel or for directives, except the nowait clause, with identical meanings and restrictions."   □

Example 9. The following code is the combined version of Example 7.

```
1 #pragma omp parallel sections
2 {
3   #pragma omp section
4   {
5     for ( int i←0 ; i<n ; i++)
6         yᵢ←2*xᵢ*xᵢ;
7   }
8   #pragma omp section
9   {
```

```
10      for(int i←0;i<n;i++)
11        z_i←sin(x_i);
12    }
13 }
```

This code is obviously more compact through the usage of the combined parallel
sections construct.   △

Master and Synchronization Constructs

In shared memory parallel programming, there is often the situation that all threads
may access the same memory location but the updates of this memory location
must be synchronized in some way. We will later see that these constructs are
necessary to achieve the closure property of the adjoint source-to-source transformation.

master Construct

OpenMP 3.1 Citation 24. p. 67
"The master construct specifies a structured block that is executed by the master
thread of the team.

> **#pragma omp master** *new-line*
> *structured-block*

Other threads in the team do not execute the associated structured block. There is
no implied barrier either on entry to, or exit from, the master construct."   □

A *reduction* in parallel programming is a method where computation results
from simultaneously running instances are brought together in the sense that they
all influence a single result. A typical example for a reduction is a parallel sum,
where parts of the sum are computed in parallel and at the end these results must
again be added to form the overall result. Such a reduction can be implemented
with the help of the master construct. For an example, we refer the reader to the
section about the barrier construct.

critical Construct

In shared memory parallel programming multiple threads may access the same
shared variable by a read or a write operation. Simultaneous read and write operations of multiple threads must be avoided since they lead to inconsistent values.

One speaks about a race condition since the concurrent threads compete each other by reading from and writing to the memory. The winning thread, or better saying the thread that determines the result, is the one that performs the last write operation.

In other words the result depends on the order of the read and write operations which depend on the nondeterministic context changes between the threads currently executing the parallel region. A race condition must be prevented and can, for example, be avoided by putting the part of the code that causes the race condition into critical construct. This construct ensures that at each point in time only one thread is inside the critical section. The critical-section problem is a famous problem in computer science and can be solved by a construct satisfying the three requirements *mutual exclusion, progress,* and *bounded waiting* [74, 75]. These requirements are given by the OpenMP critical construct.

OpenMP 3.1 Citation 25 (pages 68). "The critical construct restricts execution of the associated structured block to a single thread at a time.

```
#pragma omp critical [(name)] new-line
   structured-block
```

An optional *name* may be used to identify the critical construct. All critical constructs without a name are considered to have the same unspecified name. A thread waits at the beginning of a critical region until no thread is execution a critical region with the same name. The critical construct enforces exclusive access with respect to all critical constructs with the same name in all threads, not just those threads in the current team."   □

The critical construct is used when a sequence of statements must be executed atomically. This means that the critical region is never executed by more than one thread at the same time.

Example 10. The critical region in this code protects the simultaneous stores to reference thread_result[0].

```
1 #pragma omp parallel
2 {
3    ...
4    while( i ≤ ub) {
5       j←0;
6       y←0.;
7       while(j<n) {
8          y+←A_{i*n+j}*x_j;
```

```
9            j←j+1;
10       }
11       #pragma omp critical
12       {
13           thread_result₀←sin(thread_result₀)*sin(y);
14       }
15       i←i+1;
16     }
17 }
```

The data decomposition is here only indicated by dots. A thread local result is computed by using the private variable y. Afterwards, all results from the team of threads flow into a computation where the reference thread_result[0] is successively updated by the assignment inside the critical construct.   △

barrier Construct

OpenMP 3.1 Citation 26 (pages 70). "The barrier construct specifies an explicit barrier at the point at which the construct appears.

```
#pragma omp barrier new-line
```

Each barrier region must be ecncountered by all threads in a team or by none at all. The sequence of worksharing regions and barrier regions encountered must be the same for every thread in a team."   □

Example 11. In case that a reduction is necessary, one can use the master construct as displayed in the current code. The dots hide the part where a data decomposition takes place.

```
1 #pragma omp parallel
2 {
3     ...
4     while( i ≤ ub) {
5         j←0;
6         while(j<n) {
7             y←y*sin(A_{i*n+j}*x_j)*cos(A_{i*n+j}*x_j);
8             j←j+1;
9         }
10        i←i+1;
```

```
11    }
12    thread_result_{tid}←y;
13    #pragma omp barrier
14    #pragma omp master
15    {
16       i←1;
17       while(i<p) {
18          thread_result_0←thread_result_0*thread_result_i;
19          i←i+1;
20       }
21    }
22 }
```

A loop computes the thread local result for the reference thread_result[tid], where tid is the unique thread number. Subsequently, the barrier construct defines a meeting point in the parallel region that all threads have to reach before the execution of any thread may continue. This is necessary to ensure that all threads have finished their work before the reduction operation summarizes the results of each thread. After all threads have encountered the barrier, the master thread enters the corresponding construct and executes the reduction.   △

atomic Construct

The atomic construct can as well as the critical construct be used for avoiding a race condition. The difference between these constructs is that the atomic construct is only defined for the consecutive assignment instead for a whole subsequence of statements.

OpenMP 3.1 Citation 27 (pages 73-77). "The atomic construct ensures that a specific storage location is accessed atomically, rather than exposing it to the possibility of multiple, simultaneous reading and writing threads that may result in indeterminate values.

```
#pragma omp atomic [ read | write | update | capture ]
    new-line
    expression-stmt
```

or

```
#pragma omp atomic capture new-line
    structured-block
```

where *expression-stmt* is an expression statement with one of the following forms:
[...]
If clause is update or not present:

   x *binop= expr* ;

[...]
and where *structured-block* is a structured block with one of the following forms:
[...]
The atomic construct with the update clause forces an atomic update of the location designated by x using the designated operator or intrinsic. Note that when no clause is present, the semantics are equivalent to atomic update. Only the read and write of the location designated by x are performed mutually atomically. The evaluation of *expr* or *expr_list* need not be atomic with respect to the read or write of the location designated by x. No task scheduling points are allowed between the read and the write of the location designated by x."   □

We only consider the atomic construct with no additional clauses. This is actually the definition of the atomic construct until OpenMP 3.0. The additional clauses were first defined with publication of version 3.1 and will be neglected in this work.

Example 12. Similar as in Example 10, the current code contains an assignment where the memory location thread_result[0] is consecutively assigned to by all threads.

```
 1 #pragma omp parallel
 2 {
 3    ...
 4    while( i ≤ ub) {
 5      j←0;
 6      y←0.;
 7      while(j<n) {
 8        y+←A_{i*n+j}*x_j;
 9        j←j+1;
10      }
11    #pragma omp atomic
12      thread_result_0+←y;
13      i←i+1;
14    }
15 }
```

The atomic construct synchronizes this race condition and the critical reference used in a nondeterministic way. $\triangle$

OpenMP Data Environment

We briefly introduced the memory model of OpenMP at the beginning of this section by stating that all variables in a OpenMP code are either private or shared. To complete this statement the current subsection introduces the descriptions from the OpenMP document.

OpenMP 3.1 Citation 28 (Data-sharing attribute rules for variables referenced in a construct, p. 84ff). "Specifying a variable on a firstprivate, lastprivate, or reduction clause of an enclosed construct causes an implicit reference to the variable in the enclosing construct. Such implicit references are also subject to the data-sharing attribute rules outlined in this section.

Certain variables and objects have predetermined data-sharing attributes as follows:

- Variables appearing in threadprivate directives are threadprivate.

- [...]

- The loop iteration variable(s) in the associated *for-loop(s)* of a for or parallel for construct is (are) private.

- [...]

Rules for variables with implicitly determined data-sharing attributes are as follows:

- In a parallel or task construct, the data-sharing attributes of these variables are determined by the default clause, if present.

- In a parallel construct, if no default clause is present, these variables are shared.

- For constructs other than task, if no default clause is present, these variables inherit their data-sharing attributes from the enclosing context.

- In a task construct, if no default clause is present, a variable that in the enclosing context is determined to be shared by all implicit tasks bound to the current team is shared.

- In a task construct, if no default clause is present, a variable whose data-sharing attribute is not determined by the rules above is firstprivate.

Additional restrictions [...]"   □

threadprivate Directive (OpenMP 3.1 Section 2.9.2)

The first pragma introduced here is the threadprivate directive. This directive defines static data to be replicated for each thread.

OpenMP 3.1 Citation 29 (p. 88). "The threadprivate directive specifies that variables are replicated, with each thread having its own copy. The syntax of the threadprivate directive is as follows:

`#pragma omp threadprivate` (*list*)   *new-line*

Each copy of a threadprivate variable is initialized once, in the manner specified by the program, but at an unspecified point in the program prior to the first reference to that copy. [...] The values of data in the threadprivate variables of non-initial threads are guaranteed to persist between two consecutive active parallel regions only if all the following conditions hold:

- Neither parallel region is nested inside another explicit parallel region.

- The number of threads used to execute both parallel regions is the same.

[...] A threadprivate directive for file-scope variables must appear outside any definition or declaration, and must lexically precede all references to any of the variables in its list."   □

private clause

To declare thread local data one can use the private clause.

OpenMP 3.1 Citation 30 (p. 96). "The private clause declares one or more list items to be private to a task. The syntax of the private clause is as follows:

`private` (*list*)

Each task that references a list item that appears in a private clause in any statement in the construct receives a new list item whose language-specific attributes are derived from the original list item. Inside the construct, all references to the original list item are replaced by references to the new list item. In the rest of the region, it is unspecified whether references are to the new list item or the original list item. Therefore, if an attempt is made to reference the original item, its value after the region is also unspecified. If a task does not reference a list item that appears in a private clause, it is unspecified whether that task receives a new list item.

The value and/or allocation status of the original list item will change only:

1. if accessed and modified via pointer,

2. if possibly accessed in the region but outside of the construct, or

3. as a side effect of directives or clauses.

List items [...] may also appear in a private clause in an enclosed parallel [...] construct."   □

firstprivate clause

To declare thread local data that is initialized with the value of the global instance one can use the firstprivate clause.

OpenMP 3.1 Citation 31 (p. 98). "The firstprivate clause declares one or more list items to be private to a task, and initializes each of them with the value that the corresponding original item has when the construct is encountered. The syntax of the firstprivate clause is as follows:

firstprivate (*list*)

The firstprivate clause provides a superset of the functionality provided by the private clause. [...]
In addition, the new list item is initialized from the original list item before the construct. The initialization of the new list item is done once for each task that references the list item in any statement in the construct. The initialization is done prior to the execution of the construct.

For a firstprivate clause on a parallel or task construct, the initial value of the new list item is the value of the original list item that exists immediately prior to the construct in the task region where the construct is encountered. For a firstprivate clause on a worksharing construct, the initial value of the new list item for each implicit task of the threads that execute the worksharing construct is the value of the original list item that exists in the implicit task immediately prior to the point in time that the worksharing construct is encountered.

To avoid race conditions, concurrent updates of the original list item must be synchronized with the read of the original list item that occurs as a result of the firstprivate clause. If a list item appears in both firstprivate and lastprivate clauses, the update required for lastprivate occurs after all the initializations for firstprivate."   □

lastprivate clause

OpenMP 3.1 Citation 32 (p. 101). "The lastprivate clause declares one or more list items to be private to an implicit task, and causes the corresponding original list item to be updated after the end of the region.

The syntax of the lastprivate clause is as follows:

l a s t p r i v a t e (*list*)

The lastprivate clause provides a superset of the functionality provided by the private clause.

A list item that appears in a lastprivate clause is subject to the private clause semantics [...] In addition, when a lastprivate clause appears on the directive that identifies a worksharing construct, the value of each new list item from the sequentially last iteration of the associated loops, or the lexically last section construct, is assigned to the original list item.

The original list item becomes defined at the end of the construct if there is an implicit barrier at that point. To avoid race conditions, concurrent reads or updates of the original list item must by synchronized with the update of the original list item that occurs as a result of the lastprivate clause.

If the lastprivate clause is used on a construct to which nowait is applied, accesses to the original list item may create a race condition. To avoid this, synchronization must be inserted to ensure that the sequentially last iteration or lexically last section construct has stored and flushed that list item."   □

reduction clause

OpenMP 3.1 Citation 33 (p. 103). "The reduction clause specifies an operator and one or more list items. For each list item, a private copy is created in each implicit task, and is initialized appropriately for the operator. After the end of the region, the original list item is updated with the values of the private copies using the specified operator.
The syntax of the clause reduction is:

    r e d u c t i o n (*operator:list*)

The following table lists the operators that are valid and their initialization values. The actual initialization value depends on the data type of the reduction list item.

Operator	Initialization value
+	0
*	1
-	0
⋮	⋮

[...] A private copy of each list item is created, one for each implicit task, as if the private clause had been used. The private copy is then initialized to the initialization value for the operator, as specified above. At the end of the region for which the reduction clause was specified, the original list item is updated by combining its original value with the final value of each of the private copies, using the operator specified. (The partial results of a subtraction reduction are added to form the final value)."   □

This was a brief introduction into the OpenMP 3.1 standard and the original document with several hundred pages contains much more information. However, we cited these parts from the original document that are of importance for this work. The next section describes related work.

1.4 Related Work

An early approach of coupling AD with OpenMP can be found in [10]. The authors consider a given derivative code Q' that implements the tangent-linear model in *vector mode*. The forward mode in vector mode can be defined as

$$\mathbf{y}^{(1)} = F^{(1)}(\mathbf{x}, \mathbf{x}^{(1)}) \equiv \nabla F(\mathbf{x}) \cdot A^{(1)}$$

in difference to our definition $\mathbf{y}^{(1)} = F^{(1)}(\mathbf{x}, \mathbf{x}^{(1)}) \equiv \nabla F(\mathbf{x}) \cdot \mathbf{x}^{(1)}$ in (1.10). This means in vector mode the derivative code does not compute a Jacobian vector product as we defined it in (1.10) but instead the derivative code computes a Jacobian matrix product.

The forward mode in non vector mode augments each assignment in the original code by another assignment that computes one directional derivative value that corresponds to the original assignment. The vector mode on the other side augments the original assignment not by an assignment but by a loop that computes the whole gradient of the original assignment. We do not cover the vector mode in this work since it is connected with the consumption of a big amount of memory, especially in the adjoint case. However, since the computation of the individual gradient components are independent of one another the corresponding loop is parallelizable. The authors of [10] suggest to use the OpenMP loop construct for declaring the loops of the tangent-linear vector mode as parallelizable. To avoid side effects during the execution of the statements coming from the original code Q, they recommend to use the master construct. This implies the need for synchronization with the barrier directive.

To prevent the synchronization overhead of the previous approach, [11] describes an extension to get better scaling properties. This paper suggests to avoid the master construct to compute the original statements by the master, but instead to compute the original statements redundantly by the whole team of threads. Obviously, this requires the usage of additional thread local memory such that the different computations do not interfere with each other. The synchronization is prevented but paid with the price of redundant computations and an higher memory usage. Another extension, recommended by [11], is to use preprocessed loop bounding scheduling. This is possible here as all loops corresponding to a gradient computation have the exact same number of iterations.

In case that the original code Q already contains OpenMP directives, the above approach can still be applied by using nested parallelism. This is discussed in [12] and leads to two levels of parallel execution. Each thread attending the parallel execution on the first level creates another team of threads, each time it encounters an OpenMP #pragma parallel directive on the second level of the parallel execution. The implicit barrier at the end of each parallel region suggests that this approach causes a lot of synchronization overhead on the second level. However, since the authors only present a potential speedup formula instead of runtime results, we can only speculate that the overhead of this approach is a major factor.

The papers mentioned above mainly focus on the tangent-linear vector mode. The main advantage of the vector mode is that it provides the whole Jacobian matrix after one evaluation of the derivative code. Due to the high amount of memory necessary for the execution, we do not use the vector mode of AD. This fact immediately prohibits the above approaches of computing the derivatives with an OpenMP loop.

Our work differs from the approaches above in a way that we do not assume to have a derivative code given that should be tuned by using OpenMP. Instead, we assume that an original code Q is given that contains an OpenMP parallel region P. We examine how to achieve the generation of the tangent-linear code and the adjoint code of P while assuming that the derivative code of Q without the parallel region P is known. Therefore, we focus of transforming P correctly into the corresponding derivative code depending which AD mode we apply.

As mentioned in the AD introduction, besides the source transformation method there is also the overloading method to achieve derivative values. In case that the overloading method should be applied to a code containing a parallel region P then P can be adjusted such that certain information is provided during runtime about the parallel execution. This approach was introduced in [6, 48]. The adjustment of P is that a firstprivate clause is used to inform the AD overloading tool ADOL-C[13] at runtime that a parallel execution is underway. Since the overloading method is a runtime solution, the OpenMP pragma information cannot be exploited without any adjustments of the pragmas in the original code.

Applications of AD by source transformation applied to Fortran code are described in [34, 35, 27, 26, 39, 28]. The source transformation in [27] uses OpenMP 1.1. Nevertheless, it is not mentioned what OpenMP constructs are known to the AD tool and [34] states that adjoint support routines were written by hand to offer a basic support of OpenMP.

The computation of derivatives in tangent-linear and in adjoint model while using distributed memory parallel programming with MPI is examined in [71, 72, 77]. In order to prove the correctness of MPI adjoint code, [58] proposes a framework where the different possible interleavings of a parallel execution are considered. This work was taken as a starting point for studies how to prove the correctness of the source transformation results. It turns out that the distributed memory model has a fundamental advantage compared to the shared memory model. When considering all the possible interleavings of a possible parallel execution, the usage of the distributed memory model reduces the number of possibilities considerably. This originates from the fact that each process only sees its own local memory. Hence, a combination of statements that are from two different processes can be executed in arbitrary order. The result is the same since the store operations to the distributed memory locations cannot interfere each other.

This dissertation is about shared memory parallel programming and therefore each and every possible interleaving must be examined to reveal whether there is a

---

[13]https://projects.coin-or.org/ADOL-C

possible problem in the parallel execution or not. Hence, we will keep the number of possibilities small by reducing the set of language statements to a subset that covers most of the occurring numerical kernels.

1.5 Contributions

This dissertation makes the following contributions:

1. **A methodology to show the correctness of a parallel execution.** The formalism of this methodology is based on the approach of using interleavings as a mathematical abstraction for a parallel execution. We present a formalism that is applicable to a parallel region with a sequence of statements. This parallel region is assumed to be executed by a team of threads. We purposely keep this formalism on an abstract level in order to generally enable the possibility to apply this formalism to different parallel programming models. Our formalism is sophisticated enough to cover the effects of specific constructs of a parallel programming API such as synchronization. Despite the fact that we only consider shared memory parallel programming, our formalism can also be used for a distributed memory (MPI) or for hybrid machine architectures (GPU, FPGA, Intel Phi).

2. **Source transformation rules for applying algorithmic differentiation to pure parallel regions which comprise of SPL code.** A parallel region is a structured code block that is surrounded by curly brackets and a preceding compiler pragma that declares the code block as concurrently executable. We provide source transformation rules for such parallel regions where the code block is given in a certain language called *SPL*. This language is a simplification of C/C++ because otherwise one would have to define source transformation rules for all possible statements of C/C++. Therefore, we reduce the number of possible statements to a manageable size by defining this customized language. *SPL* is an extension of *SL* which is the input and output language of the AD source transformation tool dcc that we introduced in Section 1.2.2.

Until now the source transformation rules implemented in dcc were only verified by a comparison between the computed results. A certain derivative value is computed twice, once by the derivative code that is provided by dcc, and once by an implementation using finite differences. In contrast to this, this dissertation presents a proof that our source transformation result in fact implements the tangent-linear or the adjoint model. With these source transformation rules at hand everyone can implement her/his own AD tool that is capable of generating higher derivative codes.

3. A formal proof that our AD source transformation has the closure property. We assume the code that serves as input for our source transformation is correct in the sense that it can be executed concurrently without any race conditions or side effects. The closure property is fulfilled if we apply the source transformation to a given input code P and the result is a code that is contained in SPL and in addition it must be ensured that the parallel execution of this code is correct, for example, without any race conditions. In case of the tangent-linear source transformation we will see that this property is fulfilled without any restrictions to the input code except that correctness must be given. The adjoint source transformation on the other hand fulfills the closure property only for an input code which fulfills the so called exclusive read property. This property is fulfilled by a parallel region P if it is ensured that during the execution of P each memory location that is used on a right-hand side of a floating-point assignment is used by one thread exclusively and not by multiple threads concurrently.

The formal proofs which show the above properties differentiate between possible combinations of statements. In case of the tangent-linear code we have to consider 10 cases, the proof of the adjoint code contains 36 different cases.

4. A static program analysis for recognizing the exclusive read property of a given SPL code. If P does not fulfill the exclusive read property, we have to use synchronization methods for the reverse section of the adjoint code in order to allow a correct concurrent execution. One could be conservative by defining that the adjoint source transformation introduces for each adjoint assignment a synchronization mechanism. This provides a correct adjoint code but from a certain number of threads on, the execution will be sequential since all threads are busy with waiting on each other. This conservatism is expected to be very expensive in terms of runtime performance. Therefore, we present a static program analysis that provides the information whether or not synchronization is necessary for a certain adjoint assignment. The static program analysis that we introduce is called the exclusive read analysis (ERA).

We present experimental results where we compare the runtime results of an adjoint code obtained without using the ERA with the results of an adjoint code created using the ERA. These results show clearly that a static analysis of the original code is crucial. The adjoint code generated with help of information from the ERA is in average three times faster than the conservative adjoint code. The runtime improvements lie between a factor of 1.5 and 7.74 what means that the least improvement is 50% and the biggest provides a code that is almost eight times faster than the adjoint code obtained without using ERA. This improvement can also expressed in million floating-point operations per second (MFLOPS). For example, the execution of the adjoint code with 32 threads without using ERA achieves 4622 MFLOPS, the execution of the code with using ERA reaches 16378 MFLOPS which is an improvement factor of 3.5.

5. **Implementation of the defined source transformation rules in a tool called** SPLc. The source transformation rules from Chapter 2 and Section 4.2 have been implemented in SPLc (see Appendix A). This means that SPLc implements the source transformations that are necessary to fulfill the closure property. This allows to generate arbitrary higher derivative codes. SPLc allows a source transformation interactively on the console or the user can provide a filename that contains *SPL* code.

To give the reader an impression of the codes sizes that SPLc generates, we take an OpenMP code, called 'plain-parallel', as example. This example code is contained together with the derivative codes in Appendix B.1. The source has 33 lines of code and is referred to as original code. The corresponding first-order tangent-linear code has 77 lines, the first-order adjoint code has 148 lines of code. The reapplication of SPLc to the first derivative code provides the second derivative code. The second-order tangent-linear code has 291 lines of code, the second-order adjoint code in forward-over-reverse mode has 395 lines of code, and the second-order adjoint code in reverse-over-forward has 507 lines of code.

6. **Extensive tests for showing scalability and the successful application..** We developed a test suite to show the scalability of this work. This test suite consists of seven different OpenMP codes starting from a code with only a parallel region, then codes with synchronization constructs of OpenMP and finally a test case where we examine the properties of the second-order derivative code. In each test case we show the results in terms of runtime, MFLOPS, and the stack sizes used during the adjoint computation. We use two compilers, the Intel compiler and the g++ compiler to produce a binary executable. In addition, we use first the compiler without any code optimization. Afterwards, we show the impact when using the code optimization of the Intel and the g++ compiler. We show that these compilers supply very different results in terms of speedup and MFLOPS.

Another result is that the adjoint code is not well suited for the typical code optimization provided by the Intel and the g++ compiler. The experimental results of some adjoint codes reveal that the speedup value is higher than the one from the original code. This fact shows that it is substantial to use concurrency for computing adjoint values.

To show the application of our approach we implemented the least-squares problem and the constrained optimization problem which we introduced in this chapter.

The corresponding first and second derivative codes are provided by SPLc.

7. Rules for applying AD by source transformation to OpenMP directives inside a parallel region. This contribution is the continuation of contribution 2 where we only assume to have a pure parallel region as input code for our source transformation. Instead of a pure parallel region this contribution presents transformation rules for OpenMP constructs which are frequently used in practice. We show how the synchronization constructs of OpenMP can be transformed. This part will give us the missing part to show the closure property of our source transformation. Afterwards, we show the transformation rules for the worksharing and the data-sharing constructs of the OpenMP 3.1 standard. In addition to just defining a source transformation rule we have to show that certain properties inside the resulting code holds. For example, in accordance to the OpenMP 3.1 document an OpenMP barrier construct must be encountered by all threads or by none at all. In case that the derivative code contains a barrier this property must be shown formally.

1.6 Outline of the Thesis

The structure of this dissertation is as follows. Section 2.1 introduces the notation that we use throughout this work. In Section 2.2, we introduce the *SPL* language that serves as input and output language for the source transformation rules which are presented in Section 2.3. The fact that these source transformations produce the tangent-linear or the adjoint model of a given input code P is shown in Section 2.4. The input code P is assumed to be a code block enclosed by a compiler pragma that declares the code block as concurrently executable. The code where the parallel region P is included is denoted by Q. Despite the fact that we only consider the source code transformation of P and not the transformation of the code Q, Section 2.3.3 describes how the derivative code of the parallel region P fits into the derivative code of Q.

Chapter 3 proposes a static program analysis for providing the information whether the exclusive read property could be verified or not. After motivating the approach with an example and defining the formalism of the control flow of *SPL* code in Section 3.2, we present a methodology how a static interval analysis examines the possible value range of integer variables. This approach is subsequently extended such that the value ranges of the integer variables are described by expressions. These expressions are represented by directed acyclic graphs. This approach is established from Section 3.3 until Section 3.5. Section 3.6 concludes this chapter by displaying how an obtained fixed point of intervals of directed acyclic graphs can be exploited to verify the exclusive read property.

The transition from a general parallel region P that does not contain further pragmas inside, to an OpenMP parallel region which contains certain OpenMP pragmas inside is performed in Chapter 4. Section 4.2 discusses the transformation of the synchronization constructs while these constructs enable us to show the closure property of the adjoint transformation of *SPL* code without any restrictions to the input code (Section 4.2.5). The source transformation of worksharing constructs such as the loop construct or the sections construct is displayed in Section 4.3. Finally, we show transformation rules for data-sharing clauses such as private or firstprivate in Section 4.4.

The experimental results of our implementation are discussed in Chapter 5. We present a test suite containing several code examples in Section 5.1. For each example we illustrate the runtime of the original code and its corresponding first-order tangent-linear and adjoint code. In Section 5.2, we compare the runtime values of a second derivative code. Subsequently, we examine in Section 5.3 the impact of the exclusive read analysis. The runtimes of the implementations of the nonlinear least-squares problem and the nonlinear constrained optimization problem from Chapter 1 are examined in Section 5.4 and Section 5.5. The concluding Chapter 6 summarizes this dissertation and describes possible future work.

2 Source Transformation of Pure Parallel Regions

In this chapter, we will first describe the notation that we use for describing a parallel code region and its execution. This work is about the source transformation of pragma-based parallel regions. Therefore, we will define the *SPL* language that is used as input and as output language for our source transformation. After we have defined the source transformation rules, we will show that the resulting code represents the AD model as defined in the introduction chapter. The *SPL* language is shaped such that it comprises one compiler directive in form of a pragma that defines an associated code region as concurrently executable. We will denote this kind of parallel region as a *pure* parallel region.

A pure parallel region is a sequence of statements surrounded by curly brackets and a preceding parallel pragma such as, for example,

```
#pragma omp parallel
{
    s₁ ; s₂ ; ··· ; sq ;
}
```

Listing 2.1: A pure parallel region.

where the appearance of the code inside of the parallel region is irrelevant at this point. The term 'pure' means in this context that there are no additional pragmas inside the parallel region. However, we try to stay on a level that allows to apply the results of this chapter to a non-OpenMP parallel region.

2.1 Formalism and Notation

In general, we assume a parallel region P to have a structure as shown in Listing 2.1 where a sequence of q statements are surrounded by a pragma that declares these statements as parallelizable. A sequence of statements is indicated by S. There are potentially subsequent statements below S, which occur, for example, when there is a conditional statement or a loop statement. We display these subsequent statements with S'. The sequence that is on the first level does not have any parent sequences and is referred by S_0. By writing $P = S_0$ we indicate that the parallel region P has the sequence S_0 on its first level. When we want to express a special structure of the parallel region, we use sequences of statements for better readability. For example, consider the structure of P in

```
#pragma omp parallel
{
    S₁
    s    ;
    S₂
}
```

where the parallel region consists of two sequences of statements S_1, S_2 and a statement s that separates these sequences from one another.

To show the correctness of the source transformations in this work, we need a mathematical abstraction for the concurrent execution of the parallel regions. This abstraction is based on the usage of interleaved statements and was proposed by Ben-Ari in [5, Chapter 2], whereby this approach goes back to [52, Chapter 2]. No matter how many physical processors are available, we only consider one processor and p different threads. Each thread holds a sequence of q atomic statements. An atomic statement means that this statement cannot be divided into smaller instructions.

The parallel execution consist of pq statements (p threads times q statements). The execution of these pq statements takes place on one processor. Thus, the statements have to be merged or interleaved by a scheduler into one sequence. This sequence is called *interleaving* and it represents the order in which the processor encounters the individual statements. An important assumption is that the strategy of the scheduler is unknown and every permutation of the pq statements is possible. The interleaving is empty at the beginning of the parallel execution and all threads have a queue with q statements waiting for the scheduler to pick them up. Let us consider a possible status of an execution with three threads after the scheduler has already taken six statements. One possible appearance of the interleaving is shown in Figure 2.1.

We denote a statement with s or with s_i when the statement is the i-th statement in a sequence of statements $(s_1; s_2; \ldots; s_q)$. Each thread has an unique identifier number starting with zero and ending with $p - 1$ because p threads are participating the execution. We differ between threads such that we note the different identifier numbers. For example, thread 0 is the thread that is identified with the number zero. Thread t is a thread where the identifier number is arbitrarily chosen, which means that t can be a number from zero to $p - 1$. The fact that statement

s is executed by thread t is denoted by s^t. Consider, for example, the case that a possible interleaving contains

$$s_3^1 s_5^0.$$

This means that the scheduler first took the third statement of the thread 1 and subsequently it chose, for whatever reason, the fifth statement of thread 0.

An atomic statement is one that is executed by a thread t without any context change to a different thread t'. In the literature there is often the explanation of an atomic statement as a statement that cannot be divided into smaller instructions. However, given an atomic statement s, it can in general be divided into several assembly instructions. In [52, Section 2.2] it is shown that a concurrent sequence of statements where each statement has at most one *critical reference* can be assumed to be atomic. And most important, it is shown that the behavior of the result is the same as if the sequence is compiled to a machine architecture with an atomic load and store.

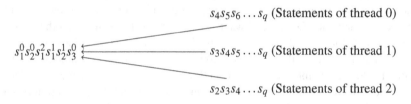

Figure 2.1: This figure shows one possible status of the program execution after processing six statements with three threads. Each thread has q statements $(s_1; \ldots; s_q)$ at the beginning of the execution. Then the scheduler chooses some statements from the three threads and puts them to the interleaving. After the scheduler has taken six statements, the status of the execution can appear as in this figure. At this point in the execution, the first three statements of thread 0 have already been executed. The next time the scheduler decides to choose thread 0 to be the execution thread, statement s_4^0 is put into the interleaving. Thread 1 is waiting for statement s_3^1 being put into the interleaving. For thread 2, this holds analogously for statement s_2^2. Because all statements have a fixed order inside their respective executing thread, statement s_{i+1}^t cannot be executed before statement s_i^t. In the figure we see that statement s_2^0 occurs before statement s_3^0.

In OpenMP one can define certain assignments to be atomic by using the directive atomic. We do not mean this atomicity when we write about an atomic statement. Nevertheless, the atomic construct can be associated with a statement s such that s is executed atomically as we will see in Chapter 4. This is very helpful in case that statement s contains a critical reference. We cite the definition of a critical reference that Ben Ari presents in his textbook.

Definition 34 ([5], p.27). An occurrence of a variable v is defined to be a *critical reference*,

1. if it is assigned to in one thread and has an occurrence in another thread, or

2. if it has an occurrence in an expression in one thread and is assigned to in another.

A program satisfies the *limited-critical-reference* (LCR) restriction if each statement contains at most one critical reference.   □

The author Ben-Ari presents an example as justification for the abstraction with interleavings of atomic statements, see [5, p. 16]. The example describes the following situation where two threads want to write to the memory location on a multiprocessor machine. One thread wants to write the value 1 and the other thread wants to write the value 2. Thus, both threads write different bit values. The question is what the result of the store operation is. The value might be undefined when both bit values are combined per logical OR operation. In this case we would have the value 3 inside the memory location. This would reveal the abstraction as unqualified, but in practice this case does not occur because the memory hardware writes values not bitwise but cell wise. A cell is, for example, 32 bit on a 80386 processor or 64 bit on modern processors. The bits that the threads want to write are contained in the same cell. Therefore, the memory hardware completes one access before the other takes effect. In this case the memory hardware performs the atomicity and interleaving.

For our approach it is important to understand that the atomic statements of each individual thread are interleaved arbitrarily. For example, let us consider a parallel region P that consists of the sequence of statements $S = (s_1; \ldots; s_q)$. The atomic statements considered during the execution are in case of thread 0 $(s_1^0; \ldots; s_q^0)$, for thread 1 they are $(s_1^1; \ldots; s_q^1)$, and so on. The permutations where the order does not reflect the order of the statements inside of a particular thread are not valid. For example, an interleaving beginning with $s_1^1 s_2^0 s_2^1 s_1^0 s_1^2$ is not possible, since the second statement of thread 0 is executed before the first statement s_1^0 of thread 0 appears in the interleaving.

The set of variables used in a parallel region P is denoted by VAR_P, the set of variables occurring in statement s is denoted by VAR_s. Most of the time we will discuss about comparing references instead of comparing variables. Therefore, we need a mapping from the set of variables to the memory address space. This is done by the operator &:

$$\& : VAR_P \rightarrow MemLoc \subset \mathbb{N}.$$

The following example uses a statement where multiple threads increment a shared variable to clarify the term of a critical reference and the LCR property.

Example 13. Definition 34 can be explained by a source statement s represented by

$$s = \quad n \leftarrow n + 1 \quad ,$$

where n is a shared variable. This example is similar to the example in [5, p. 27]. Depending on the set of statements contained in the assembly language of a given architecture, this statement can be compiled into an atomic increment instruction or into a sequence of assembly statements whose semantic are the same as the source statement. What one can expect when considering Figure 2.1 is that the number of interleavings is different when we consider, on the one hand, a program consisting of one atomic increment instruction, or on the other hand, a program that consists of a sequence of assembly statements. The behavior can be different opposed to the case that the sequence of assembly statements would be evaluated atomically. Nevertheless, the abstraction of atomic source statements can be used when each statement fulfills the LCR property. To clarify this, we split the statement s into[1]

$$
\begin{aligned}
s_1 &= \quad a \leftarrow n \\
s_2 &= \quad n \leftarrow a + 1 \quad ,
\end{aligned}
$$

where a is a private (thread local) variable.

We assume that the statement s and the statements $s_1 ; s_2$ are executed by thread t and by thread t'. According to statement s, thread t executes at runtime the SPMD instance

$$s^t = \quad n^t \leftarrow n^t + 1$$

and thread t' executes

$$s^{t'} = \quad n^{t'} \leftarrow n^{t'} + 1$$

---

[1]Please note that there is an error in [5, p. 27] that we reported the author. It should be similar as presented here as pointed out on the errata website (http://www.weizmann.ac.il/sci-tea/benari/books/pcdp2-errata.html).

where we use a superscript index to indicate the thread that accesses the variable n. The memory locations of n concerning thread t is $\&n^t$, thread t' uses $\&n^{t'}$. Since n is a shared variable, the reference of both threads is the same, namely $\&n^t = \&n^{t'}$.

When we consider the reference $\&n$ occurring in statement s on the left-hand side, we conclude that this reference is critical. It is critical by Definition 34 (1) because it occurs in thread t as left-hand side reference and in thread t' as right-hand side reference. Let us now consider the reference $\&n$ on the right-hand side of statement s. It is critical by Definition 34 (2) since it is read by thread t and is assigned to in thread t'. Therefore, s has two critical references and does not fulfill the LCR property.

Now, we consider the code where we split s into s_1 and s_2. This code is semantically equivalent to s but differs according to the number of critical references.

$$
\begin{aligned}
s_1^t &= a^t \leftarrow n^t \\
s_2^t &= n^t \leftarrow a^t + 1
\end{aligned}
\qquad\qquad
\begin{aligned}
s_1^{t'} &= a^{t'} \leftarrow n^{t'} \\
s_2^{t'} &= n^{t'} \leftarrow a^{t'} + 1
\end{aligned}
$$

Variable a is a thread local variable is therefore noncritical. Statement s_1^t reads the shared reference $\&n^t$ and assigns it to the thread local reference $\&a^t$. But because of $\&n^t = \&n^{t'}$, this reference is also assigned to in thread t' as $\&n^{t'}$ in $s_2^{t'}$. Therefore, reference $\&n^t$ is a critical reference in statement s_1 due to Definition 34 (2). The reference $\&n^t$ is assigned to in s_2^t and is read in $s_1^{t'}$. Thus, the occurrence of variable n in s_2 is also critical because of (1) in Definition 34. In summary, s_1 and s_2 both fulfill the LCR property whereby s does not. This means that the approach of interleavings of atomic statements can be applied to the code $s_1; s_2$ but not to s.

Just to be clear, the LCR property does not ensure the correct evaluation of the sequence, it only ensures that this abstraction shows the same behavior as an architecture with an atomic load and store. If we abstract $s_1; s_2$ by interleavings, then we get different results for n due to the fact that n is critical. The following table shows that only two of the possible six interleavings result in the correct result of two increment operations which results in being $n + 2$. The remaining four interleavings all end up with setting n to a value of $n + 1$. The column titled with a^0 shows the value of variable a that belongs to thread 0, whereby a^1 belongs to thread 1.

Interleaving	a^0	a^1	n (after execution)	Correct
$s_0^0; s_1^0; s_0^1; s_1^1;$	n	$n+1$	$n+2$	true
$s_0^0; s_0^1; s_1^0; s_1^1;$	n	n	$n+1$	false
$s_0^1; s_0^0; s_1^0; s_1^1;$	n	n	$n+1$	false
$s_0^0; s_0^1; s_1^1; s_1^0;$	n	n	$n+1$	false
$s_0^1; s_0^0; s_1^1; s_1^0;$	n	n	$n+1$	false
$s_0^1; s_1^1; s_0^0; s_1^0;$	$n+1$	n	$n+2$	true

In contrast to the above, let us reconsider the statement s and the possible interleavings

$$s^0; s^1$$

and

$$s^1; s^0 .$$

Both interleavings lead to n having the value $n+2$, which is the correct result but shows that in this case our mathematical abstraction would be wrong, since the associated interleaved assembly instructions could lead to the result $n+1$.

The above table shows why the runtime situation of a code where a critical reference is involved is called a race condition in the domain of shared memory parallel programming. The finishing value in the critical reference depends on the thread that performs the last store to the reference. $\triangle$

If we consider synchronization methods in parallel programming, then certain interleavings are not possible. Therefore, we define an order of interleavings. Let us suppose that two statements s_i and s_j are executed by the threads t and t'. The order of s_i^t and $s_j^{t'}$ inside of I is denoted by

$$s_i^t \prec s_j^{t'}$$

when s_i^t is evaluated prior to $s_j^{t'}$, or in other words, in the interleaved sequence s_i^t is before $s_j^{t'}$. Please note that, when considering statements executed by two different threads, this order is independent of the order of statements that each thread executes. For example let s_1 and s_8 be two statements inside of a sequence of statements $S = (s_1; s_2; \ldots; s_8)$. Suppose that S is executed by two threads identified by 0 and 1. If we consider only thread 0, then it can only hold

$$s_1^0 \prec s_8^0$$

because $s_8^0 \prec s_1^0$ is a contradiction to the order of $(s_1; s_2; \ldots; s_8)$. In case that we consider different threads, for example 0 and 1, then both

$$s_1^0 \prec s_8^1$$

and

$$s_8^0 \prec s_1^1$$

are possible because thread 0 could execute s_8 prior as thread 1 executes s_1.

The following definition describes the set of possible interleavings for a given parallel region P with q statements that is executed with p threads. This set contains all the possible interleavings and it may possibly have infinite size, for example, when at least one thread executes an infinite loop.

Definition 35 (set of possible interleavings). Suppose a parallel region P with q statements, this is $P = S$ with $S = (s_1; \ldots; s_q)$, $q \in \mathbb{N}$, and P is executed by p threads. The set of possible interleavings is denoted by

$$\mathscr{I}(P, q, p)$$

□

Lemma 36. *Suppose a parallel region P is executed by p threads and consists of the sequence S with q straight-line code statements. Then the set of possible interleavings is finite and is defined by*

$$\mathscr{I}(P, q, p) = \left\{ I \,|\, I = (s_i^t)_{i \in 1, \ldots, q}^{t \in \{0, \ldots, p-1\}}, \text{ with } s_i^t \prec s_j^t \right\}$$

where $s \in S$, $i, j \in \{1, \ldots, q\}$, $t \in \{0, \ldots, p-1\}$. I is a sequence of instances of statements in execution. This sequence has a length of $p \cdot q$ instances. The restriction $s_i^t \prec s_j^t$ means that the ordering of the statements from S concerning the same thread must be preserved. The size of $\mathscr{I}(P, q, p)$ is

$$|\mathscr{I}(P, q, p)| := \frac{(q \cdot p)!}{(q!)^p}$$

Proof. We show this by induction over q, the number of straight-line code statements. Assume we only have one statement in the parallel region. The parallel region is executed by p threads resulting in p statements inside of a possible interleaving. The p statements can be scheduled in any order. Therefore, we have $p!$ possible interleavings.

Assuming that the number of possible interleavings for $(q-1)$ statements in a parallel region executed by p threads is

$$\frac{((q-1)\cdot p)!}{((q-1)!)^p},\qquad(2.1)$$

then we have to show that the set of possible interleavings is

$$\frac{(q\cdot p)!}{(q!)^p}$$

when the number of statements in the parallel region is q. A parallel region with q statements executed by p threads results in $q\cdot p$ statements. We can illustrate this by an urn problem with qp balls where q balls have one of p colors. Therefore, we have qp possibilities to choose the first statement in the interleaving and $qp-1$ possibilities to choose the second statement, and so forth. For the p-th statement we have $qp-p+1$ possibilities. We concern about the statements coming from a group of p threads or as shown with the urn problem; the fact that we have p different colors must be taken into account. Hence, the number of possibilities to choose the first p statements is

$$\frac{qp\cdot(qp-1)\cdot\ldots\cdot(qp-p+1)}{\underbrace{q\cdot q\cdot\ldots\cdot q}_{p\text{-times}}}.$$

According to the prerequisites, the number of possibilities to choose the remaining $(q-1)p$ statements is (2.1). Therefore, the number of interleavings is

$$\frac{qp\cdot(qp-1)\cdot\ldots\cdot(qp-p+1)}{\underbrace{q\cdot q\cdot\ldots\cdot q}_{p\text{-times}}}\cdot\frac{((q-1)\cdot p)!}{((q-1)!)^p}=\frac{(q\cdot p)!}{(q!)^p}$$

$\square$

To get familiar with the above formalism, we present in the following several examples where we apply our notation to some OpenMP code examples. These examples contain different OpenMP constructs as the barrier or the master construct.

However, these examples serve only as proof of concept and from the next section on, we consider parallel regions such that these regions only consist of code and not further pragmas. The barrier and master constructs are topic of Chapter 4.

The OpenMP standard provides a loop construct for sharing loop iterations among a group of threads, see Section 1.3. To see the difference between a regular loop statement and a worksharing loop, let us consider the following example where a loop consist of three iterations.

Example 14 (Loop Worksharing Construct of OpenMP). For simplicity we define a group of threads of size two and we assume that all the statements s_1, s_2, s_3 in the parallel region are straight-line code statements. In OpenMP this can be realized by defining omp_set_num_threads(2) before the execution encounters the parallel region.

```
omp_set_num_threads(2);
#pragma omp parallel
{
    s₁
    for( i←0; i <3; i←i+1)
        s₂
    s₃
}
```

$$s_1^0 s_1^1 \overset{s_2 s_2 s_2 s_3 \text{ (statements of thread 0)}}{\underset{s_2 s_2 s_2 s_3 \text{ (statements of thread 1)}}{\Longleftarrow}}$$

Above we see that thread 0 and thread 1 have three instances of statement s_2 waiting for the scheduler to be put into execution. The first statement of each thread (s_1^0 and s_1^1) is already executed at the current point in time. Next, we show the same code structure but with a preceding worksharing construct.

```
omp_set_num_threads(2);
#pragma omp parallel
{
    s₁
    #pragma omp for
    for( i←0; i <3; i←i+1)
        s₂
    s₃
}
```

$$s_1^0 s_1^1 \Longleftarrow \begin{array}{l} s_2 s_2 s_3 \ \text{(statements of thread 0)} \\ \\ s_2 s_3 \ \text{(statements of thread 1)} \end{array}$$

The worksharing construct instructs the runtime environment to distribute all the loop iterations among the two threads. Therefore, the three iterations are distributed to the two threads in chunks of a certain size. This size can be influenced per schedule clause, but we do not consider this possibility here. Without the schedule clause, the standard procedure is static which means that the chunks are approximately equal in size. For our example the chunks are such that thread 0 gets two instances of s_2 and thread 1 gets the remaining third instance. △

Example 15 (Barrier Synchronizing). In OpenMP 3.1 citation 26 (page 45), we saw that one can define an explicit barrier inside an OpenMP parallel region. What this means according to the possible interleavings can be understood when viewing all possible interleavings of two straight-line code statements $(s_0; s_1)$ that are executed by two threads. Let us assume that there is a barrier between s_0 and s_1 to make sure that each thread has finished its evaluation of s_0 before continuing with s_1. As a code snippet this appears as follows.

```
...
s0
#pragma omp barrier
s1
...
```

The set $\mathscr{I}((s_0; s_1), 2, 2)$ of possible interleavings consists of

$$s_0^0 s_0^1 s_1^0 s_1^1$$
$$s_0^1 s_0^0 s_1^0 s_1^1$$
$$s_0^0 s_0^1 s_1^1 s_1^0$$
$$s_0^1 s_0^0 s_1^1 s_1^0$$

Due to Lemma 36, we know that the number of interleavings is

$$\frac{(2 \cdot 2)!}{(2!)^2} = \frac{24}{4} = 6 \,,$$

but the interleavings

$$s_0^1 s_1^1 s_0^0 s_1^0 \tag{2.2}$$

$$s_0^0 s_1^0 s_0^1 s_1^1. \tag{2.3}$$

are not possible because s_1 is executed before the other thread has finished s_0 and this is contradiction to the definition of a barrier which ensures that all threads have finished their statements preceding the barrier. Formally, we express this by writing $s_1^1 \prec s_0^0$ for the situation in (2.2) and analogously $s_1^0 \prec s_0^1$ for (2.3). This shows that synchronization reduces the number of possible interleavings. In practice this fact may lead to a lower performance at runtime. $\triangle$

Example 16 (Master Synchronizing). In Section 1.3 we saw that one can define a region that is only executed by the master thread. This is the thread with the identification number 0 in the current group of threads. Consider the following code as an example where we again have two straight-line code statements.

```
...
s0
#pragma omp master
{
    s1
}
...
```

The set $\mathscr{I}((s_0, s_1), 2, 2)$ of possible interleavings consists of

$$s_0^0 s_0^1 s_1^0$$
$$s_0^1 s_0^0 s_1^0$$
$$s_0^0 s_1^0 s_0^1. \tag{2.4}$$

As mentioned in OpenMP 3.1 [63, p. 68], the master construct does not have an implicit barrier. This is reflected in (2.4) where thread 1 executes s_0 after the master thread finishes s_1. This is formally expressed by $s_1^0 \prec s_0^1$ and would not be possible in case that an implicit barrier would conclude the block connected to the master construct. $\triangle$

Example 17 (Single Synchronizing). The OpenMP standard provides a single construct described in OpenMP 3.1 citation 20 (page 40), which specifies an associated code region to be executed only by one thread. All the other threads wait at an implicit barrier at the end of the single construct.

> . . .
>
> s_0
> **#pragma omp** single
> s_1
>
> . . .

We assume that these two straight-line code statements are executed by two threads. Therefore, the set $\mathscr{I}((s_0,s_1),2,2)$ of possible interleavings consists of

$$s_0^0 s_0^1 s_1^0$$
$$s_0^1 s_0^0 s_1^0$$
$$s_0^0 s_0^1 s_1^1$$
$$s_0^1 s_0^0 s_1^1.$$

Here, all possibilities are present where a certain thread enters the single block. The OpenMP 3.1 mentions on page 52 that the method for choosing the thread is implementation-defined. In practice, an useful implementation is to let the first thread that reaches the single block enter it.

We have a total number of four interleavings for the two statements executed by two threads. This slightly indicates that using the single directive is preferable to the master directive with only three interleavings for this example code.

The number of possible interleavings can even be higher if one uses the nowait clause together with the single construct. The nowait clause suppresses the implicit barrier after the statement s_1. Obviously, this can only be done if there are not data dependencies from s_1 to any following statements. The usage of the nowait clause increases the number of possible interleavings by

$$s_0^0 s_1^0 s_0^1$$
$$s_0^1 s_1^1 s_0^0 \quad ,$$

where the cases $s_1^0 \prec s_0^1$ and $s_1^1 \prec s_0^0$ are only possible due to the missing implicit barrier after the statement s_1. $\triangle$

Considering a certain statement s, we classify s by being a *straight-line code* statement or a *control flow* statement. A control flow statement influences the control flow of the program, for example, when a statement tests a boolean expression and, depending on the result, it continues with different parts of the program. An example for a straight-line code statement is an assignment. The set of possible straight-line code statements is denoted by SLC, the set of control flow statements is referred to as $CFSTMT$.

For a parallel region P, the set of statements contained in P is denoted by $STMTS$. The mapping

$$LABEL : STMTS \rightarrow \mathbb{N}$$

maps a statement of P to an unique label. The assignment

$$l : y \leftarrow \phi(x_1, \ldots, x_n),$$

for example, is labeled with l where $l \in \mathbb{N}$. This label can be obtained by

$$LABEL(\quad l : y \leftarrow \phi(x_1, \ldots, x_n) \quad) = l \in \mathbb{N}.$$

In addition, each memory reference $\&y$, $\&x_{i=1,\ldots,n}$ has an unique label as well.

In the next section we will define a source code transformation which supplies again code as a parallel region. In order to prove that the outcome of the source transformation can be executed concurrently, we need to consider the references that can occur during the parallel execution. Therefore, we define the following mappings where the actual definition is done as soon we have introduced the language SPL. We denote the power set of a set $A = \{a, b, c\}$ with $\mathscr{P}$. Therefore, it holds

$$\mathscr{P}(A) := \{U | U \subseteq A\} = \{\emptyset, \{a\}, \{b\}, \{c\}, \{a,b\}, \{b,c\}, \{a,b,c\}\}.$$

The set of left-hand side references of a statement is given by the mapping

$$LHSREF : STMTS \rightarrow \mathscr{P}(MemLoc).$$

Analogously, we define the mapping that supplies the right-hand side references as

$$RHSREF : STMTS \rightarrow \mathscr{P}(MemLoc).$$

Conditional expressions that influence the control flow arise in conditional statements as well as in loop statements. These expressions are indicated by b. Therefore, to map a statement to the set of references occurring in a conditional expression, we use

$$CONDREF : STMTS \rightarrow \mathscr{P}(MemLoc).$$

For example, let us consider two instances of a statement s that are executed by thread t and thread t'. To express that the left-hand side reference in thread t does not occur on the right-hand side of t', we note

$$LHSREF(s^t) \cap RHSREF\left(s^{t'}\right) = \emptyset \quad .$$

As an abbreviation for the joined sets of all possible references of s^t we use

$$REF(s^t) = LHSREF(s^t) \cup RHSREF(s^t) \cup CONDREF(s^t) \quad .$$

The set of integer variables of P is denoted by INT_P, the set of floating-point variables $FLOAT_P$. A comma separated list of variables is denoted by $list$ while $v \in list$ means that v occurs in the comma separated list despite the fact that this is actually not a set of variables. For example, the notion $FLOAT_{list}$ stands for the set of floating-point variables occurring in the comma separated $list$.

A parallel region P that serves as input for a source transformation is assumed to be correct in the sense that for the same given input values all possible interleavings result in the same output values. This behavior must be proved for P' that is the result of the source transformation of P. We will prove this by showing the absence of any data dependencies inside of the set of possible interleavings $\mathscr{I}(P', q, p)$ while assuming that P has no data dependencies. In other words, we show the correctness of code P' by assuming that the parallel region P is correct. More formally, this means we choose an arbitrary interleaving $I \in \mathscr{I}(P', q, p)$. Each consecutive pair of statements $(s_i^t, s_j^{t'}) \in I$ with $t \neq t'$ and $i, j \in \{1, \ldots, q\}$ must be considered whether there is a data dependence or not. Since we know that P does not have any data dependencies we only have to show that our source transformation does not introduce a new data dependence. But we will see that the adjoint source transformation in certain cases introduces a new data dependence that leads to a race condition at runtime. This race condition must be resolved appropriately as shown later in this work.

In this context it must be guaranteed that assignments as well as evaluations of conditions in control statements do not contain any critical references. Otherwise, a race condition is given at runtime which leads to a nondeterministic result. We want to define the absence of a critical reference as *noncritical*. According to Definition 34 we define a critical parallel region as follows.

Definition 37 (A critical parallel region). Suppose a parallel region P containing a sequence of statements $S = (s_1; \ldots; s_q)$. In the following $s_i^t, s_j^{t'} \in S$ are two statements, executed by thread t and t' with $t, t' \in \{0, \ldots, p-1\}$, and $i, j \in \{1, \ldots, q\}$.

P is *critical* if there exists a pair $(s_i^t, s_j^{t'})$ of consecutive statements with $t \neq t'$ that is part of an interleaving $I \in \mathscr{I}(P, q, p)$ and for this consecutive pair of statements in I holds

$$\&v \in LHSREF(s_i^t) \quad \wedge \quad \&v \in REF\left(s_j^{t'}\right). \tag{2.5}$$

The memory reference $\&v$ is called critical reference. $\qquad \square$

The condition (2.5) can be expressed in words by saying that statement s_i^t performs a store operation to a memory location that is simultaneously used by thread t' in $s_j^{t'}$. Please note that neither the term critical parallel region nor critical reference $\&v$ imply that the concurrent execution is wrong. A nondeterministic behavior only occurs when there is the situation as described in Definition 37 but besides the interleaving $I \in \mathscr{I}(P, q, p)$ with $(s_i^t, s_j^{t'}) \in I$ there exists another interleaving $I' \in \mathscr{I}(P, q, p)$ with $(s_j^{t'}, s_i^t) \in I'$.

Obviously, we do not want to achieve a critical parallel region P' when transforming the source code of a *noncritical* parallel region P. P is called noncritical if for all consecutive pairs of statements $(s_i^t, s_j^{t'})$, executed by the two different threads t and t', and contained in an arbitrary interleaving I out of the set of possible sets $\mathscr{I}(P, q, p)$, the following holds. An occurrence of reference $\&v$ on the left-hand side of s_i^t implies that this reference may not be contained in $s_j^{t'}$.

Lemma 38 (A noncritical parallel region). *Consider again the parallel region P from Definition 37 and the two statements $s_i^t, s_j^{t'}$. P is noncritical if and only if*

$$\forall I \in \mathscr{I}(P, q, p) \quad \forall (s_i^t; s_j^{t'}) \in I, t \neq t' :$$
$$\&v \in LHSREF(s_i^t) \Longrightarrow \&v \notin REF\left(s_j^{t'}\right), \tag{2.6}$$

Proof. In terms of Definition 37 P is critical if

$$\exists I \in \mathscr{I}(P, q, p) \quad \exists (s_i^t; s_j^{t'}) \in I, t \neq t' :$$
$$\&v \in LHSREF(s_i^t) \wedge \&v \in REF\left(s_j^{t'}\right)$$

When we negate the condition we can rewrite this into a formula as an implication:

$$\neg \left(\&v \in LHSREF(s_i^t) \wedge \&v \in REF\left(s_j^{t'}\right) \right)$$

$$\Leftrightarrow \quad \neg\, \&v \in LHSREF(s_i^t) \vee \neg\, \&v \in REF\left(s_j^{t'}\right)$$

$$\Leftrightarrow \quad \neg\, \&v \in LHSREF(s_i^t) \vee \&v \notin REF\left(s_j^{t'}\right)$$

$$\Leftrightarrow \quad \&v \in LHSREF(s_i^t) \Rightarrow \&v \notin REF\left(s_j^{t'}\right) \quad .$$

Therefore, we conclude that P is noncritical in case of

$$\forall I \in \mathscr{I}(P,q,p) \quad \forall (s_i^t; s_j^{t'}) \in I, t \neq t' :$$

$$\&v \in LHSREF(s_i^t) \Rightarrow \&v \notin REF\left(s_j^{t'}\right)$$

$$\square$$

Example 18. We show that the code in Example 4 on page 35 is noncritical. According to Lemma 38, we have to consider all possible pairs $(s_i^t, s_j^{t'})$ of consecutive statements in all possible interleavings $I \in \mathscr{I}(P,q,p)$. Since the only shared variables are the pointers x and y, we only consider the two floating-point assignments in line 17 and line 18:

```
17      y_i ← 2 * x_i * x_i ;
18      x_i ← 0 . ;
```

We label these two statements with the corresponding line number in the code sample. Therefore, we consider pairs of possible combinations of s_{17} and s_{18}. An important fact is that i has an unique value range due to the data decomposition that takes place before the loop.

1. Case $(s_{17}^t, s_{18}^{t'})$:
   Since $s_{18}^{t'}$ has a constant on its right-hand side there is no confict with shared memory locations.

2. Case $(s_{18}^t, s_{17}^{t'})$:
   Since i has different values in thread t and in thread t' we conclude that $\&x_i^t \neq \&y_i^{t'}$ and $\&x_i^t \neq \&x_i^{t'}$ holds. Please note that this even holds for x and y pointing to the same address.

3. Case $(s_{17}^t, s_{17}^{t'})$:
   $\&y_i^t \notin \{\&y_i^{t'}, \&x_i^{t'}\}$ is achieved when reasoning analogous to case 2.

4. Case $(s_{18}^t, s_{18}^{t'})$:

Analogously, we obtain $\&x_i^t \neq \&x_i^{t'}$ because i has different values in thread t and in t'.

Since not one of the above combinations does lead to a critical reference, we conclude that the parallel region in Example 4 is noncritical. $\qquad$ $\triangle$

Example 19. We show that the code in Example 5 is critical. As in the previous example, we consider only the floating-point statements

5      $y_i \leftarrow 2 * x_i * x_i$;
6      $x_i \leftarrow 0$. ;

since these statements contain the shared references x and y. In constrast to Example 4, i is not unique since there is no data decomposition preceding the loop. Instead, i has the same value range in all threads. This indicates that both statements are critical what can be shown by considering the combinations where two different threads execute the statements simultaneously. In our context, we show this with help of the possible interleavings $I \in \mathscr{I}(P, q, p)$.

1. Case $(s_5^t, s_6^{t'})$:

   s_5^t has the left-hand side reference $\&y_i^t$. This reference is only used in $s_6^{t'}$ if the pointer x and y point to the same memory location ($\&y_i^t = \&x_i^{t'}$). Is this the case then $\&x_i^{t'}$ is a critical reference.

2. Case $(s_6^t, s_5^{t'})$:

   The left-hand side of s_6^t, namely $\&x_i^t$, is contained in $s_5^{t'}$ since thread t and t' uses the same offset value i. Therefore, $\&x_i^t$ is a critical reference.

3. Case $(s_5^t, s_5^{t'})$:

   $\&y_i^t$ is critical since it occurs on the left-hand side in s_5^t and also in $s_5^{t'}$.

4. Case $(s_6^t, s_6^{t'})$:

   Even this situtation reveals a critical reference $\&x_i^t$ but there is no wrong effect visible since both threads assign the constant value zero to the same memory location.

This shows that the parallel region in Example 5 is critical. $\qquad$ $\triangle$

In order to define a source transformation we have to determine what the input code and what the output code is. In this work, we define the code inside a parallel region P to be in language *SPL*, which stands for *simple parallel language*. *SPL* is introduced in the next section. The code contained in P is assumed to be noncritical and therefore correct in the sense that its concurrent execution does not have any race conditions or side effects.

We will define a source transformation in Section 2.3 such that the output code is again in *SPL* what provides the possibility of a reapplication of the source transformation to its own output. For the reapplication it must again be ensured that the code is correct. This leads to the following definition of the closure property.

Definition 39 (Closure property of a source transformation). Suppose that $P \in SPL$ is a noncritical parallel region. A source transformation $\alpha(P)$ fulfills the *closure property* if and only if

1. $\alpha : SPL \to SPL$ holds such that the code $\alpha(P)$ is again in *SPL*, and

2. the code obtained by $\alpha(P)$ is noncritical. $\qquad\qquad\qquad\qquad\square$

In the following will introduce the *SPL* language and its syntax. Subsequently, Section 2.3 presents the source transformations that provide the tangent-linear, and the adjoint model of a given parallel region P written in *SPL* code.

2.2 *SPL* - A Simple Language for Parallel Regions

In this section we introduce a language called *SPL* that is on the one side simple compared to the coverage of C/C++ but on the other side sophisticated enough to cover most of the numerical kernels. Later in this work when we switch from just focusing on pure parallel region to general OpenMP parallel regions, we extend *SPL* step by step with constructs defined in the OpenMP 3.1 standard. These extensions will be called $SPLOMP^1$, $SPLOMP^2$, and so forth. With *SPL* we found the balance between defining a language that is sufficient enough to cover occurring codes in practices while making the verification proof of the source transformation not too difficult. The problem is that each possible statement must be considered as a possible input statement for the transformation and with the interleaving approach that is used to verify the correct parallel execution we examine combinations of possible statements. This indicates that with a full language support of C or C++, it is error-prone or almost infeasible to verify all of the possible outcomes.

SPL provides support of stack operations, where the stack is a data structure implemented as usual with a last in, first out data scheme (LIFO). The common methods push(), pop(), and top() can be used to insert, remove or access the data on the top of the stack. *SPL* is an extension of *SL* [57, ch. 4] which is the input and output language of dcc, our derivative code compiler. One extension is that statements associated with a stack operation are represented by an abstract statement instead of using a static C/C++ array as stack implementation. This allows later a customized implementation of these stacks. The second extension is that we allow assignments shaped like $y \mathrel{+\!\!\leftarrow} e$ instead of $y \leftarrow y + e$, where y is a scalar variable and e is an expression. This extension becomes clear as soon as we know that we need the atomic construct to achieve the closure property of the adjoint source transformation of a general OpenMP parallel region. Until OpenMP 3.0, the atomic construct was only defined for assignments shaped as $y \mathrel{+\!\!\leftarrow} e$ and we do not want to exclude potential codes that are written in an OpenMP version lower than 3.1.

Category	Range	Meta Symbols
Boolean values	$\{true, false\}$	B
Unsigned integer numbers	$\mathbb{N}_0 := \mathbb{N} \cup \{0\}$	N
Integer numbers	$\mathbb{Z}$	Z
Floating-point numbers	$\mathbb{R}$	R
(Unsigned) Integer variables	$\{a, b, c\}$	a
Derivative variables	$DerivVar := \left\{ x^{(1)}, x_{(1)}, x^{(2)}_{(1)}, \ldots \right\}$	$x^{(1)}, x_{(1)}, x^{(2)}_{(1)}$
Floating-point variables	$\{x_1, x_n, y, z, A_{i*j+n}\} \cup DerivVar$	$x, y, x^{(1)}, x_{(1)}, x^{(2)}_{(1)}$
Boolean expressions	$\{B, a < Z, a <= b, \ldots\}$	b
Integer expressions	$\{Z, a, e_1 + e_2, e_1 * e_2, e_1 \% e_2,$, omp_get_thread_num(), $STACK_i.top(), STACK_c.top()\}$	e
Floating-point expressions	$\{R, x, x_1 * x_2, x_1 + x_2, \sin(x) \ldots\}$	$\phi(x_1, \ldots, x_n)$

Table 2.1: This table lists the possible terminal symbols of the language *SPL*. A derivative variable is assumed to be uniquely associated with one variable, for example $x^{(1)}$ is tangent-linear component associated with x. We want to display left-hand side variables with y and right-hand side variables with x

Table 2.1 gives an overview of the set of terminal symbols, contained in *SPL*. Please note that we have to code variables, as for example $x^{(1)}$ or $x_{(1)}$, in some way that they are representable only with ASCII characters. Our AD tool does this by defining a prefix 't1_' or 'a1_'. This means a variable called a1_x is the derivative variable associated with x. Nevertheless, we want to remain this notation as implementation defined and we will further write $x_{(1)}$ instead of a1_x. Definition 40 presents the syntax of *SPL*.

A parallel region $P \in SPL$ consists of a sequence of declarations D and a sequence of statements S. The syntax is

$$P : \quad \#\text{pragma omp parallel} \quad \{ \quad D \quad S \quad \} \quad ,$$

with $D = (d_1; d_2; \ldots; d_r)$, $S = (s_1; s_2; \ldots; s_q)$, where $r, q \in \mathbb{N}$. Please note that we use the syntax of the OpenMP parallel construct despite the fact that we want to consider general parallel regions. This is just to avoid adjustments when we extend *SPL* in Chapter 4 by OpenMP directives. The omp can just be omitted or replaced by a different API name such as in #pragma acc parallel for the OpenACC syntax. Any declaration in D can be of the following kind:

$$d_i \in \begin{cases} \text{unsigned int } a \\ \text{unsigned int } a \leftarrow N \text{ , where } N \text{ is an integer number.} \\ \text{int } a \\ \text{int } a \leftarrow Z \text{ , where } Z \text{ is an integer number.} \\ \text{double } x \\ \text{double } x \leftarrow R \text{ , where } R \text{ is a floating-point number.} \end{cases}$$

As mentioned briefly in the previous section, the straight-line code statements are contained in the set *SLC* whereby the possible control flow statements are collected in *CFSTMT*. According to *SPL*, we define the set *CFSTMT* consisting of one conditional statement and one loop statement.

- if (b) $\{$ S' $\}$ $\in CFSTMT$,

- while (b) $\{$ S' $\}$ $\in CFSTMT$
  b is a boolean expression or a test that checks whether or not a certain element is on top of the stack STACK_c. S' is a nested sequence of statements.

The set *SLC* includes the following statements:

- $\text{STACK}_{(c|f|i)}.\text{push}(v) \in SLC$, v for storing some data v with $v \in \mathbb{N} \cup VAR_P$. $\text{STACK}_{(c|f|i)}$ is a thread local data structure.

- $\text{STACK}_{(c|f|i)}.\text{pop}() \in SLC$ for removing an element from the top of the stack,

- $a \leftarrow e \in SLC, a \in INT_P$, e is an expression of integer variables, a call to $\text{STACK}_i.\text{top}()$, or a call to a runtime library providing an integer value such as omp_get_thread_num() or omp_get_num_threads().

- $y \leftarrow \phi(x_1, \dots, x_n) \in SLC$, $y, x_{1,\dots,n} \in FLOAT_P$ are floating-point variables, ϕ is an intrinsic function as sin or cos, or an expression of floating-point variables. $\text{STACK}_f.\text{top}()$ is used to regain values from the floating-point value stack.

- $y \mathbin{+\!\!\!\leftarrow} \phi(x_1, \dots, x_n) \in SLC$
  As the floating-point assignment before, but the assigned value is the value of the evaluation of the right-hand side plus the value of the left-hand side reference before the assignment. We call this kind of statement an *incremental assignment*.

Definition 40 (Syntax of *SPL*). We denote P as the parallel region, D as a sequence of declarations, and S as a sequence of statements. A statement is referred to as s. We define the set V of non-terminal symbols as $V := \{P, D, d, S, s\}$. The context-free grammar $\mathscr{G}_{SPL} = (V, \Sigma, R, P)$ produces the language *SPL*. The set of terminal symbols Σ represents the set of ASCII[2] characters. R is the set of production rules and is defined as:

$$
\begin{aligned}
R := \{ \\
P \quad &: \quad \text{\#pragma omp parallel} \quad \{ \quad D \quad S \quad \} \\
D \quad &: \quad \varepsilon \\
&\mid \quad D \quad d \\
d \quad &: \quad \text{unsigned int } a; \\
&\mid \quad \text{int } a; \\
&\mid \quad \text{double } x; \\
&\mid \quad \text{unsigned int } a \leftarrow N; \\
&\mid \quad \text{int } a \leftarrow Z; \\
&\mid \quad \text{double } x \leftarrow R; \\
S \quad &: \quad s \\
&\mid \quad S \quad s \\
s \quad &: \quad \text{STACK}_i.\text{push}(a); \\
&\mid \quad \text{STACK}_f.\text{push}(x); \\
&\mid \quad \text{STACK}_c.\text{push}(N); \\
&\mid \quad \text{STACK}_i.\text{pop}(); \\
&\mid \quad \text{STACK}_f.\text{pop}(); \\
&\mid \quad \text{STACK}_c.\text{pop}(); \\
&\mid \quad a \leftarrow e; \\
&\mid \quad y \leftarrow \phi(x_1, \dots, x_n); \\
&\mid \quad y \mathbin{+\!\!\!\leftarrow} \phi(x_1, \dots, x_n);
\end{aligned}
\tag{2.7}
$$

---

[2] American Standard Code for Information Interchange

$$\mid \quad \text{if } (b) \ \{ \ S \ \}$$
$$\mid \quad \text{while } (b) \ \{ \ S \ \}$$
$$\}$$

P is the start symbol and ε is the empty word. $\qquad\qquad\qquad$ □

The following example shows the code from Example 4 rewritten into *SPL* code.

Example 20. In Example 4 we saw a typical pattern in parallel programming where we partition a data set into chunks and let each thread process a certain chunk. This code is shown here expressed as *SPL* code.

```
1  #pragma omp parallel
2  {
3      int i ← 0;
4      int tid ← 0;
5      int lb ← 0;
6      int ub ← 0;
7      int chunk_size ← 0;
8      tid ← omp_get_thread_num();
9      p ← omp_get_num_threads();
10     chunk_size ← n/p;
11     i ← chunk_size*p;
12     if(i ≠ n)
13     { chunk_size ← chunk_size+1; }
14     lb ← tid*chunk_size;
15     ub ← (tid+1)*chunk_size -1;
16     if(ub ≥ n)
17     { ub ← n-1; }
18     i ← lb;
19     while(i ≤ ub) {
20         yᵢ ← 2*xᵢ*xᵢ;
21         xᵢ ← 0;
22         i ← i+1;
23     }
24 }
```

The difference between this code and the one from Example 4 is quite small. Instead of the counting loop the current implementation uses a while-loop for processing the array. $\qquad\qquad\qquad\qquad\qquad\qquad\qquad\qquad\qquad\qquad\qquad$ △

Since we will very often speak about instances of assignments that are part of a concurrent execution, we display an example to present the notation. Let us consider two assignments that are executed by thread t and thread t'. We note this by writing

$$s_i^t = y_i^t \leftarrow \phi(x_{i,1}^t, \ldots, x_{i,n}^t)$$

and

$$s_j^{t'} = y_j^{t'} \leftarrow \phi(x_{j,1}^{t'}, \ldots, x_{j,n}^{t'}),$$

which means that both right-hand sides consist of an intrinsic function ϕ with n arguments. The intrinsic functions ϕ represents some intrinsic function not necessarily the same. The evaluated value of ϕ is assigned to the left-hand side reference y. The identifier x and y should only be considered as symbol names for declaring on which side of an assignment these symbols occur. The y_i^t does not need to be connected with $y_j^{t'}$ just because they have the same name. These kind of relation is expressed by an equation that defines the memory locations of two references as equal. For example, to express that the left-hand side reference of s_i^t also occurs on the right-hand side of $s_j^{t'}$, we write

$$\&y_i^t = \&x_{j,k}^{t'} \quad , k \in \{1, \ldots, n\}.$$

Each read access of a memory reference is assumed to be defined in the sense that each memory reference is initialized by an assignment before the first read access happens.

We introduced the mappings of a statement to its left-hand side, right-hand side, and conditional references in the previous section. Now, after defining *SPL*, we show the actual definitions of these mappings. Let us consider an instance s^t of statement $s \in STMTS$ that is executed by thread t. The set that contains the left-hand side reference is defined as

$$LHSREF(s^t) := \begin{cases} \{\&y^t\} & \text{if} \quad s = y \leftarrow \phi(x_1, \ldots, x_n) \\ \emptyset & \text{otherwise} \end{cases}$$

Analogously, we define the set of right-hand side references as

$$RHSREF(s^t) := \begin{cases} \{\&x_1^t, \&x_2^t, \ldots, \&x_n^t\} & \text{if} \quad s = y \leftarrow \phi(x_1, \ldots, x_n) \\ \emptyset & \text{otherwise} \end{cases} .$$

The references occurring in conditional statements, for example, $s = \text{if } (b) \{ S' \}$

as well as in loop statements $s =$ while (b) $\{$ S' $\}$ are contained in

$$CONDREF(s^t) := \begin{cases} \{\&v \mid v \in VAR_b\} & \text{if} \quad s \in CFSTMT_P, \text{ and } b \text{ is} \\ & \text{test expression in } s \\ \emptyset & \text{otherwise} \end{cases}.$$

In short, we regard a parallel region as a sequence of statements. The statements in this sequence can be integer or floating-point assignments, stack operations, branching, or they are loop statements. The straight-line code statements as well as the evaluation of a test expression in a control flow statement are considered as atomic.

The next section introduces the fundamental source transformations that we want to use throughout this work. These source transformations provide an implementation of the tangent-linear or the adjoint model as shown in the upcoming section as well.

2.3 AD Source Transformation of *SPL* Code

Algorithmic Differentiation by source transformation is basically a transformation of the assignments occurring in the input code. Each floating-point variable in the input code is associated with a derivative component that stores the derivative information of this variable. In Section 1.2, we have familiarized ourselves with the tangent-linear model and the adjoint model. The derivative component of variable x is indicated by a subscript or a superscript index in the generated code. The transformation in tangent-linear mode introduces a variable $x^{(1)}$ for the derivative information of x. In adjoint mode the generated code uses variable $x_{(1)}$ to store the derivative information of x.

In this section we will define a source transformation τ that provides the tangent-linear model and two source transformations σ and ρ, which serve for obtaining the adjoint model of a given *SPL* code. We assume that our input code is a parallel region P written in *SPL*. In this work we will only focus on how to transform this parallel region P into its derivative code pendant. The code outside of P is assumed to be in a form that an external AD tool is capable of providing the different derivative models. Therefore, we assume that the source transformations for the code before and after the parallel region P is given by an external tool and we only have to ensure that the transformation of P fits into the overall derivative code. The source transformation will be defined by rules for all the possible *SPL* statements. Afterwards, we show the correctness of these transformations by proving that the outcome implements in fact the tangent-linear or the adjoint model, respectively. Hereafter we show that the transformations have the closure property under certain restrictions to the input code.

In this work we assume that variables are uniquely associated with exactly one reference. We do not consider alias effects occurring when more than one symbol represents a certain memory location. In AD literature there is the term of *associated by name*, which means that each program variable x is uniquely associated with a variable x'. Variable x' represents the derivative information of variable x. From now on we will use names as $x^{(1)}$ or $x_{(1)}$ instead of x' for the derivative component of variable x. The '(1)' indicates that this is a first derivative component. Each floating-point variable $x \in FLOAT_P$ has an uniquely associated variable $x^{(1)}$ in case of the tangent-linear transformation. The adjoint transformation uniquely associates a variable x with $x_{(1)}$.

Suppose that the code of the parallel region P is contained in a code Q. The code of Q without the code of P is denoted by $Q \setminus P$. We assume that the derivative code of $Q \setminus P$ is given. The interface between the derivative codes of $Q \setminus P$ and the derivative code of P must be defined. This interface can be expressed by an equivalence between memory references.

We define that the tangent-linear model of Q fulfills the following equivalence considering the variables u and v and their tangent-linear associates $u^{(1)}$ and $v^{(1)}$.

$$\&u = \&v \Longleftrightarrow \&u^{(1)} = \&v^{(1)} \tag{2.8}$$

Analogously, we define that the adjoint model of Q fulfills

$$\&u = \&v \Longleftrightarrow \&u_{(1)} = \&v_{(1)} \tag{2.9}$$

where u, v are variables and $u_{(1)}, v_{(1)}$ are the corresponding adjoint variables. These properties define the interface between the derivative code of Q and the derivative code that we will achieve by the source transformations of this work. The following sections will introduce the AD source transformations of *SPL* code for achieving the tangent-linear and the adjoint model of a given code.

2.3.1 Tangent-Linear Model of *SPL* - Transformation $\tau(P)$

In this section we define the source transformation τ to obtain the tangent-linear model of a code written in *SPL* code. Suppose the mathematical function $F : \mathbb{R}^n \to \mathbb{R}^m$ is implemented by the parallel region $P \in SPL$. The source transformation $\tau : SPL \to SPL$ takes P as input code and provides $P' = \tau(P)$ as output. P is assumed to be

$$P = \#\text{pragma omp parallel} \quad \{ \quad D \quad S \quad \} \quad .$$

τ is then defined by:

$$\tau(P) := \#\text{pragma omp parallel} \quad \{ \quad \tau(D) \quad \tau(S) \quad \} \quad , \tag{2.10}$$

$$\tau(D) := \tau((d_1;\ldots;d_r)) := (\tau(d_1);\ldots;\tau(d_r)), \text{ where } D = (d_1;\ldots;d_r) \tag{2.11}$$

$$\tau(\text{unsigned int } a) := \text{unsigned int } a; \tag{2.12}$$

$$\tau(\text{int } a) := \text{int } a; \tag{2.13}$$

$$\tau(\text{double } x) := \begin{cases} \text{double } x; \\ \text{double } x^{(1)}; \end{cases} \tag{2.14}$$

$$\tau(\text{double } x \leftarrow R) := \begin{cases} \text{double } x \leftarrow R; \\ \text{double } x^{(1)} \leftarrow 0.; \end{cases} \tag{2.15}$$

$$\tau(S) := \tau((s_1;\ldots;s_q)) := (\tau(s_1);\ldots;\tau(s_q)), \text{ where } S = (s_1;\ldots;s_q) \tag{2.16}$$

$$\tau(y \leftarrow \phi(x_1,\ldots,x_n)) := \begin{cases} y^{(1)} \leftarrow \sum_{k=1}^{n} \phi_{x_k}(x_1,\ldots,x_n) \cdot x_k^{(1)}; \\ y \leftarrow \phi(x_1,\ldots,x_n); \end{cases} \tag{2.17}$$

$$\tau(y \mathrel{+\!\!\leftarrow} \phi(x_1,\ldots,x_n)) := \begin{cases} y^{(1)} \mathrel{+\!\!\leftarrow} \sum_{k=1}^{n} \phi_{x_k}(x_1,\ldots,x_n) \cdot x_k^{(1)}; \\ y \mathrel{+\!\!\leftarrow} \phi(x_1,\ldots,x_n); \end{cases} \tag{2.18}$$

$$\tau\left(y \leftarrow \text{STACK}_f.\text{top}()\right) := \begin{cases} y^{(1)} \leftarrow \text{STACK}_f^{(1)}.\text{top}(); \\ y \leftarrow \text{STACK}_f.\text{top}(); \end{cases} \tag{2.19}$$

$$\tau(\text{if } (b) \ \{ \ S' \ \}) := \text{if } (b) \ \{ \ \tau(S') \ \} \tag{2.20}$$

$$\tau(\text{while } (b) \ \{ \ S' \ \}) := \text{while } (b) \ \{ \ \tau(S') \ \} \tag{2.21}$$

$$\tau(a \leftarrow e) := a \leftarrow e \tag{2.22}$$

$$\tau(\text{STACK}_f.\text{push}(x)) := \begin{cases} \text{STACK}_f^{(1)}.\text{push}(x^{(1)}); \\ \text{STACK}_f.\text{push}(x); \end{cases} , \tag{2.23}$$

$$\tau(\text{STACK}_{(i|c)}.\text{push}(v)) := \text{STACK}_{(i|c)}.\text{push}(v); \qquad , v \in \mathbb{N} \cup \mathit{INT}_P \tag{2.24}$$

$$\tau(\text{STACK}_f.\text{pop}()) := \begin{cases} \text{STACK}_f^{(1)}.\text{pop}() \\ \text{STACK}_f.\text{pop}() \end{cases} \tag{2.25}$$

$$\tau(\text{STACK}_{(i|c)}.\text{pop}()) := \text{STACK}_{(i|c)}.\text{pop}() \tag{2.26}$$

Remark. In (2.17) we used the partial derivative of $\phi(x_1,\ldots,x_n)$ with respect to x_k with $k = 1,2,\ldots,n$. This can also be rewritten as

$$y^{(1)} \leftarrow \langle \nabla\phi(\mathbf{x}), \mathbf{x}^{(1)} \rangle, \tag{2.27}$$

where $\langle \nabla \phi(\mathbf{x}), \mathbf{x}^{(1)} \rangle$ is the inner product $\nabla \phi(\mathbf{x}) \cdot \mathbf{x}^{(1)}$, and $\mathbf{x}, \mathbf{x}^{(1)} \in \mathbb{R}^n$.

In [29] as well as in [57] the authors assume that the numerical program is given by a single assignment code (SAC) and they show the corresponding tangent-linear model [57, ch. 2]. In our case, we have Similarly, we have to show that the result of the transformation of $\tau(P)$ implements the tangent-linear model of F.

The following proposition shows that the source transformation $\tau(P)$ provides the tangent-linear model of P.

Proposition 41. *Given a parallel region code $P \in SPL$ that implements a mathematical function $F : \mathbb{R}^n \to \mathbb{R}^m$, then the source transformation $\tau(P)$ implements $F^{(1)}$.*

Proof. We have to show that $\tau(P)$ implements the tangent-linear model $F^{(1)}$ of F with

$$F^{(1)}(\mathbf{x}, \mathbf{x}^{(1)}) \equiv \nabla F(\mathbf{x}) \cdot \mathbf{x}^{(1)} \in \mathbb{R}^m.$$

Let us assume that P consists of q statements and P is processed by p threads. The possible set of interleavings is then $\mathscr{I}(P, q, p)$. Since P implements F, the computation terminates, and the length of an interleaving $I \in \mathscr{I}(P, q, p)$ is finite. We consider an arbitrary interleaving $I \in \mathscr{I}(P, q, p)$ of length $r < \infty$. In case of P containing only straight-line code statements, r is equal to pq, but in general P contains conditional statements and loop statements as well. This means that the computation of the parallel region P can be expressed by an interleaving I with r instances of atomic statements in execution:

$$I = \left(s_{i,1}^t; s_{i,2}^t; \ldots; s_{i,r}^t \right) \quad , i \in \{1, \ldots, q\}, t \in \{0, \ldots, p-1\} \quad .$$

This means the sequence of r statements provides the m output values of $F(\mathbf{x})$. Since the interleaving does not only contain assignments but also statements where a test expression is evaluated, we define r' to be the number of assignments contained in the interleaving I where $r' \leq r$. In general there are more than m assignments involved in the computation of the m output values, therefore it holds $r' \geq m$.

We conclude that there is a sequence of r' floating-point assignments that determines the m output values $F(\mathbf{x})$. Suppose that the following sequence of assignments $y_{1,\ldots,r'} \in SLC$ reflects the computation of the outputs. The association of what variable y_j with $j \in \{1, \ldots, r'\}$ defines which output value is determined by the implementation. The computation expressed in r' assignments can be expressed as a SAC:

$$y_1 \leftarrow \phi^1(x_1,\ldots,x_n)$$
$$y_2 \leftarrow \phi^2(x_1,\ldots,x_n)$$
$$\vdots$$
$$y_{r'} \leftarrow \phi^{r'}(x_1,\ldots,x_n),$$

The $\phi^{\{1,\ldots,r'\}}$ are intrinsic scalar functions.

In the following we express the value propagation from the input values to the output values by a program state transformation. Each program state consists of $n + r'$ states where the first n states represent the n input values. The remaining r' states represent the results of the r' assignments $y_{1,2,\ldots,r'}$. The first program state is the one where only the first n input values are set. The program state transformation is performed by $\Phi^i : \mathbb{R}^{n+r'} \to \mathbb{R}^{n+r'}$ with

$$(\Phi^i(\mathbf{z}))_j := \begin{cases} \phi^i(x_1,x_2,\ldots,x_n) & \text{, if } i = j \\ \mathbf{z}_j & \text{, otherwise} \end{cases}, \qquad (2.28)$$

where $j = 1,\ldots,n+r', \mathbf{z} = (x_1,\ldots,x_n,y_1,\ldots,y_{r'}) \in \mathbb{R}^{n+r'}$. The function Φ^i takes the current program state $\mathbf{z}_{i-1}$ and provides the new program state $\mathbf{z}_i$ where the state only differs in the i-th component. Hence, the computation displayed as a sequence of state transformations is

$$\mathbf{z}_1 \leftarrow \Phi^1(\mathbf{z}_0)$$
$$\mathbf{z}_2 \leftarrow \Phi^2(\mathbf{z}_1)$$
$$\vdots$$
$$\mathbf{z}_{r'} \leftarrow \Phi^{r'}(\mathbf{z}_{r'-1}).$$

The state $\mathbf{z}_0$ is the initial state where only the input values are set. The final state $\mathbf{z}_{r'}$ reflects all the output values after executing the whole computation. Therefore, all the output values of $F(\mathbf{x})$ are contained in the vector $\mathbf{z}_{r'}$ what we denote by $F(\mathbf{x}) \subset \mathbf{z}_{r'}$. All the intermediate states of the computation are indicated by the sequence $\mathbf{z}_1,\mathbf{z}_2,\ldots,\mathbf{z}_{r'-1}$. When $\Phi^i \circ \Phi^{i-1}$ displays the consecutive application of Φ^i after Φ^{i-1} we can write the computation as consecutive application of $\Phi^{r'},\ldots,\Phi^1$ and this is referred to as $\mathscr{F}(\mathbf{z}_0)$. It holds

$$F(\mathbf{x}) \subset \mathbf{z}_{r'} \equiv \mathscr{F}(\mathbf{z}_0) \tag{2.29}$$

$$\equiv \Phi^{r'} \circ \ldots \circ \Phi^1(\mathbf{z}_0) \tag{2.30}$$

$$\equiv \Phi^{r'}(\Phi^{r'-1}(\ldots \Phi^1(\mathbf{z}_0)\ldots)).$$

We perform now the transition from the interleaving I of the execution of P to the execution of $\tau(P)$. The transformation rules (2.20) and (2.21) show that the control flow of P is inherited to $\tau(P)$ through the identity mapping of the corresponding test expressions. Therefore, the exact same order of computations $y_{1,\ldots,r'}$ exists also in the set of possible interleavings but augmented with derivative computations.
We choose the existing interleaving $I' \in \mathscr{I}(\tau(P), q', p)$ with

$$I' = \left(\tau\left(s_{i,1}^t\right) ; \tau\left(s_{i,2}^t\right) ; \ldots ; \tau\left(s_{i,r}^t\right) \right)$$

where $i \in \{1,\ldots,q\}, t \in \{0,\ldots,p-1\}$, that represents the same order of computations of $y_{1,\ldots,r'}$ as in the interleaving I. The number of statements q' in the tangent-linear model is obviously bigger than the number of statements q in the parallel region P. By applying the source transformation rule (2.17) to I', we obtain

$$y_1^{(1)} \leftarrow \left\langle \nabla\phi^1(x_1,\ldots,x_n), \left(x_1^{(1)},\ldots,x_n^{(1)}\right)^T \right\rangle$$

$$y_1 \leftarrow \phi^1(x_1,\ldots,x_n)$$

$$y_2^{(1)} \leftarrow \left\langle \nabla\phi^2(x_1,\ldots,x_n), \left(x_1^{(1)},\ldots,x_n^{(1)}\right)^T \right\rangle$$

$$y_2 \leftarrow \phi^2(x_1,\ldots,x_n)$$

$$\vdots$$

$$y_{r'}^{(1)} \leftarrow \left\langle \nabla\phi^{r'}(x_1,\ldots,x_n), \left(x_1^{(1)},\ldots,x_n^{(1)}\right)^T \right\rangle$$

$$y_{r'} \leftarrow \phi^{r'}(x_1,\ldots,x_n).$$

where

$$\left(x_1^{(1)},\ldots,x_n^{(1)}\right)^T = \begin{pmatrix} x_1^{(1)} \\ x_2^{(1)} \\ \vdots \\ x_n^{(1)} \end{pmatrix}$$

In addition, we used the first-order tangent-linear projection as defined in Definition 7.

For expressing this by a sequence of state transformations we need, besides the $\mathbf{z}$ vector, a state vector that reflects the current state of the derivative values. This is performed by the vector

$$\mathbf{z}^{(1)} = (x_1^{(1)}, \ldots, x_n^{(1)}, y_1^{(1)}, \ldots, y_{r'}^{(1)}) \in \mathbb{R}^{n+r'}.$$

$\mathbf{z}_0^{(1)}$ represents the program state before the execution and $\mathbf{z}_i^{(1)}$ is the program state after the evaluation the i-th tangent-linear projection. The execution of the tangent-linear code can be expressed by a program state transformation as

$$\mathbf{z}_1^{(1)} \leftarrow \left\langle \nabla \Phi^1(\mathbf{z}_0), \mathbf{z}_0^{(1)} \right\rangle$$

$$\mathbf{z}_1 \leftarrow \Phi^1(\mathbf{z}_0)$$

$$\mathbf{z}_2^{(1)} \leftarrow \left\langle \nabla \Phi^2(\mathbf{z}_1), \mathbf{z}_1^{(1)} \right\rangle$$

$$\mathbf{z}_2 \leftarrow \Phi^2(\mathbf{z}_1)$$

$$\vdots$$

$$\mathbf{z}_{r'}^{(1)} \leftarrow \left\langle \nabla \Phi^{r'}(\mathbf{z}_{r'-1}), \mathbf{z}_{r'-1}^{(1)} \right\rangle$$

$$\mathbf{z}_{r'} \leftarrow \Phi^{r'}(\mathbf{z}_{r'-1}).$$

The resulting vector $\mathbf{z}_q^{(1)}$ can be expressed as a chain of matrix-matrix multiplications and a matrix-vector multiplication.

$$\mathbf{z}_{r'}^{(1)} \equiv \left\langle \nabla \Phi^{r'}(\mathbf{z}_{r'-1}), \mathbf{z}_{r'-1}^{(1)} \right\rangle$$

$$\equiv \left\langle \nabla \Phi^{r'}(\mathbf{z}_{r'-1}), \left\langle \nabla \Phi^{r'-1}(\mathbf{z}_{r'-2}), \mathbf{z}_{r'-2}^{(1)} \right\rangle \right\rangle$$

$$\equiv \ldots$$

$$\equiv \nabla \Phi^{r'}(\mathbf{z}_{r'-1}) \cdot \nabla \Phi^{r'-1}(\mathbf{z}_{r'-2}) \cdot \ldots \cdot \nabla \Phi^1(\mathbf{z}_0) \cdot \mathbf{z}_0^{(1)}$$

With $\mathbf{z}_i \equiv \Phi^i \circ \Phi^{i-1} \circ \ldots \circ \Phi^1(\mathbf{z}_0)$ we can replace the Jacobian arguments by a consecutive application of the function Φ.

$$\mathbf{z}_{r'}^{(1)} \equiv \left\langle \nabla \Phi^{r'}(\Phi^{r'-1} \circ \ldots \circ \Phi^1(\mathbf{z}_0)), \ldots, \nabla \Phi^1(\mathbf{z}_0), \mathbf{z}_0^{(1)} \right\rangle$$

The matrix product

$$A = \nabla \Phi^{r'}(\mathbf{z}_{r'-1}) \cdot \nabla \Phi^{r'-1}(\mathbf{z}_{r'-2}) \cdot \ldots \cdot \nabla \Phi^1(\mathbf{z}_0)$$

results in a matrix $A \in \mathbb{R}^{(n+r') \times (n+r')}$ where we only consider the lines which define the m output values. Without loss of generality, we assume that the last m assignments $y_{r'-m+1}, y_{r'-m+2}, \ldots, y_{r'}$ define the m output values. Therefore, we consider a submatrix inside of A, namely the last m lines and the first n columns. This submatrix contain exactly the components of the Jacobian matrix $\nabla F(\mathbf{x})$. The first n components in $\mathbf{z}_0^{(1)}$ are $x_1^{(1)}, \ldots, x_n^{(1)}$. Thus, the submatrix and the first n components of $\mathbf{z}_0^{(1)}$ define the tangent-linear values $y_{r'-m+1}^{(1)}, y_{r'-m+2}^{(1)}, \ldots, y_{r'}^{(1)}$ where

$$
\begin{pmatrix}
y_{r'-m+1}^{(1)} \\
y_{r'-m+2}^{(1)} \\
\vdots \\
y_{r'}^{(1)}
\end{pmatrix}
\equiv \nabla F(\mathbf{x}) \cdot \mathbf{x}^{(1)}.
$$

$\square$

Example 21. We apply the source transformation τ to the code from Example 20. The resulting code is the following.

```
 1 #pragma omp parallel
 2 {
 3    int i ← 0;
 4    int tid ← 0;
 5    int lb ← 0;
 6    int ub ← 0;
 7    int chunk_size ← 0;
 8    tid ← omp_get_thread_num();
 9    p ← omp_get_num_threads();
10    chunk_size ← n/p;
11    i ← chunk_size*p;
12    if(i ≠ n)
13    { chunk_size ← chunk_size+1; }
14    lb ← tid*chunk_size;
15    ub ← (tid+1)*chunk_size-1;
16    if(ub ≥ n)
17    { ub ← n-1; }
18    i ← lb;
19    while(i ≤ ub) {
20        y_i^(1) ← 2*x_i*x_i^(1)+2*x_i*x_i^(1);
21        y_i ← 2*x_i*x_i;
22        x_i^(1) ← 0;
```

```
23      xᵢ ← 0;
24      i ← i+1;
25    }
26  }
```

The derivative code looks quite similar to the input code except for the two additional assignments in line 20 and line 22. The assignment in line 20 represents the tangent-linear model of the assignment in line 21. Analogously, the tangent-linear model of the assignment in line 23 is shown in line 22. $\triangle$

We finish this section with a lemma that shows that the source transformation $\tau(P)$ fulfills the interface requirements that we formulated in (2.8). This equivalence will play an important role when we prove that the parallel region of the tangent-linear model is noncritical.

Lemma 42. *Suppose $u, v \in FLOAT_P$ are two variables occuring in a parallel region $P \in SPL$, and $\tau(P)$ is the tangent-linear model of P. Then it holds for the variables $u, v, u^{(1)}, v^{(1)} \in \tau(P)$:*

$$\&u = \&v \Longleftrightarrow \&u^{(1)} = \&v^{(1)} \tag{2.31}$$

Proof. In case that u, v are defined outside of P, the equivalence is given because we determined the interface in such a way, see (2.8). Otherwise, the definitions of u, v are contained in P and the association by name is achieved by the transformation rules (2.14) and (2.15). Assume that a variable $v \in FLOAT_{\tau(P)}$ is defined in the sequence of definitions D contained in $\tau(P)$. This definition is unique since in *SPL* we only allow definitions on the scope level of the parallel region and not below as it is possible in C/C++. However, the definition of v inside of the tangent-linear code is always connected with the definition of the corresponding derivative component $v^{(1)}$. Thus, the equivalence in (2.31) is true. $\square$

This section introduced the source transformation $\tau(P)$ of a given parallel region P. $\tau(P)$ represents the tangent-linear model of P as we showed in the proof of Proposition 41. The next section introduces the source transformation $\sigma(P)$ that provides the adjoint model of P.

2.3.2 Adjoint Model of *SPL* - Transformation $\sigma(P)$

Since the adjoint model requires a reversal of the dataflow, the adjoint transformation is not as straightforward as the tangent-linear transformation was. In fact, we use two transformations, σ and ρ. In addition, we need a separation of the code into sequences of *SLC* and *CFSTMT* statements. Consider, for example, the sequence $(s_1; \ldots; s_j; \ldots; s_q)$, where statement s_j is a control flow statement $(s_j \in CFSTMT)$ and all other statements are assignments or stack operations and therefore contained in *SLC*. In this case, we split the transformation and apply the transformation to $(s_1; \ldots; s_{j-1})$, followed by the transformation of the control flow statement s_j and finally the transformation of $(s_{j+1}; \ldots; s_q)$.

The source transformations $\sigma : SPL \to SPL$ and $\rho : SPL \to SPL$ are used for different parts of the output code. Transformation σ provides the forward section of the adjoint code and the transformation ρ is used to obtain the reverse section. We define the source transformations by using the non-terminal and terminal symbols of $\mathscr{G}_{SPL}$ from (2.7). The σ transformation is defined on all levels of our grammar $\mathscr{G}_{SPL}$ with the non-terminal symbol P as its start symbol. The transformation ρ on the other hand is not defined for the non-terminal symbols P and D but rather starts on the level of a sequence of statements S. Suppose an input code $P \in SPL$ is given with

$$P = \#\text{pragma omp parallel} \quad \{ \quad D \quad S \quad \}$$

We obtain the adjoint model of P by transforming P with σ:

$$\sigma(P) := \begin{cases} \#\text{pragma omp parallel} \\ \{ \\ \quad \sigma(D) \\ \quad \sigma(S) \\ \quad \text{while} \left(\text{not STACK}_{(1)c}.\text{empty()} \right) \{ \rho(S) \} \\ \} \end{cases} \tag{2.32}$$

The transformation of the declaration part D is performed by

$$\sigma(D) := \sigma((d_1; \ldots; d_r)) := (\sigma(d_1); \ldots; \sigma(d_r)) \tag{2.33}$$

$$\sigma(\text{unsigned int } a) := \text{unsigned int } a; \tag{2.34}$$

$$\sigma(\text{int } a) := \text{int } a; \tag{2.35}$$

$$\sigma(\text{double } x) := \begin{cases} \text{double } x; \\ \text{double } x_{(1)} \leftarrow 0.; \end{cases} \tag{2.36}$$

$$\sigma(\text{double } x \leftarrow R) := \begin{cases} \text{double } x \leftarrow R; \\ \text{double } x_{(1)} \leftarrow 0.; \end{cases} \quad (2.37)$$

The forward section or augmented forward section of the adjoint code is provided by the transformation $\sigma(S)$ in (2.32). As you will see, the forward section is
quite similar to the code in P, but augmented by statements for storing the control
flow of the current execution and by statements for storing values that are being
overwritten. The data structure for storing these values is the $\text{STACK}_{(1)c}$ in case of
the control flow, floating-point values are stored in $\text{STACK}_{(1)f}$, and integer values
are put into $\text{STACK}_{(1)i}$. The reversal of the control flow during the reverse section
is done by consuming all the labels contained in $\text{STACK}_{(1)c}$. The floating-point as
well as the integer values that are overwritten during the forward section must be
stored in order to allow the reverse section to regain these values. In (2.32), the
reverse section is represented as

$$\text{while} \left(\text{not } \text{STACK}_{(1)c}.\text{empty}() \right) \{ \rho(S) \},$$

where the loop is responsible for consuming all the labels from the control flow
stack $\text{STACK}_{(1)c}$. Inside the loop, $\rho(S)$ emits a sequence of branching statements
in order to execute the adjoint statements that correspond to the current label.

For transforming a sequence of statements, we distinguish between the cases
where the whole sequence is *SLC* code or not. In case that there is a conditional
statement or a loop statement, the sequence is split up into parts:

1. If $S = (s_1; \ldots; s_j; \ldots; s_q)$ with

$$(s_1; \ldots; s_{j-1}) \in SLC,$$
$$s_j \in CFSTMT, \text{ and}$$
$$(s_{j+1}; \ldots; s_q) \in SLC$$

This means there is a control flow statement enclosed by straight-line code
sequences. The transformation is defined through:

$$\sigma\left((s_1; \ldots; s_q) \right) := \begin{cases} \sigma\left((s_1; \ldots; s_{j-1}) \right) \\ \sigma(s_j) \\ \sigma\left((s_{j+1}; \ldots; s_q) \right) \end{cases} \quad (2.38)$$

Please note that in (2.38) the arguments of

$$\sigma\left((s_1; \ldots; s_{j-1}) \right)$$

and
$$\sigma\left((s_{j+1};\ldots;s_q)\right)$$
are provided as a sequence unlike the argument of $\sigma(s_j)$ that is a statement. This fact is important even when only one statement is succeeding the control flow statement. For instance, in case of $j+1=q$, only one *SLC* statement follows the control flow statement. Nevertheless, this final statement is first transformed as sequence with $\sigma((s_q))$ and through the recursive definition, σ is applied to the single statement per $\sigma(s_q)$.

2. If $S = (s_1;\ldots;s_q) \in SLC$, this means there is no control flow statement present:

$$\sigma\left((s_1;\ldots;s_q)\right) := \begin{cases} \text{STACK}_{(1)c}.\text{push}(l); \\ \sigma(s_1); \\ \vdots \\ \sigma(s_q); \end{cases} \quad , \quad (2.39)$$

where $l = LABEL(s_1)$.

The individual statement transformations are:

$$\sigma\left(y \leftarrow \phi(x_1,\ldots,x_n)\right) := \begin{cases} \text{STACK}_{(1)f}.\text{push}(y); \\ y \leftarrow \phi(x_1,\ldots,x_n); \end{cases} \quad (2.40)$$

$$\sigma\left(y \mathrel{+\!\!+\!\!\leftarrow} \phi(x_1,\ldots,x_n)\right) := \begin{cases} \text{STACK}_{(1)f}.\text{push}(y); \\ y \mathrel{+\!\!+\!\!\leftarrow} \phi(x_1,\ldots,x_n); \end{cases} \quad (2.41)$$

$$\sigma\left(y \leftarrow \text{STACK}_f.\text{top}()\right) := \begin{cases} \text{STACK}_{(1)f}.\text{push}(y); \\ y \leftarrow \text{STACK}_f.\text{top}(); \end{cases} \quad (2.42)$$

$$\sigma\left(\text{if } (b) \ \{ \ S' \ \}\right) := \text{if } (b) \ \{ \ \sigma(S') \ \} \quad (2.43)$$

$$\sigma\left(\text{while } (b) \ \{ \ S' \ \}\right) := \text{while } (b) \ \{ \ \sigma(S') \ \} \quad (2.44)$$

$$\sigma\left(a \leftarrow e\right) := \begin{cases} \text{STACK}_{(1)i}.\text{push}(a); \\ a \leftarrow e \end{cases} \quad (2.45)$$

$$\sigma\left(\text{STACK}_{(f|i|c)}.\text{push}(v)\right) := \text{STACK}_{(f|i|c)}.\text{push}(v) \quad , v \in \mathbb{N} \cup VAR_P \quad (2.46)$$

$$\sigma\left(\text{STACK}_{(f|i|c)}.\text{pop}()\right) := \text{STACK}_{(f|i|c)}.\text{pop}() \quad (2.47)$$

Let us now switch to the transformation ρ that generates the code for computing the adjoint values. As with the transformation σ, we have two different cases depending on whether or not the code contains control flow statements.

1. If $S = (s_1; \ldots; s_j; \ldots; s_q)$, with

$$(s_1; \ldots; s_{j-1}) \in SLC,$$
$$s_j \in CFSTMT, \text{ and}$$
$$(s_{j+1}; \ldots; s_q) \in SLC :$$

$$\rho\Big((s_1; \ldots; s_q) \Big) := \begin{cases} \rho\Big((s_1; \ldots; s_{j-1}) \Big); \\ \rho(s_j); \\ \rho\Big((s_{j+1}; \ldots; s_q) \Big); \end{cases} \tag{2.48}$$

2. If $S = (s_1; \ldots; s_q) \in SLC$:

$$\rho\Big((s_1; \ldots; s_q) \Big) := \begin{cases} & \text{if } \Big(\text{STACK}_{(1)c}.\text{top}() = l \Big) \ \{ \\ & \quad \text{STACK}_{(1)c}.\text{pop}(); \\ & \quad \rho(s_q); \\ & \quad \vdots \\ & \quad \rho(s_1); \\ & \} \end{cases} \tag{2.49}$$

where $l = LABEL(s_1)$. The reader should notice the reversed order of the statements[3] as opposed to the order used in the argument.

We conclude the definition of the transformation for the reverse section with the

---

[3]These statements are actually assignments or stack operations due to the *SLC* characteristics of this sequence.

rules for individual statements.

$$\rho\left(y \leftarrow \phi(x_1,\ldots,x_n)\right)$$

$$:= \begin{cases} y \leftarrow \text{STACK}_{(1)f}.\text{top}(); \\ \text{STACK}_{(1)f}.\text{pop}(); \\ x_{(1)1} +\!\!\leftarrow \phi_{x_1}(x_1,\ldots,x_n) \cdot y_{(1)}; \\ \vdots \\ x_{(1)n} +\!\!\leftarrow \phi_{x_n}(x_1,\ldots,x_n) \cdot y_{(1)}; \\ y_{(1)} \leftarrow 0; \end{cases} \qquad (2.50)$$

$$\rho\left(y +\!\!\leftarrow \phi(x_1,\ldots,x_n)\right)$$

$$:= \begin{cases} y \leftarrow \text{STACK}_{(1)f}.\text{top}(); \\ \text{STACK}_{(1)f}.\text{pop}(); \\ x_{(1)1} +\!\!\leftarrow \phi_{x_1}(x_1,\ldots,x_n) \cdot y_{(1)}; \\ \vdots \\ x_{(1)n} +\!\!\leftarrow \phi_{x_n}(x_1,\ldots,x_n) \cdot y_{(1)}; \end{cases} \qquad (2.51)$$

$$\rho\left(y \leftarrow \text{STACK}_f.\text{top}()\right)$$

$$:= \begin{cases} y \leftarrow \text{STACK}_{(1)f}.\text{top}(); \\ \text{STACK}_{(1)f}.\text{pop}(); \\ y_{(1)} \leftarrow 0; \end{cases} \qquad (2.52)$$

$$\rho\left(\text{if } (b) \; \{ \; S' \; \} \right) := \rho(S') \qquad (2.53)$$

$$\rho\left(\text{while } (b) \; \{ \; S' \; \} \right) := \rho(S') \qquad (2.54)$$

$$\rho\left(a \leftarrow e\right) := \begin{cases} a \leftarrow \text{STACK}_{(1)i}.\text{top}(); \\ \text{STACK}_{(1)i}.\text{pop}(); \end{cases} \qquad (2.55)$$

$$\rho\left(\text{STACK}_{(f|i|c)}.\text{push}(c)\right) := \varepsilon \qquad (2.56)$$

$$\rho\left(\text{STACK}_{(f|i|c)}.\text{pop}()\right) := \varepsilon, \qquad (2.57)$$

With ε we denote the empty word, which means that the rules (2.55), (2.56), and (2.57) do not emit any code. The next proposition shows that the above source transformation provides code that implements the adjoint model.

Proposition 43. *Given a parallel region code $P \in SPL$ that implements a mathematical function $F : \mathbb{R}^n \to \mathbb{R}^m$, then the source transformation $\sigma(P)$ implements the adjoint model $F_{(1)}$.*

Proof. As in the proof of Proposition 41, we assume that P consists of q statements and P is processed by p threads. We consider an arbitrary interleaving $I \in \mathscr{I}(P,q,p)$ of length $r < \infty$ with

$$I = \left(s_{i,1}^t ; s_{i,2}^t ; \ldots ; s_{i,r}^t \right)$$

where $i \in \{1,\ldots,q\}$, and $t \in \{0,\ldots,p-1\}$. This computation supplies the m output values of $F(\mathbf{x})$ with $r' < r$ assignments. These r' assignments can be expressed as a SAC:

$$y_1 \leftarrow \phi^1(x_1,\ldots,x_n)$$
$$y_2 \leftarrow \phi^2(x_1,\ldots,x_n)$$
$$\vdots$$
$$y_{r'} \leftarrow \phi^{r'}(x_1,\ldots,x_n) \quad .$$

Each assignment belongs to a certain basic block contained in the parallel region P. We assume without loss of generality that each assignment is contained in a different basic block. Therefore, each assignment is tracked by a push operation to the control flow stack during the forward section. In practice this usually is not the case, and we only need one push operation per basic block to trace the flow of control.

The forward section emitted by $\sigma(S)$ looks like:

$$\text{STACK}_{(1)c}.\text{push}(1);$$

$$\text{STACK}_{(1)f}.\text{push}(y_1);$$

$$y_1 \leftarrow \phi^1(x_1,\dots,x_n);$$

$$\text{STACK}_{(1)c}.\text{push}(2);$$

$$\text{STACK}_{(1)f}.\text{push}(y_2);$$

$$y_2 \leftarrow \phi^2(x_1,\dots,x_n);$$

$$\vdots$$

$$\text{STACK}_{(1)c}.\text{push}(r');$$

$$\text{STACK}_{(1)f}.\text{push}(y_{r'});$$

$$y_{r'} \leftarrow \phi^{r'}(x_1,\dots,x_n);$$

The stack for the control flow contains at the end of the forward section the labels $r', r'-1,\dots,1$ (r' at the top of the stack) and the floating-point stack contains the values $y_{r'}, y_{r'-1},\dots,y_1$ (y'_r at the top of the stack). The reverse section looks as follows.

$$\text{while (not STACK}_{(1)c}.\text{empty())} \{$$

$$\text{if (STACK}_{(1)c}.\text{top()} = 1) \{$$

$$\text{STACK}_{(1)c}.\text{pop()};$$

$$y_1 \leftarrow \text{STACK}_{(1)f}.\text{top()};$$

$$\text{STACK}_{(1)f}.\text{pop()};$$

$$x_{(1)1} \mathrel{+\!\!\leftarrow} \phi_{x_1}(x_1,\dots,x_n)\cdot y_{(1)1};$$

$$\vdots$$

$$x_{(1)n} \mathrel{+\!\!\leftarrow} \phi_{x_n}(x_1,\dots,x_n)\cdot y_{(1)1};$$

$$y_{(1)1} \leftarrow 0;$$

$$\}$$

$$\text{if (STACK}_{(1)c}.\text{top()} = 2) \{$$

$$\text{STACK}_{(1)c}.\text{pop()};$$

$$y_2 \leftarrow \text{STACK}_{(1)f}.\text{top}();$$

$$\text{STACK}_{(1)f}.\text{pop}();$$

$$x_{(1)1} \mathrel{+\!\!\leftarrow} \phi_{x_1}(x_1,\ldots,x_n) \cdot y_{(1)2};$$

$$\vdots$$

$$x_{(1)n} \mathrel{+\!\!\leftarrow} \phi_{x_n}(x_1,\ldots,x_n) \cdot y_{(1)2};$$

$$y_{(1)2} \leftarrow 0;$$

    }

$$\vdots$$

if $(\text{STACK}_{(1)c}.\text{top}() = r')$ {

    $\text{STACK}_{(1)c}.\text{pop}();$

    $y_q \leftarrow \text{STACK}_{(1)f}.\text{top}();$

    $\text{STACK}_{(1)f}.\text{pop}();$

    $x_{(1)1} \mathrel{+\!\!\leftarrow} \phi_{x_1}(x_1,\ldots,x_n) \cdot y_{(1)r'};$

$$\vdots$$

    $x_{(1)n} \mathrel{+\!\!\leftarrow} \phi_{x_n}(x_1,\ldots,x_n) \cdot y_{(1)r'};$

    $y_{(1)r'} \leftarrow 0;$

    }

}

The n adjoint assignments in branch i with $x_{(1)1},\ldots,x_{(1)n}$ on their left-hand side can be interpreted as a product of the transposed gradient of $\phi(x_1,\ldots,x_n)$ with the scalar $y_{(1)i}$. Therefore, we rewrite the n assignments into one assignment using a vector notation. We obtain

while (not $\text{STACK}_{(1)c}.\text{empty}()$) {

    if $(\text{STACK}_{(1)c}.\text{top}() = 1)$ {

        $\text{STACK}_{(1)c}.\text{pop}();$

        $y_1 \leftarrow \text{STACK}_{(1)f}.\text{top}();$

        $\text{STACK}_{(1)f}.\text{pop}();$

$$
\begin{pmatrix} x_{(1)1} \\ x_{(1)2} \\ \vdots \\ x_{(1)n} \end{pmatrix} \leftarrow \begin{pmatrix} x_{(1)1} \\ x_{(1)2} \\ \vdots \\ x_{(1)n} \end{pmatrix} + (\nabla \phi(x_1, \ldots, x_n))^T \cdot y_{(1)1}
$$

$y_{(1)1} \leftarrow 0;$

}

if $(\text{STACK}_{(1)c}.\text{top}() = 2)$ {

   $\text{STACK}_{(1)c}.\text{pop}();$

   $y_2 \leftarrow \text{STACK}_{(1)f}.\text{top}();$

   $\text{STACK}_{(1)f}.\text{pop}();$

$$
\begin{pmatrix} x_{(1)1} \\ x_{(1)2} \\ \vdots \\ x_{(1)n} \end{pmatrix} \leftarrow \begin{pmatrix} x_{(1)1} \\ x_{(1)2} \\ \vdots \\ x_{(1)n} \end{pmatrix} + (\nabla \phi(x_1, \ldots, x_n))^T \cdot y_{(1)2}
$$

   $y_{(1)2} \leftarrow 0;$

}

$\vdots$

if $(\text{STACK}_{(1)c}.\text{top}() = r')$ {

   $\text{STACK}_{(1)c}.\text{pop}();$

   $y_{r'} \leftarrow \text{STACK}_{(1)f}.\text{top}();$

   $\text{STACK}_{(1)f}.\text{pop}();$

$$
\begin{pmatrix} x_{(1)1} \\ x_{(1)2} \\ \vdots \\ x_{(1)n} \end{pmatrix} \leftarrow \begin{pmatrix} x_{(1)1} \\ x_{(1)2} \\ \vdots \\ x_{(1)n} \end{pmatrix} + (\nabla \phi(x_1, \ldots, x_n))^T \cdot y_{(1)r'}
$$

   $y_{(1)r'} \leftarrow 0;$

}

}

As in the proof of Proposition 41, we use the extended function Φ to express the program state propagation along the execution of the code. Each

$$\mathbf{z} = (x_1, x_2, \ldots, x_n, y_1, \ldots, y_{r'})^T \in \mathbb{R}^{n+r'}$$

begins with n components that represent the input variables of P. Subsequently, the r' components reflect the results of the r' assignments. The first assignment defines the value of the $(n+1)$-th component in $\mathbf{z}$, the second assignment the $(n+2)$-th component and so forth. In vector $\mathbf{z}_0$, only the first n components are initialized with the input values of P. Vector $\mathbf{z}_1$ differs from $\mathbf{z}_0$ only in the $(n+1)$-th component as it contains the result of the first assignment y_1. Vector $\mathbf{z}_{r'}$ on the other hand contains all the results from the r' assignments and represents the end-state of the execution of P.

$$\mathbf{z}_1 \leftarrow \Phi^1(\mathbf{z}_0)$$
$$\mathbf{z}_2 \leftarrow \Phi^2(\mathbf{z}_1)$$
$$\vdots$$
$$\mathbf{z}_{r'} \leftarrow \Phi^{r'}(\mathbf{z}_{r'-1}) \tag{2.58}$$

The forward section ends up in the same program state as the original code, but in addition the stacks contain the following information where the topmost element on the stack is the leftmost element in the notation with a tuple:

$$\text{STACK}_{(1)c} = (r', r'-1, \ldots, 2, 1)$$
$$\text{STACK}_{(1)f} = (y_{r'}, y_{r'-1}, \ldots, y_2, y_1).$$

This means r' is the label on the top of the control flow stack. Therefore, the test expression that is valid during the first iteration of the reverse section, is the one that corresponds to the r'-th assignment. The second iteration of the reverse section enters the branch where the test for $r'-1$ is valid because this label is on top of the stack $\text{STACK}_{(1)c}$, and so forth. Therefore, we can express the computation of the reverse section by the following program transformations where we again use the notation with the transposed Jacobian.

$$\mathbf{z}_{(1)r'-1} \leftarrow \left(\nabla\Phi^{r'}(\mathbf{z}_{r'-1})\right)^T \cdot \mathbf{z}_{(1)r'} \tag{2.59}$$
$$\mathbf{z}_{(1)r'-2} \leftarrow \left(\nabla\Phi^{r'-1}(\mathbf{z}_{r'-2})\right)^T \cdot \mathbf{z}_{(1)r'-1}$$
$$\vdots$$
$$\mathbf{z}_{(1)0} \leftarrow \left(\nabla\Phi^1(\mathbf{z}_0)\right)^T \cdot \mathbf{z}_{(1)1}$$

Please note that the transposed Jacobian in the i-th adjoint assignment is evaluated with the program state $\mathbf{z}_{r'-i} \equiv \Phi^{r'-i} \circ \Phi^{r'-i-1} \ldots \circ \Phi^1(\mathbf{z}_0)$, but the current program state at this adjoint assignment is actually $\mathbf{z}_{r'-i+1}$. For example, when the execution reaches the end of the forward section and the assignment (2.58) is performed, this leads to the program state $\mathbf{z}_{r'}$. Afterwards, the reverse section starts with executing (2.59). There, the current program state is still $\mathbf{z}_{r'}$ not $\mathbf{z}_{r'-1}$. The two states differ only at the component for $y_{r'}$. Hence, we restore the program state $\mathbf{z}_{r'-1}$ with the help of $\text{STACK}_{(1)f}$.

By recursively replacing the adjoint program states we obtain

$$\mathbf{z}_{(1)0} \equiv (\nabla\Phi^1(\mathbf{z}_0))^T \cdot (\nabla\Phi^2(\mathbf{z}_1))^T \cdot \ldots \cdot (\nabla\Phi^{r'}(\mathbf{z}_{r'-1}))^T \cdot \mathbf{z}_{(1)r'}. \tag{2.60}$$

The chain of matrix multiplications can be evaluated where the result is denoted by A:

$$A \equiv (\nabla\Phi^1(\mathbf{z}_0))^T \cdot (\nabla\Phi^2(\mathbf{z}_1))^T \cdot \ldots \cdot (\nabla\Phi^{r'}(\mathbf{z}_{r'-1}))^T.$$

Without loss of generality we assume that the m values $y_{r'-m+1}, y_{r'-m+2}, \ldots, y_{r'}$ determine the output values of the computation. We extract a submatrix of A by only considering the first n lines and the last m columns of A. This submatrix represents the values of the Jacobian $\nabla F(\mathbf{x})$ at point $\mathbf{x}$. The last m values in $\mathbf{z}_{(1)r'}$, namely $y_{(1)r',r'-m+1}, y_{(1)r',r'-m+2}, \ldots, y_{(1)r',r'}$ are the initial adjoint values defined by the user of the adjoint code, see (2.59). The first n values in $\mathbf{z}_{(1)0}$, namely $x_{(1)0,1}, y_{(1)0,2}, \ldots, y_{(1)0,n}$ are the adjoint values that the adjoint code provides as output, see (2.60). Therefore, we can rewrite the adjoint computation as the following matrix-vector product

$$\begin{pmatrix} x_{(1)0,1} \\ y_{(1)0,2} \\ \vdots \\ y_{(1)0,n} \end{pmatrix} \equiv \nabla F(\mathbf{x})^T \cdot \begin{pmatrix} y_{(1)r',r'-m+1} \\ y_{(1)r',r'-m+2} \\ \vdots \\ y_{(1)r',r'} \end{pmatrix}$$

$$\square$$

To get familiar with the adjoint source transformation, we show the adjoint model of same example code that we used for explaining the tangent-linear source transformation.

Example 22. This examples shows the application of the reverse mode to the code of Example 20. The transformation by σ is as follows, where indexes of statements corresponds to line numbers in Example 20:

$$\sigma(D) \stackrel{(2.33),(2.35)}{=} \begin{cases} \sigma(d_3) = \text{int i} \leftarrow 0; \\ \sigma(d_4) = \text{int tid} \leftarrow 0; \\ \sigma(d_5) = \text{int lb} \leftarrow 0; \\ \sigma(d_6) = \text{int ub} \leftarrow 0; \\ \sigma(d_7) = \text{int chunk\_size} \leftarrow 0; \end{cases} \tag{2.61}$$

$$\sigma(S) \stackrel{(2.38)}{=} \begin{cases} \sigma((s_8; s_9; s_{10}; s_{11})) \\ \sigma(s_{12}) \\ \sigma((s_{14}; s_{15})) \\ \sigma(s_{16}) \\ \sigma((s_{18})) \\ \sigma(s_{19}) \end{cases} \tag{2.62}$$

We avoid to blow up the example and show only the first two transformations that represent a transformation of a *SLC* code and a statement from *CFSTMT*. The transformation of the *SLC* sequence $(s_8; s_9; s_{10}; s_{11})$ is:

$$\sigma((s_8; s_9; s_{10}; s_{11})) \stackrel{(2.39)}{=} \begin{cases} \text{STACK}_{(1)c}.\text{push}(8); \\ \sigma(s_8); \\ \sigma(s_9); \\ \sigma(s_{10}); \\ \sigma(s_{11}); \end{cases}$$

$$\stackrel{(2.45)}{=} \begin{cases} \text{STACK}_{(1)c}.\text{push}(8); \\ \text{STACK}_{(1)i}.\text{push}(\text{tid}) \\ \text{tid} \leftarrow \text{omp\_get\_thread\_num}(); \\ \text{STACK}_{(1)i}.\text{push}(p) \\ p \leftarrow \text{omp\_get\_num\_threads}(); \\ \text{STACK}_{(1)i}.\text{push}(\text{chunk\_size}) \\ \text{chunk\_size} \leftarrow n/p; \\ \text{STACK}_{(1)i}.\text{push}(i) \\ i \leftarrow \text{chunk\_size}*p; \end{cases}$$

The control flow statement s_{12} is:

$$\sigma(s_{12}) \overset{(2.43)}{=} \text{if } (i \neq n) \; \{ \; \sigma((s_{13})) \; \}$$

$$\overset{(2.39)}{=} \text{if } (i \neq n) \; \{$$
$$\qquad \text{STACK}_{(1)c}.\text{push}(13);$$
$$\qquad \sigma(s_{13});$$
$$\}$$

$$\overset{(2.45)}{=} \text{if } (i \neq n) \; \{$$
$$\qquad \text{STACK}_{(1)c}.\text{push}(13);$$
$$\qquad \text{STACK}_{(1)i}.\text{push}(chunk\_size)$$
$$\qquad chunk\_size \leftarrow chunk\_size + 1;$$
$$\}$$

The remaining transformation steps are left to the reader. The forward section emitted by $\sigma(S)$ looks as follows

```
1  #pragma omp parallel
2  {
3     int i ← 0;
4     int tid ← 0;
5     int lb ← 0;
6     int ub ← 0;
7     int chunk_size ← 0;
8  STACK_c . push (8) ;
9  STACK_i . push ( tid ) ;
10    tid ← omp_get_thread_num ( ) ;
11 STACK_i . push ( p ) ;
12    p ← omp_get_num_threads ( ) ;
13 STACK_i . push ( chunk_size ) ;
14    chunk_size ← n/p ;
15 STACK_i . push ( i ) ;
16    i ← chunk_size*p ;
17    if ( i ≠ n )
18    {
19    STACK_c . push (13) ;
20    STACK_i . push ( chunk_size ) ;
21       chunk_size ← chunk_size+1;
```

```
22   }
23 STACK_c . push ( 14 ) ;
24 STACK_i . push ( lb ) ;
25   lb  ←  tid * chunk_size ;
26 STACK_i . push ( ub ) ;
27   ub  ←  ( tid +1 )* chunk_size −1;
28   if ( ub  ≥  n )  {
29   STACK_c . push ( 17 ) ;
30   STACK_i . push ( ub ) ;
31     ub  ←  n −1;
32   }
33 STACK_c . push ( 18 ) ;
34 STACK_i . push ( i ) ;
35   i  ←  lb ;
36   while ( i  ≤  ub )  {
37   STACK_c . push ( 20 ) ;
38   STACK_f . push ( y_i ) ;
39     y_i  ←  2* x_i * x_i ;
40   STACK_f . push ( x_i ) ;
41     x_i  ←  0;
42   STACK_i . push ( i ) ;
43     i  ←  i +1;
44   }
```

Listing 2.2: The first half of the parallel region consists of the forward section of the adjoint code where we take the code from Example 20 as input.

The reverse section of the adjoint code is supplied by transformation ρ:

$$
\rho(S) \overset{(2.48)}{=}
\begin{cases}
\rho\left((s_8; s_9; s_{10}; s_{11})\right) \\
\rho(s_{12}) \\
\rho\left((s_{14}, s_{15})\right) \\
\rho(s_{16}) \\
\rho\left((s_{18})\right) \\
\rho(s_{19})
\end{cases}
$$

$$\rho\big((s_8;s_9;s_{10};s_{11})\big) \overset{(2.49)}{=} \begin{cases} \text{if } \Big(\text{STACK}_{(1)c}.\text{top}() = 8\Big) \{ \\ \qquad \text{STACK}_{(1)c}.\text{pop}(); \\ \qquad \rho(s_{11}); \\ \qquad \rho(s_{10}); \\ \qquad \rho(s_9); \\ \qquad \rho(s_8); \\ \} \end{cases}$$

$$\overset{(2.55)}{=} \begin{cases} \text{if } \Big(\text{STACK}_{(1)c}.\text{top}() = 8\Big) \{ \\ \qquad \text{STACK}_{(1)c}.\text{pop}(); \\ \qquad \text{i} \leftarrow \text{STACK}_{(1)i}.\text{top}(); \\ \qquad \text{STACK}_{(1)i}.\text{pop}(); \\ \qquad \text{chunk\_size} \leftarrow \text{STACK}_{(1)i}.\text{top}(); \\ \qquad \text{STACK}_{(1)i}.\text{pop}(); \\ \qquad \text{p} \leftarrow \text{STACK}_{(1)i}.\text{top}(); \\ \qquad \text{STACK}_{(1)i}.\text{pop}(); \\ \qquad \text{tid} \leftarrow \text{STACK}_{(1)i}.\text{top}(); \\ \qquad \text{STACK}_{(1)i}.\text{pop}(); \\ \} \end{cases}$$

Statement s_{19} represents a loop and is therefore transformed by rule (2.54):

$$\rho(s_{19}) \overset{(2.54)}{=} \rho\big((s_{20},s_{21},s_{22})\big) \overset{(2.49)}{=} \begin{cases} \text{if } \Big(\text{STACK}_{(1)c}.\text{top}() = 20\Big) \{ \\ \qquad \text{STACK}_{(1)c}.\text{pop}(); \\ \qquad \rho(s_{22}); \\ \qquad \rho(s_{21}); \\ \qquad \rho(s_{20}); \\ \} \end{cases}$$

We transform $\rho(s_{22})$ as the previous integer assignments with (2.55). More interesting are the transformations $\rho(s_{20})$ and $\rho(s_{21})$ since the arguments are floating-point assignments and therefore being transformed by (2.50).

$$
\rho(s_{20}) \overset{(2.50)}{=}
\begin{cases}
y_i \leftarrow \text{STACKf.top}(); \\
\text{STACKf.pop}(); \\
x_{(1)i}+ \leftarrow 2 * x_i * y_{(1)i}; \\
x_{(1)i}+ \leftarrow 2 * x_i * y_{(1)i}; \\
y_{(1)i} \leftarrow 0;
\end{cases}
$$

$$
\rho(s_{21}) \overset{(2.50)}{=}
\begin{cases}
x_i \leftarrow \text{STACKf.top}(); \\
\text{STACKf.pop}(); \\
x_{(1)i} \leftarrow 0;
\end{cases}
$$

The remaining transformations are left to the reader and we present the continuation of the adjoint code by displaying the complete reverse section.

```
46    while( notSTACKc.empty() ) {
47      if(STACKc.top()  =  8 ) {
48    STACKc.pop();
49       i ←STACKi.top();
50    STACKi.pop();
51       chunk_size ←STACKi.top();
52    STACKi.pop();
53       p ←STACKi.top();
54    STACKi.pop();
55       tid ←STACKi.top();
56    STACKi.pop();
57    }
58      if(STACKc.top()  =  13 ) {
59    STACKc.pop();
60       chunk_size ←STACKi.top();
61    STACKi.pop();
62    }
63      if(STACKc.top()  =  14 ) {
64    STACKc.pop();
65       ub ←STACKi.top();
66    STACKi.pop();
67       lb ←STACKi.top();
68    STACKi.pop();
69    }
70      if(STACKc.top()  =  17 ) {
```

```
71    STACK_c . pop ( ) ;
72       ub ←STACK_i . top ( ) ;
73    STACK_i . pop ( ) ;
74    }
75    if (STACK_c . top ( )   =   18 ) {
76    STACK_c . pop ( ) ;
77       i ←STACK_i . top ( ) ;
78    STACK_i . pop ( ) ;
79    }
80    if (STACK_c . top ( )   =   20 ) {
81    STACK_c . pop ( ) ;
82       i ←STACK_i . top ( ) ;
83    STACK_i . pop ( ) ;
84       x_i ←STACK_f . top ( ) ;
85    STACK_f . pop ( ) ;
86       x_(1)i ← 0 ;
87       y_i ←STACK_f . top ( ) ;
88    STACK_f . pop ( ) ;
89       x_(1)i ++← 2*x_i*y_(1)i ;
90       x_(1)i ++← 2*x_i*y_(1)i ;
91       y_(1)i ← 0 ;
92    }
93    } /* End of while loop */
94 } /* End of parallel region */
```

Listing 2.3: This code shows the second half of the parallel region, namely the reverse section of the adjoint code.

The complete adjoint code consists of a parallel region with first Listing 2.2 as forward section and afterwards follows Listing 2.3 which represents the reverse section. Δ

The following lemma shows that the $\sigma(P)$ transformation fulfills the interface requirements that we defined in (2.9). The requirement was that if two memory references are equal it must be ensured that their adjoint associates also have the same reference.

Lemma 44. *Let $u, v \in FLOAT_P$ be two variables of a parallel region $P \in SPL$ and $\sigma(P)$ is the adjoint model of P. Then it holds for the variables u, v, $u_{(1)}$,*

$v_{(1)} \in \sigma(P)$:

$$\&u = \&v \iff \&u_{(1)} = \&v_{(1)} \tag{2.63}$$

Proof. In the case that u and v are defined outside of P, then this holds due to (2.9). Otherwise, the variable is defined inside of P and therefore transformed by rule (2.36) or (2.37). This definition is unique inside the parallel region $\sigma(P)$ since we only allow one sequence of declarations inside the parallel region. As the transformations (2.36) and (2.37) show, the declaration of v is always connected with the declaration of $v_{(1)}$ inside of $\sigma(P)$. Hence, the equivalence is fulfilled. $\square$

Before we finish this section, we introduce a notation that will improve the readability when we prove the closure property of the adjoint source transformation. During the proof we will examine different combinations of possible statements. For example, we have pairs of instances of statements in execution where one statement is from the forward section of thread t and the other statement is from the reverse section of thread t'. We try to prove the correctness of the source transformation by reducing the information about the statements inside the source transformation code to the original statements. This allows to reason that the source transformation is correct because the original statement is assumed to be correct. This method requires an association of the interleaving of the original code P with the interleaving of the code obtained by $\sigma(P)$. The association is done by a notation that we introduce in the following.

Suppose the parallel region P has a sequence of statements of length q and is executed by p threads. The set of possible interleavings is $\mathscr{I}(P, q, p)$. Let us assume that q' is the number of statements in $\sigma(P)$. The set of possible interleavings of the source transformation is therefore $\mathscr{I}(\sigma(P), q', p)$.

We write $s_g \in P$ when the statement s_g is contained in P. In case that s_i is obtained by transforming s_g through σ, we write $s_i \in \sigma(s_g)$. We do not use $s_i = \sigma(s_g)$ since in general the transformation supplies a sequence of statements. Analogously, we write $s_j \in \rho(s_h)$ for the case that s_j is contained in the resulting code of applying the transformation ρ to statement s_h that stems from P.

Let us consider two consecutive statements $s_i^t; s_j^{t'}$ in $J \in \mathscr{I}(\sigma(P), q', p)$. The fact that $s_i \in \sigma(s_g)$ and $s_j \in \rho(s_h)$ is denoted by

$$(s_g^t, s_h^{t'}) \xrightarrow{(\sigma, \rho)} (s_i^t, s_j^{t'}). \tag{2.64}$$

We illustrate this situation in Figure 2.2. First, we choose an arbitrary interleaving $J \in \mathscr{I}(\sigma(P), q', p)$ and focus on two consecutive instances $(s_i^t; s_j^{t'})$ of statements in execution. This is shown in the lower half of the figure. These two statements

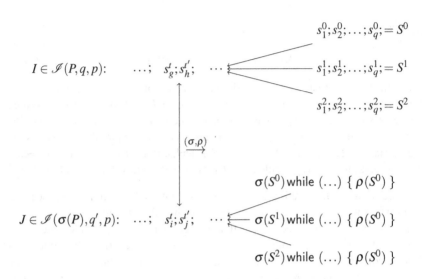

Figure 2.2: We define a notation that connects the interleaving of the original code P and its adjoint source transformation $\mathscr{I}(\sigma(P), q', p)$.

arised out of the transformations $\sigma(s_g)$ and $\rho(s_h)$. Since s_g and s_h are contained in P, there exists an interleaving $I \in \mathscr{I}(P, q, p)$ that contains the consecutive pair $(s_g^t; s_h^{t'})$. This pair is shown in the upper half of Figure 2.2. To describe this relation we write this as shown in (2.64).

Please note the important fact that we have to consider four possible combinations of instances inside the interleaving of the adjoint source transformation, namely $\mathscr{I}(\sigma(P), q', p)$. Either statement s_i^t or $s_j^{t'}$ may be part of the forward section or from the reverse section. When we combine these possibilities, we achieve four different cases that need to be considered. For each of the four possible combinations, we define in (2.65) an association to describe the connection between the statements in the derivative code and the statements from the original code.

$$
\begin{aligned}
(s_g^t, s_h^{t'}) \xrightarrow{(\sigma, \sigma)} (s_i^t, s_j^{t'}) &\text{ with } s_i \in \sigma(s_g), \text{ and } s_j \in \sigma(s_h) \\
(s_g^t, s_h^{t'}) \xrightarrow{(\sigma, \rho)} (s_i^t, s_j^{t'}) &\text{ with } s_i \in \sigma(s_g), \text{ and } s_j \in \rho(s_h) \\
(s_g^t, s_h^{t'}) \xrightarrow{(\rho, \sigma)} (s_i^t, s_j^{t'}) &\text{ with } s_i \in \rho(s_g), \text{ and } s_j \in \sigma(s_h) \\
(s_g^t, s_h^{t'}) \xrightarrow{(\rho, \rho)} (s_i^t, s_j^{t'}) &\text{ with } s_i \in \rho(s_g), \text{ and } s_j \in \rho(s_h)
\end{aligned}
\tag{2.65}
$$

This section introduced the adjoint source transformation $\sigma(P)$ that takes a parallel region P as input and provides the adjoint model of P as output. In order to implement the reverse section, we defined another source transformation $\rho\,(S)$ that takes a sequence of statements S as input and provides the corresponding adjoint statements as output. The fact that the code provided by the transformation represents the adjoint model of P was proven in Proposition 43. The following section explains how the code that we obtain by the source transformations fits into the derivative code of Q which is the code where the parallel region P is embedded.

2.3.3 *SPL* Code Inside of C/C++ Code

In Section 2.3.1 and Section 2.3.2, we defined source transformations for *SPL* code. This *SPL* code is contained in a parallel region P whereby P is embedded in C/C++ code referred to as Q. The current section explains how the code that we obtain by using the source transformations fits into the tangent-linear or the adjoint model of Q. Suppose the code Q has the structure shown in Listing 2.4 containing a parallel region that is referred to as P.

```
S₀
#pragma omp parallel
{
    S₁
}
S₂
```

Listing 2.4: Example Structure of a code Q with an OpenMP parallel region P.

There are three sequences of statements where S_0 and S_2 are C/C++ code and S_1 is *SPL* code. We assume that the source transformations $\tau(\,)$, $\sigma(\,)$, and $\rho()$ are well-defined, when S_0 and S_2 serve as input codes. The actual definition does not play a role in this work, since we only focus on the source transformation of a parallel region. However, we have to ensure that the interface between our source transformation result and the remaining derivative code fits together. This interface is described in (2.8) and (2.9) on page 86.

Suppose that the tangent-linear models of S_0 and S_2 are given by $\tau(S_0)$ and $\tau(S_2)$. The next listing shows the transition of the original structure of Q to the one in $\tau(Q)$. The tangent-linear model of Q inherits the structure from Q to its derivative pendant. This corresponds to the fact that the dataflow keeps the same in the tangent-linear model. Our contribution to $\tau(Q)$ is the parallel region with $\tau(S_1)$ in it.

S_0
#pragma omp parallel
{
    S_1
}
S_2

$\xrightarrow{\tau(Q)}$

$\tau(S_0)$
#pragma omp parallel
{
    $\tau(S_1)$
}
$\tau(S_2)$

Original structure of Q.   The tangent-linear model of Q.

The usage of the adjoint model $\sigma(P)$ needs to be explained in more detail. The adjoint model computes the derivative values with the help of the forward section and the reverse section. First of all, the forward section is executed containing the code of the original code augmented by statements for storing values before they potentially are overwritten. The augmented code also traces the flow of control. After the termination of the forward section, the reverse section follows with the reversal of the control flow and the calculation of the adjoint values. The rigorous separation of the forward section and the reverse section is called *split reversal*. This scheme generally consumes a lot of memory. A remedy for this drawback is the *joint reversal* scheme.

In our case, the joint reversal scheme can be applied to the parallel region P. This means that the forward section of Q does not contain the augmented forward section of P but instead the original code of P. Before the execution of the original code P begins, all input values of P are stored, for example, by writing a *checkpoint* [47, 30, 78]. A checkpoint, taken at a certain point in the program execution, contains all values that are necessary to allow a later restart of the computation at this point. The checkpoint that stores all input data of P during the forward section is restored in the reverse section of Q just before the execution of the adjoint code of P starts. Hence, it is ensured that the adjoint code of P has the exact same input data as the execution of P had during the forward section of Q. In case that the input data of P is constant or is not changed after P we do not need to store anything. The above method leads to the following structure of the forward section of $\sigma(Q)$.

$\sigma(S_0)$
#include "store_checkpoint.inc"
#pragma omp parallel
{
    S_1
}
$\sigma(S_2)$

As in the tangent-linear case, we assume that $\sigma(S_0)$ and $\sigma(S_2)$ are given. The forward section $\sigma(S_0)$ follows the creation of a checkpoint which is illustrated by including the code store_checkpoint.inc from an external file. Sequence S_1 in the parallel region is present as it was in the original code Q. The reverse section of Q looks as follows.

```
ρ(S₂)
#include "restore_checkpoint.inc"
#pragma omp parallel
{
    σ(S₁)
    while( not STACK₍₁₎c.empty() ) { ρ(S₁) }
}
ρ(S₀).
```

The codes $\rho(S_0)$ and $\rho(S_2)$ appear in reversed order compared to the forward section. The checkpoint containing the input data of P is restored by including the external code file restore_checkpoint.inc. The reason for restoring the input data becomes clear while considering the code inside the parallel region. First, the augmented forward section $\sigma(S_1)$ is preceding the reverse section $\rho(P)$. This joint computation of the adjoint values inside of the reverse section of Q is called joint reversal. To summarize, we illustrate the transformation of Q with the following listing.

```
S₀
#pragma omp parallel
{
    S₁
}
S₂
```

$\xrightarrow{\ \sigma(Q)\ }$

```
σ(S₀)
#include "store_checkpoint.inc"
#pragma omp parallel
{
    S₁
}
σ(S₂)
ρ(S₂)
#include "restore_checkpoint.inc"
#pragma omp parallel
{
    σ(S₁)
    while( not STACK₍₁₎c.empty() )
    { ρ(S₁) }
}
ρ(S₀)
```

Original program Q.   The adjoint model of Q.

More details about the split reversal and the joint reversal scheme can be found in [29, 57]. The next section shows that if P fulfills certain conditions, the code obtained by applying the source transformations from this chapter to P, can in fact be executed in parallel.

2.4 Closure of the Source Transformation

The closure property as defined in Definition 39 is crucial for generating higher derivative codes because it allows to reapply the transformation on its own output and ensures that the outcome is executable concurrently. In Section 2.4.1, we show that the tangent-linear source transformation $\tau(P)$ fulfills the closure property. Afterwards, we present a proof that shows that the adjoint source transformation $\sigma(P)$ only fulfills the closure property if the input code for the transformation fulfills the so called *exclusive read property*. A general assumption in this section is that a parallel region $P \in SPL$ is given that is noncritical.

2.4.1 Closure Property of $\tau(P)$

Section 2.3.1 defined the tangent-linear source transformation $\tau(P)$ that expects a parallel region P as input code and provides the tangent-linear model of P as shown in Proposition 41. The rules for the source transformation $\tau(P)$ reveal that the resulting code is contained in the language *SPL*. To achieve the closure property for $\tau(P)$ we have to show that the derivative code is noncritical.

Proposition 45. *Let $P \in SPL$ be a noncritical parallel region, then $\tau(P)$ is noncritical.*

Proof. Suppose that P contains the sequence $S = (s_1; \ldots; s_q)$. We denote the statements from S as original statements. Since P is noncritical, all statements $s \in S$ are noncritical. Hence, it holds that

$$\forall \left(s_g^t, s_h^{t'}\right) \in I \in \mathscr{I}(P, q, p), t \neq t' :$$

$$\&v \in LHSREF\left(s_g^t\right) \implies \&v \notin REF\left(s_h^{t'}\right).$$

We choose an arbitrary interleaving $J \in \mathscr{I}(\tau(P), q, p)$ that represents a parallel execution of the tangent-linear model of P. In the following we will examine the different possibilities of occurring pairs in J and show that these pairs are noncritical. Let us consider $\left(s_i^t; s_j^{t'}\right) \in J$ where t and t' are two different threads. Suitable to this pair, there exists an interleaving that represents the parallel execution of P and contains the pair of consecutive statements that are the original statements of s_i^t and s_j^t. Choose $I \in \mathscr{I}(P, q, p)$ such that $(s_g^t; s_h^{t'}) \in I$ with $s_i \in \tau(s_g)$ and $s_j \in \tau(s_h)$.

We assume that $\&v \in LHSREF(s_i^t) \neq \emptyset$ what implies that s_i is an assignment. We have to show that the reference $\&v$ cannot occur in $s_j^{t'}$ no matter which kind of statement s_j is.

Case $s_i = a \leftarrow e$ (s_i is an integer assignment): We only consider the cases where the integer reference $\&a_i^t$ may occur in $s_j^{t'}$.

1. $s_j = STACK_i.push(a)$

2. $s_j = a \leftarrow e$

3. $s_j = $ if (b) $\{$ S $\}$ or $s_j = $ while (b) $\{$ S $\}$

The first two cases for s_j lead to the exact same shape of s_h since these statements are transformed by the identity mapping. In case of the branch or loop statement the important part is the test condition b which is also transformed by the identity. Therefore, the assumption that $\&a^t$ occurs in $s_j^{t'}$ implies that this also hold for the pair $(s_g^t; s_h^{t'}) \in I$ which leads to the claim that P is critical.

Case $s_i = y_i \leftarrow \phi(x_{i,1}, \ldots, x_{i,n})$ (s_i is a floating-point assignment): The reference $\&y_i^t$ may occur in $s_j^{t'}$ in the following cases.

1. $s_j = STACK_f.push(x)$

2. $s_j = y \leftarrow \phi(x_1, \ldots, x_n)$ or $s_j = y +\!\!\leftarrow \phi(x_1, \ldots, x_n)$

3. $s_j = $ if (b) $\{$ S $\}$ or $s_j = $ while (b) $\{$ S $\}$

4. $s_j = y^{(1)} \leftarrow \sum_{k=1}^n \phi_{x_k}(x_1, \ldots, x_n) \cdot x_k^{(1)}$

5. $s_j = y^{(1)} +\!\!\leftarrow \sum_{k=1}^n \phi_{x_k}(x_1, \ldots, x_n) \cdot x_k^{(1)}$

For the first three cases we conclude analogously as we did in the previous case. s_i as well as s_j have the same shape as s_g and s_h have in the original code. Hence, we conclude that these cases are noncritical. Therefore, let us assume that s_j is an assignment where a derivative value is computed and we assume further that $\&y_i^t$ is contained in $s_j^{t'}$. The original statement s_h has the shape

$$y_h \leftarrow \phi(x_{h,1}, \ldots, x_{h,n})$$

or

$$y_h \mathrel{+\!\!\leftarrow} \phi(x_{h,1},\ldots,x_{h,n}).$$

The fact that $\&y_i^t$ is part of the derivative computation means for the left-hand side reference $\&y_g^t$ of s_g^t that this reference is also contained in $s_h^{t'}$. As P is noncritical, this assumption cannot be true.

Case $s_i = y_i^{(1)} \leftarrow \sum_{k=1}^n \phi_{x_{i,k}}(x_{i,1},\ldots,x_{i,n}) \cdot x_{i,k}^{(1)}$: The only case where $\&y_i^{(1)}$ can occur in $s_j^{t'}$ is that s_j is a tangent-linear assignment as well. Therefore, the original statements s_g and s_h are

$$s_g = y_g \leftarrow \phi(x_{g,1},\ldots,x_{g,n})$$

and

$$s_h = y_h \leftarrow \phi(x_{h,1},\ldots,x_{h,n}).$$

The claim that $\&y_i^{(1)t}$ occurs in $s_j^{t'}$ means for the original statements that $\&y_g^{(1)t}$ occurs in $s_h^{t'}$. This shows again a contradiction to the assumption that P is noncritical.

Case $s_i = y_i^{(1)} \mathrel{+\!\!\leftarrow} \sum_{k=1}^n \phi_{x_{i,k}}(x_{i,1},\ldots,x_{i,n}) \cdot x_{i,k}^{(1)}$: Analogously as shown in the previous case. $\square$

Lemma 46. *Suppose $P \in SPL$ is a noncritical parallel region, then the source transformation $\tau(P)$ is closed.*

Proof. We know from Proposition 45 that $\tau(P)$ is noncritical. Since P does not contain any OpenMP constructs besides the parallel directive, the tangent-linear source transformation is closed. $\square$

This section showed the very pleasing fact that our source transformation that provides the tangent-linear model of a parallel region P fulfills the closure property in case that P is noncritical. The coming section will introduce the exclusive read property and we show that the closure property for the adjoint transformation is only given if the original code P has the exclusive read property.

2.4.2 Closure Property of $\sigma(P)$ and the Exclusive Read Property

After showing that the tangent-linear source transformation fulfills the closure property in case that the input code is noncritical, we try to show the same for the adjoint source transformation $\sigma(P)$. To achieve this we perform similar steps as in the previous section and we consider the possible combinations of pairs in the interleaving that represents the parallel execution of the adjoint model. To anticipate it, a certain combination will reveal that the closure property is only given in case that the input code P is noncritical and in addition never reads a memory location concurrently. The latter requirement is different to the tangent-linear transformation where the noncritical requirement was sufficient.

In the introduction we saw that a race condition during runtime may lead to a nondeterministic behavior. Such a race condition appears, for example, when two threads write to the same memory location at the same time. In contrast to this, a simultaneous read of the same memory location does not change the memory location and is therefore noncritical no matter how many threads read this certain memory location. Thus, a concurrent read is anything but an exception in shared-memory parallel programming.

The property of a parallel region P that a parallel execution never leads to a situation where a memory location is read concurrently by more than one thread is denoted as the *exclusive read property*. Speaking in terms of an interleaving, the exclusive read property can be expressed as follows. We consider all possible interleavings $\mathscr{I}(P, q, p)$ of a parallel execution of parallel region $P \in SPL$ with q statements and p threads. We focus on an arbitrary consecutive pair of statements $(s_i^t; s_j^{t'})$ contained in an arbitrary interleaving $I \in \mathscr{I}(P, q, p)$. The exclusive read property is fulfilled if and only if each occurrence of the reference $\&v$ on the right-hand side of s_i^t leads to the fact that this reference does not occur on the right-hand side of $s_j^{t'}$. More formal, we write:

$$\forall I \in \mathscr{I}(P, q, p) \quad \forall (s_i^t; s_j^{t'}) \in I, t \neq t' :$$

$$\&v \in RHSREF(s_i^t) \implies \&v \notin RHSREF\left(s_j^{t'}\right). \tag{2.66}$$

In other words there is no combination in an interleaving possible such that two statements have the same memory reference on a right-hand side of an assignment. Please note that we exclude the case where a memory reference is read by a test expression of a branch or a loop statement. In case that (2.66) holds, we can be sure that it never happens that two threads use the same memory location simultaneously due to an evaluation of a right-hand side of an assignment. As indicated, this property of P plays a crucial role when the adjoint model of P should be achieved.

In practice, the exclusive read property is a strong restriction. In fact, in shared memory parallel programming it is very often the case that multiple threads read from the same memory location $\&v$. As we will see in the following proof, by applying the adjoint source transformation $\sigma(P)$, this leads to code that contains a race condition where the critical reference is $\&v_{(1)}$. In other words, the simultaneous read operation in the original code leads to a simultaneous write operation in the adjoint code.

Proposition 47. *Suppose that $P \in SPL$ is noncritical and fulfills the exclusive read property (2.66) then $\sigma(P)$ is noncritical.*

Proof. P consists of the sequence of statements $S = (s_1; \ldots; s_q)$ where the individual statements in S are referred to as original statements. P is noncritical what is equivalent with the fact that all statements $s \in S$ are noncritical. Therefore, it holds:

$$\forall I \in \mathscr{I}(P,q,p) \quad \forall (s_g^t; s_h^{t'}) \in I, t \neq t' :$$

$$\&v \in LHSREF\left(s_g^t\right) \implies \&v \notin REF\left(s_h^{t'}\right)$$

Suppose that $J \in \mathscr{I}(\sigma(P), q', p)$ is an arbitrary interleaving that represents the parallel execution of $\sigma(P)$, the adjoint model of P. With q' we denote the number of statements in the source transformation result $\sigma(P)$. We consider $(s_i^t; s_j^{t'}) \in J$ which is a pair of consecutive statements executed by the threads t and t' where $t \neq t'$. The statements s_i and s_j are from the sequence $(s_1, s_2, \ldots, s_{q'})$ which means that i and j are indexes out of $\{1, \ldots, q'\}$ and not necessarily different. The original statements corresponding to s_i and s_j are referred to as s_g and s_h. Thus, g and h have values between 1 and q.

There is an interleaving $I \in \mathscr{I}(P, q, p)$ that contains $(s_g^t; s_h^{t'})$. This fact is important as we trace back the absence of a critical reference in J to the absence of a critical reference in I. We illustrate the correlation between the interleavings I and J in Figure 2.2. We use the notation (2.65) to make clear what kind of transformations have been used to achieve $(s_i^t; s_j^{t'})$. For example,

$$\left(s_g^t, s_h^{t'}\right) \xrightarrow{(\sigma, \rho)} \left(s_i^t, s_j^{t'}\right)$$

means that s_i^t is a statement from the forward section ($s_i = \sigma(s_g)$) and $s_j^{t'}$ is a statement from the reverse section ($s_j = \rho(s_h)$). The pair $(s_g^t, s_h^{t'})$ is from the interleaving I that represents a parallel execution of P, and the pair $(s_i^t, s_j^{t'})$ is from the interleaving J that represent a parallel execution of $\sigma(P)$.

A parallel execution of $\sigma(P)$ means that each thread executes its part of the adjoint model of P independently. At any time, some threads execute the forward section while other threads execute the reverse section. Thus, we cannot be sure in which phase thread t and thread t' are when we consider the pair $(s_i^t; s_j^{t'})$. There are four possible combinations for s_i^t and $s_j^{t'}$ being either in the forward section or in the reverse section. Considering one of these four combinations, the statements themselves have a certain kind of shape what lead to further case distinctions. Obviously, this results in many possible combinations for $(s_i^t; s_j^{t'})$.

The reasoning method will be the same for all possible cases. We assume that s_i is an assignment and therefore s_i performs a store operation to the memory location referenced by the left-hand side reference of s_i^t. This reference can be $\&a_i^t$ in case of an integer assignment or $\&y_i^t$ if the assignment is a floating-point assignment. In either case this left-hand side reference may not be used in $REF\left(s_j^{t'}\right)$. Otherwise, this reference is critical. In order to show that the consecutive pair of statements $(s_i^t; s_j^{t'})$ is noncritical, we use a proof by contradiction. Therefore, we suppose that the left-hand side reference of s_i^t is also used by the instance $s_j^{t'}$. In most of the possible cases this leads to a contradiction because this would mean that P is critical as well. The only case where this does not lead to a contradiction is the case where we consider two adjoint assignments as s_i^t and $s_j^{t'}$. Thus, the impatient reader may skip the cases where at least one thread executes a forward section and jump to the case

$$\left(s_g^t, s_h^{t'}\right) \xrightarrow{(\rho,\rho)} \left(s_i^t, s_j^{t'}\right)$$

where both threads execute their reverse section code.

To recap, first we classify the cases depending on whether s_i^t and $s_j^{t'}$ are from the forward section or the reverse section. In each of these four different cases we further differ between possible shapes for s_i and s_j. We examine those combinations $(s_i^t; s_j^{t'})$ where s_i is an assignment and s_j is a statement such that the left-hand side reference s_i^t may be contained in $s_j^{t'}$. The possible combinations are illustrated in Figure 2.3, Figure 2.4, Figure 2.5, and Figure 2.6. The figures show 36 different cases in total. For example, Figure 2.3 displays the shapes of interest for s_i and s_j if we assume that s_i^t and $s_j^{t'}$ are both from the forward section.

The question may appear why the figures display cases that an integer left-hand side reference may occur in each and every possible statement $s_j^{t'}$ no matter if it is an integer statement or not. These cases review the situation where an integer reference is used as an offset for referencing floating-point elements contained in an array. For example, the expression A[i*n+j]*x[j] makes use of the integer variables i, n, and j.

In general holds that we do not consider the case where a floating-point assignment has the shape $y \mathbin{+\!\!\leftarrow} \phi(x_1, \ldots, x_n)$ since this shape is a special case of the general shape $y \leftarrow \phi(x_1, \ldots, x_n)$.

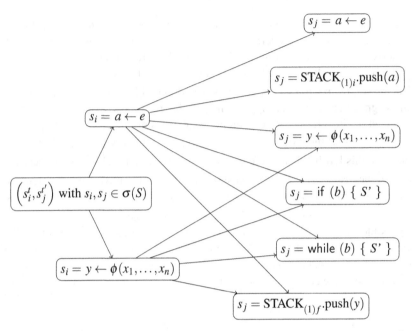

Figure 2.3: This tree shows the possibilities for the case that both statements s_i^t and $s_j^{t'}$ are assumed to be from the forward section of the adjoint code. The assumption that the left-hand side reference of s_i^t is critical allows the cases that s_i is an integer assignment or a floating-point assignment. From each of these cases we obtain another four possible cases for the shape of $s_j^{t'}$.

Case $(s_g^t, s_h^{t'}) \xrightarrow{(\sigma,\sigma)} (s_i^t, s_j^{t'})$ **(Figure 2.3):** In this case we assume that the threads t and t' both execute statements from the forward section of the adjoint model $\sigma(P)$. Suppose that the left-hand side reference of an assignment in thread t occurs also in the execution of thread t'. In terms of an instance of a statement in execution, we consider instance s_i^t that is executed by thread t and instance $s_j^{t'}$ that is executed by thread t'. We assume that the left-hand side reference &v of s_i^t occurs in instance $s_j^{t'}$ which means that

$$\&v \in LHSREF(s_i^t) \wedge \&v \in REF\left(s_j^{t'}\right) \tag{2.67}$$

is valid. Depending on the different possible shapes for s_i and s_j, we trace this assumption back to the original statements s_g and s_h where we consider the instances s_g^t and $s_h^{t'}$ from the interleaving I.

1. $s_i = a \leftarrow e$:
   From transformation rule (2.45), we know that the original statement s_g has the shape $a \leftarrow e$. Because we assume that (2.67) holds we assume that $\&a_i^t$ is in $REF\left(s_j^{t'}\right)$.

   a) $s_j = a \leftarrow e$:
      Again, with (2.45) we conclude that the corresponding statement s_h is $a \leftarrow e$. With (2.67) this leads to the fact that $\&a_i^t$ occurs either on the left- or on the right-hand side of s_j. This can be represented by the equation

      $$\&a_i^t = \&a_j^{t'} \quad \vee \quad \&a_i^t = \&v \in RHSREF\left(s_j^{t'}\right)$$

      concerning interleaving J. Since $s_i = s_g$ and $s_j = s_h$ we conclude for interleaving I that

      $$\&a_g^t = \&a_h^{t'} \quad \vee \quad \&a_g^t = \&v \in RHSREF\left(s_h^{t'}\right).$$

      This leads to a contradiction because the equation implies that P is critical.

   b) $s_j = \text{STACK}_{(1)i}.\text{push}(a)$:
      Source transformation rule (2.45) reveals that the original statement s_h is $a \leftarrow e$. Therefore, we can apply the same reasoning as in 1a.

   c) $s_j = y \leftarrow \phi(x_1,\ldots,x_n)$:
      Reference $\&a_i^t$ may appear in $s_j^{t'}$ in the shape of an offset. From rule (2.40) we conclude that $s_h = s_j$. The occurring offsets in P are transformed by the identity mapping in all our source transformation rules. The assumption (2.67) and the fact that $s_i = s_g$ and $s_j = s_h$ leads to the fact that reference $\&a_g^t$ is used as an offset in $REF\left(s_h^{t'}\right)$ where we consider interleaving I. This means that $\&a_g^t$ is a critical reference and therefore P is critical.

   d) $s_j = \text{if } (b) \ \{ \ S \ \} \text{ or } s_j = \text{while } (b) \ \{ \ S \ \}$:
      From (2.43) and (2.44) we conclude that $s_j = s_h$. If we assume that (2.67) is true then $\&a_i^t$ may occur as an offset or as part of an integer expression in the test expression $b_j^{t'}$ that is evaluated by thread t'. Both cases lead to the fact that

      $$\&a_g^t \in \{\&v_h^{t'} \mid v \in VAR_b\}$$

      holds for interleaving I. Therefore, $\&a_g^t$ is a critical reference in P.

e) $s_j = \text{STACK}_{(1)f}.\text{push}(y)$:
s_j is a result of one of the source transformation rules (2.40), (2.41), (2.42), or (2.46). In all these cases assumption (2.67) means for the interleaving I that $\&a_g^t$ is equal to a reference in $REF\left(s_h^{t'}\right)$. This cannot be the case since P is noncritical.

2. $s_i = y \leftarrow \phi(x_1, \ldots, x_n)$:
Rule (2.40) shows that the original statement s_g is $y \leftarrow \phi(x_1, \ldots, x_n)$. We assert that (2.67) is true and therefore reference $\&y_i^t$ is assumed to be contained in $REF\left(s_j^{t'}\right)$.

a) $s_j = y \leftarrow \phi(x_1, \ldots, x_n)$:
The original statement s_h has the shape $y \leftarrow \phi(x_1, \ldots, x_n)$ due to transformation rule (2.40). The claim (2.67) about interleaving J implies that $\&y_i^t$ needs to be either on the left or on the right-hand side:

$$\&y_i^t = \&y_j^{t'} \quad \vee \quad \&y_i^t = \&x_{j,k}^{t'} \text{ where } k \in \{1, \ldots, n\}.$$

Since $s_i = s_g$ and $s_j = s_h$ we conclude for interleaving I that

$$\&y_g^t = \&y_h^{t'} \quad \vee \quad \&y_g^t = \&x_{h,k}^{t'} \text{ where } k \in \{1, \ldots, n\},$$

which means that the threads t and t' either both write to $\&y_g^t$ or that one thread writes and one threads reads from the reference $\&y_g^t$. However, both possibilities implies that P is critical which is a contradiction.

b) $s_j = \text{if } (b) \{ S \}$ or $s_j = \text{while } (b) \{ S \}$:
Analogously to the reasoning in 1d, we conclude that $\&y_g^t \in REF\left(b_h^{t'}\right)$ which means that $\&y_g^t$ is a critical reference and therefore P is critical as well.

c) $s_j = \text{STACK}_{(1)f}.\text{push}(y)$:
Similar to 1e, we obtain the fact that in interleaving I the reference $\&y_g^t$ is in $REF\left(s_h^{t'}\right)$ what means that this reference from P is critical.

In all the above cases, we obtain a contradiction. Therefore, the claim (2.67) must be wrong for the case $(s_g^t, s_h^{t'}) \xrightarrow{(\sigma,\sigma)} (s_i^t, s_j^{t'})$. This shows that the parallel execution of two augmented forward sections $\sigma(S)$ in different threads cannot include a critical reference. The next case examines the concurrent execution of a forward section and a reverse section in two different threads t and t'.

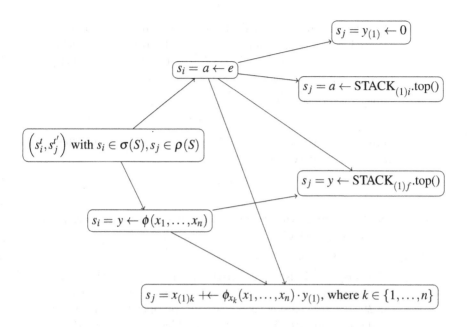

Figure 2.4: This tree shows the possibilities for the case where s_i^t is from the forward section and $s_j^{t'}$ is from the reverse section.

Case $(s_g^t, s_h^{t'}) \xrightarrow{(\sigma, \rho)} (s_i^t, s_j^{t'})$ **(Figure 2.4):**   The current case assumes that thread t executes the forward section and thread t' executes its reverse section code.

1. $s_i = a \leftarrow e$:
   We know that the original statement s_g has the shape $a \leftarrow e$ because of transformation rule (2.45). We assume that (2.67) holds. Therefore, the set $REF\left(s_j^{t'}\right)$ comprises the reference $\&a_i^t$.

   a) $s_j = y_{(1)} \leftarrow 0$:
      If we have a critical reference in J then this is only possible when reference $\&a_i^t$ is used as an offset in the left-hand side reference $\&y_{(1)j}^{t'}$. As a reminder concerning the notation, $\&y_{(1)}^{t'}$ is an adjoint reference used by thread t' that occurs in the instance $s_j^{t'}$. The original statement s_h is $y \leftarrow \phi(x_1, \ldots, x_n)$ as shown in rule (2.50). The offsets are transformed by the identity mapping. Therefore, the use of $\&a_i^t$ as an offset means for the interleaving I that $\&a_g^t$ is used as an offset in $\&y_h^{t'}$ which is the

left-hand side reference of $s_h^{t'}$. Thus, $\&a_g^t$ is critical since this reference is used in a store operation in thread t and it is read by thread t'. We infer that this is a contradiction as P is assumed to be noncritical.

b) $s_j = a \leftarrow \text{STACK}_{(1)i}.\text{top}()$:
Since (2.67) we obtain

$$\&a_i^t = \&a_j^{t'}.$$

From the transformation rule (2.55), we know that s_h has the shape $a \leftarrow e$. Thus, the equation

$$\&a_g^t = \&a_h^{t'}$$

holds for interleaving I which means that P is critical because both threads write to the same memory location.

c) $s_j = y \leftarrow \text{STACK}_{(1)f}.\text{top}()$:
Reference $\&a_i^t$ is used as an offset in referencing $\&y_i^t$. Since this offset is not changed during the source transformation, the same offset is used in P. The original statement s_h is a floating-point assignment in case that rule (2.50) or (2.51) were used to achieve s_j or s_h is $\text{STACK}_f.\text{top}()$ when (2.52) was used. However, we know that the offset is used for the left-hand side reference of s_h. For interleaving I this means that $\&a_g^t$ is critical because it is written by thread t and concurrently being read by thread t' in the context of an offset evaluation.

d) $s_j = x_{(1)1} \mathrel{+\!\!\leftarrow} \phi_{x_k}(x_1,\ldots,x_n) \cdot y_{(1)}$:
This case means as in the previous case that $\&a_i^t$ is used in $s_j^{t'}$ as an offset. The original statement s_h is a floating-point assignment due to the rule (2.50) or (2.51). The reasoning is analog to 1c since $\&a_g^t$ is once written by thread t and on the hand read by thread t'. Thus, this reference is critical.

2. $s_i = y \leftarrow \phi(x_1,\ldots,x_n)$:
The original statement s_g is $y \leftarrow \phi(x_1,\ldots,x_n)$ as a result of rule (2.40). By claiming that (2.67) is true, reference $\&y_i^t$ is assumed to be contained in $REF\left(s_j^{t'}\right)$.

a) $s_j = y \leftarrow \text{STACK}_{(1)f}.\text{top}()$:
In this case the claim (2.67) implies that

$$\&y_i^t = \&y_j^{t'}$$

Depending on whether the transformation rule (2.50), (2.51) or (2.52) was used, we obtain a floating-point assignment or $y \leftarrow \mathrm{STACK}_f.\mathrm{top}()$ as original statement s_h. We conclude that for interleaving I the equation

$$\&y_g^t = \&y_h^{t'}$$

is true. This describes a situation where two threads write concurrently to the same memory location what leads to the fact that P is critical.

b) $s_j = x_{(1)k} \,{+\!\!\leftarrow}\, \phi_{x_k}(x_1,\ldots,x_n)\cdot y_{(1)}\,, k \in \{1,\ldots,k\}$:
The left-hand side reference $\&y_i^t$ may appear on the right-hand side of $s_j^{t'}$ as an argument of the partial derivative. When we regard rule (2.50) or (2.51), we know that s_j is originated from $s_h = y \leftarrow \phi(x_1,\ldots,x_n)$. The assumption that $\&y_i^t$ occurs in the partial derivative means $\&y_i^t = \&x_{j,k}^{t'}$ where j illustrates that x is a right-hand side reference of s_j and k means the k-th argument out of $(x_1,x_2,\ldots,x_n)$. The equation $\&y_i^t = \&x_{j,k}^{t'}$ leads to the fact that for the interleaving I the equation $\&y_g^t = \&x_{h,k}^{t'}$ is true. The equality of a left-hand side reference with a right-hand side reference in different threads $(t \neq t')$ implies that P is critical.

The claim (2.67) leads to a contradiction because we require P as noncritical. This case illustrates that the situation where a left-hand side reference of the forward section in thread t is accessed in another thread t' during its reverse section cannot occur. The next case investigates if any left-hand side reference of the reverse section of thread t is accessed during the forward section of thread t'.

Case $(s_g^t, s_h^{t'}) \overset{(\rho,\sigma)}{\longrightarrow} (s_i^t, s_j^{t'})$ **(Figure 2.5):** If s_i^t is from the reverse section and $s_j^{t'}$ is from the forward section, we do not need to consider all possible assignments in the reverse section. In case that an assignment has an adjoint reference on its left-hand side, this reference cannot occur in the forward section. Therefore, Figure 2.5 shows only these cases for s_i where a left-hand side reference can possibly occur in $REF\left(s_j^{t'}\right)$.

1. $s_i = a \leftarrow \mathrm{STACK}_{(1)i}.\mathrm{top}()$:
   If s_i is a stack operation that restores an integer value then the original statement s_g is $a \leftarrow e$ due to rule (2.55). The possible cases where reference $\&a_i^t$ may occur in $REF\left(s_j^{t'}\right)$ are

   a) $s_j = a \leftarrow e$

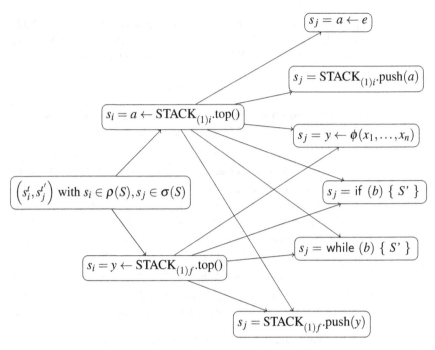

Figure 2.5: This tree shows the possibilities for the case where s_i^t is from the reverse section and $s_j^{t'}$ is from the forward section. Since no adjoint references are used inside of the forward section, we do not need the case where s_i is an adjoint assignment.

b) $s_j = \text{STACK}_{(1)i}.\text{push}(a)$

c) $s_j = y \leftarrow \phi(x_1, \ldots, x_n)$

d) $s_j = \text{if } (b) \ \{ \ S \ \} \text{ or } s_j = \text{while } (b) \ \{ \ S \ \}$

e) $s_j = \text{STACK}_{(1)f}.\text{push}(y)$

These cases for the shape of s_j are the same as displayed in Figure 2.3. Since the left-hand side reference of s_i^t is here $\&a_i^t$ as well as in Figure 2.3 we refer the reasoning in the current case to the cases

$$(s_g^t, s_h^{t'}) \overset{(\sigma, \sigma)}{\longrightarrow} (s_i^t, s_j^{t'}) \ 1a \text{ to } 1e.$$

2. $s_i = y \leftarrow \text{STACK}_{(1)f}.\text{top}()$:
The original statement s_g may have one of three possible shapes depending

on which rule was used to generate statement s_i. In case that the rules (2.50) or (2.51) were used, s_g is is $y \leftarrow \phi(x_1,\ldots,x_n)$. Otherwise, rule (2.52) displays that the original statement is $s_g = y \leftarrow \text{STACK}_f.\text{top}()$. The possible cases for the shape of s_j are

a) $s_j = y \leftarrow \phi(x_1,\ldots,x_n)$:

b) $s_j = \text{if } (b) \ \{\ S\ \} \text{ or } s_j = \text{while } (b) \ \{\ S\ \}$:

c) $s_j = \text{STACK}_{(1)f}.\text{push}(y)$:

These shapes for s_j are the same as the possible shapes in Figure 2.3. The assumption that reference $\&y_i^t$ occurs in $REF\left(s_j^{t'}\right)$ leads to the argumentation as in

$$(s_g^t, s_h^{t'}) \xrightarrow{(\sigma,\sigma)} (s_i^t, s_j^{t'}) \ 2a \text{ to } 2c.$$

that $\&y_g^t$ is critical. This contradicts to the requirement that P is noncritical.

Since we obtain a contradiction in all the possible cases we conclude that the execution of reverse section in thread t and a forward section in thread t' do not interfere each other. The next case examines whether or not an execution of a reverse section in both threads possibly interfere in some way.

Case $(s_g^t, s_h^{t'}) \xrightarrow{(\rho,\rho)} (s_i^t, s_j^{t'})$ **(Figure 2.6):** In this case, we assume that both statements s_i^t and $s_j^{t'}$ are from the reverse section. As a reminder, the reverse section has the following appearance:

$$\text{while } \left(\text{not STACK}_{(1)c}.\text{empty}()\right) \ \{\ \rho(S_0)\ \} \tag{2.68}$$

The code block $\rho(S_0)$ is structured as a sequence of branch statements as shown in (2.69).

$$\begin{aligned}
&\text{if } \left(\text{STACK}_{(1)c}.\text{top}() = 1\right) \ \{\ S_1\ \} \\
&\text{if } \left(\text{STACK}_{(1)c}.\text{top}() = 2\right) \ \{\ S_2\ \} \\
&\ \vdots \\
&\text{if } \left(\text{STACK}_{(1)c}.\text{top}() = m\right) \ \{\ S_m\ \}
\end{aligned} \tag{2.69}$$

The number of branch statements m is the number of SLC sequences that we obtain from applying rule (2.48). The test expressions in (2.68) and (2.69) are noncritical because they only read from a thread local data structure. Therefore, we only consider the sequences of statements S_1 to S_m. These sequences are the result of the source transformation rule (2.49). The pop operation which is the first statement inside the branch code block can be neglected since it only affects thread local memory. The actual statements in these sequences are obtained through the transformation rules (2.50), (2.51), (2.52), and (2.55). Hence, we consider possible combinations of statement pairs out of the following set of statements:

$$\{ \quad y \leftarrow \text{STACK}_{(1)f}.\text{top}(),$$

$$x_{(1)k} \mathrel{+\!\leftarrow} \phi_{x_k}(x_1,\ldots,x_n) \cdot y_{(1)},$$

$$y_{(1)} \leftarrow 0,$$

$$a \leftarrow \text{STACK}_{(1)i}.\text{top}() \quad \} \quad , \text{where } k \in \{1,\ldots,n\}.$$

Thus, the shape of s_i can be one of the assignments contained in this set. Depending on the kind of the left-hand side reference there are several cases where this left-hand side reference could occur in $s_j^{t'}$. The edges from s_i to s_j in Figure 2.6 corresponds with the possibility of an occurrence. To prove the absence of a critical reference by contradiction, we assume that (2.67) is true.

1. $s_i = a \leftarrow \text{STACK}_{(1)i}.\text{top}()$:
   The corresponding original statement s_g to s_i is $a \leftarrow e$ what can be recognized by rule (2.55).

   a) $s_j = a \leftarrow \text{STACK}_{(1)i}.\text{top}()$:
      As stated for s_i, due to rule (2.55) we obtain the original statement s_h with the shape $a \leftarrow e$. Claim (2.67) implies that the both left-hand side references must be the same.

      $$\&a_i^t = \&a_j^{t'}$$

      This equation means for interleaving I that

      $$\&a_g^t = \&a_h^{t'},$$

      and leads to the fact that P is critical because both threads write to the same memory location.

   b) $s_j = y_{(1)} \leftarrow 0$:
      We refer to the case

      $$(s_g^t, s_h^{t'}) \xrightarrow{(\sigma,\rho)} (s_i^t, s_j^{t'}) \; 1a \, , \text{see Figure 2.4.}$$

There, we investigate the case where the left-hand side reference a_i^t of s_i occurs in the same shape of s_j as in the current case. Thus, the reasoning is the same as shown there.

c) $s_j = y \leftarrow \text{STACK}_{(1)f}.\text{top}()$:
We conclude that this is a contradiction as shown in the case

$$(s_g^t, s_h^{t'}) \xrightarrow{(\sigma,\rho)} (s_i^t, s_j^{t'}) \; 1c \, , \text{see Figure 2.4.}$$

d) $s_j = x_{(1)k} \leftarrow\!\!\!+\!\!\!-\!\!\! \phi_{x_k}(x_1,\ldots,x_n) \cdot y_{(1)}$ for a $k \in \{1,\ldots,n\}$:
We refer to the reasoning where we covered the case

$$(s_g^t, s_h^{t'}) \xrightarrow{(\sigma,\rho)} (s_i^t, s_j^{t'}) \; 1d \, , \text{see Figure 2.4.}$$

Thus, this case leads to a contradiction.

2. $s_i = y_{(1)} \leftarrow 0$:
Source transformation rule (2.50) shows us that the original statement s_g is $y \leftarrow \phi(x_1,\ldots,x_n)$.

a) $s_j = y_{(1)} \leftarrow 0$:
The assumption (2.67) that there is a critical reference in J leads to

$$\&y_{(1)i}^t = \&y_{(1)j}^{t'}$$

From Lemma 44 we know that

$$\&u = \&v \iff \&u_{(1)} = \&v_{(1)}$$

and thus we achieve the equation

$$\&y_i^t = \&y_j^{t'} \, .$$

The original statement s_h has likewise the shape $y \leftarrow \phi(x_1,\ldots,x_n)$ due to rule (2.50). We make the transition from interleaving J to I and gain the equation

$$\&y_g^t = \&y_h^{t'}$$

which reveals a critical reference since the threads t and t' write to the same memory location.

b) $s_j = x_{(1)k} \leftarrowtail \phi_{x_k}(x_1,\ldots,x_n) \cdot y_{(1)}$ for a $k \in \{1,\ldots,n\}$:

We assume that (2.67) is true. This brings us to the fact that one of the following equations must be true:

$$\&y^t_{(1)i} = \&y^{t'}_{(1)j} \quad \vee \quad \&y^t_{(1)i} = \&x^{t'}_{(1)k,j}$$

where $\&x^{t'}_{(1)k,j}$ is a left-hand side reference from $RHSREF\left(s^{t'}_j\right)$. With Lemma 44 we infer for interleaving J that

$$\&y^t_i = \&y^{t'}_j \quad \vee \quad \&y^t_i = \&x^{t'}_{k,j}.$$

By regarding the rules (2.50) or (2.51), we conclude that the original statement is $s_h = y \leftarrow \phi(x_1,\ldots,x_n)$. For the interleaving I we obtain

$$\&y^t_g = \&y^{t'}_h \quad \vee \quad \&y^t_g = \&x^{t'}_{k,h}.$$

where $\&x^{t'}_{k,h}$ is the memory reference of the k-th argument x_k in the set $RHSREF\left(s^{t'}_h\right)$. Both equations lead to the fact that P is critical due to a write-write or a write-read conflict between the threads t and t'.

3. $s_i = y \leftarrow \text{STACK}_{(1)f}.\text{top}()$:

In this case the original statement s_g cannot be determined exactly. In case that the transformation rules (2.50) or (2.51) were used to supply s_i then the original statement s_g is $y \leftarrow \phi(x_1,\ldots,x_n)$. Otherwise, rule (2.52) displays that the original statement is $s_g = y \leftarrow \text{STACK}_f.\text{top}()$. However, since $\&y^t_i$ is the left-hand side reference in s^t_i and this reference was also a left-hand side reference in the case where we investigated the forward section and the reverse section, we can reason as shown there.

a) $s_j = y \leftarrow \text{STACK}_{(1)f}.\text{top}()$:

With the same reasoning as in the previous case

$$(s^t_g, s^{t'}_h) \xrightarrow{(\sigma,\rho)} (s^t_i, s^{t'}_j)\ 2a\ , \text{see Figure 2.4,}$$

we conclude that P is critical.

b) $s_j = x_{(1)k} \leftarrowtail \phi_{x_k}(x_1,\ldots,x_n) \cdot y_{(1)}$ for a $k \in \{1,\ldots,n\}$:

The arguments presented in

$$(s^t_g, s^{t'}_h) \xrightarrow{(\sigma,\rho)} (s^t_i, s^{t'}_j)\ 2b\ , \text{see Figure 2.4.}$$

hold also for the current case and thus we conclude that P is critical.

4. $s_i = x_{(1)k} \mathrel{+\!\!\!+\!\!\!\leftarrow} \phi_{x_k}(x_1,\ldots,x_n) \cdot y_{(1)}$ for a $k \in \{1,\ldots,n\}$:

The transformation rules (2.50) and (2.51) show us that the corresponding original statement s_g is $y \leftarrow \phi(x_1,\ldots,x_n)$. We claim that (2.67) is true and therefore the left-hand side reference of instance s_i^t, namely reference $\&x_{(1)k,i}^t \in LHSREF(s_i^t)$, occurs in the set of references $REF\left(s_j^{t'}\right)$.

a) $s_j = y_{(1)} \leftarrow 0$:

The only possibility for (2.67) being true is that

$$\&x_{(1)k,i}^t = \&y_{(1)j}^{t'} \tag{2.70}$$

is valid. With Lemma 44 we infer for interleaving J that

$$\&x_{k,i}^t = \&y_j^{t'}$$

The original statement s_h is $y \leftarrow \phi(x_1,\ldots,x_n)$ due to rule (2.50). Therefore, we conclude for interleaving I that

$$\&x_{k,g}^t = \&y_h^{t'}$$

what means that thread t uses a memory location in the context of an argument that is concurrently used in a store operation by thread t'. Therefore, $\&x_{k,g}^t$ is critical what implies that P is critical.

b) $s_j = x_{(1)l} \mathrel{+\!\!\!+\!\!\!\leftarrow} \phi_{x_l}(x_1,\ldots,x_n) \cdot y_{(1)}$ for a $l \in \{1,\ldots,n\}$:

The rule (2.50) supplies the original statement $s_h = y \leftarrow \phi(x_1,\ldots,x_n)$. In case that (2.67) is true then the left-hand side reference $\&x_{(1)k,i}^t$ occurs in $REF\left(s_j^{t'}\right)$ Therefore, $\&x_{(1)k,i}^t$ must be on the left or on the right-hand side of s_j. Hence, one of the following equations must be true.

$$\&x_{(1)k,i}^t = \&y_{(1)j}^{t'} \quad \vee \quad \&x_{(1)k,i}^t = \&x_{(1)l,j}^{t'}$$

The left equation is the same as in (2.70) and thus we achieve the contradiction stating that P is critical. Therefore, the only remaining possibility is that

$$\&x_{(1)k,i}^t = \&x_{(1)l,j}^{t'}$$

is valid. The equivalence from Lemma 44 provides the fact that in interleaving J the equation

$$\&x_{k,i}^t = \&x_{l,j}^{t'}$$

holds. The transition to the interleaving I tells us that the equation

$$\&x'_{k,g} = \&x'_{l,h} \tag{2.71}$$

is true. For the parallel execution of P this means that the threads t and t' read cocurrently from the same memory location. For the first time we do not get a contradiction because this read-read situation is not critical. Nevertheless, our original code P fulfills the exclusive read property (2.66) and therefore we achieve that this remaining equation cannot be true either.

In all cases except the last, we achieved a contradiction to the assumption that P is noncritical. Hence, we can conclude that there is no critical reference in $\sigma(P)$ as long as P fulfills the exclusive read property. We cannot relinquish the exclusive read property as we showed in the case where we achieved the equation (2.71). □

Fortunately, in shared memory parallel programming there are ways to synchronize such a situation. In Section 4.2, we will extend the language SPL such that it contains constructs for synchronizing the parallel execution of P. With these synchronization constructs we are able to provide the closure of $\sigma(P)$ for any given $P \in SPL$, as we will see in Section 4.2.5. Nevertheless, synchronization of parallel programs is often connected with a bad scaling performance since the overhead for the synchronization increases with a growing number of threads. Therefore, we only want to use synchronization when it is not avoidable. But this means that our source transformation tool must be capable of recognizing the exclusive read property during compile time. In this context, we introduce a static program analysis in Chapter 3. This static analysis provides for each floating-point assignment in P an approximation whether or not it fulfills the exclusive read property. In case the analysis cannot decide if the reference is used exclusively or not, the result is that the reference does not fulfill the exclusive read property.

2.5 Summary

This chapter started by introducing the formalism that we use not only in this chapter but also in the coming chapters. This formalism comprises the mathematical abstraction of a parallel region as an interleaving. In this approach, the concurrent instances of statements in execution are interleaved into one sequence. This sequence is called interleaving and represents one possible parallel execution. We showed in Lemma 36 that the number of possible interleavings is quite high even when we only consider a straight-line code.

We defined the terms of a critical and a noncritical parallel region and the important property called closure of a source transformation. The closure is defined in Definition 39 and it states that the source transformation takes a noncritical *SPL* code as input and produces a noncritical *SPL* code as output. The language *SPL* is defined as a subset of C/C++. This simplifies the correctness proof by restricting the set of possible statements in a parallel region. After the definition of *SPL*, we defined the source transformation $\tau(P)$ and $\sigma(P)$ which take a parallel region P as input and provides the tangent-linear ($\tau(P)$) or the adjoint model ($\sigma(P)$) as output. The formal proofs that the transformation result is in fact the corresponding model is shown in Proposition 41 and Proposition 43.

In this work, we only focus on the source transformation of a parallel region P and not the transformation of the code Q where P is embedded in. However, the derivative codes of P and Q have to fit together and we explained the corresponding interface in Section 2.3.3.

The most important section in this chapter was Section 2.4. In this section we proved that the closure property for $\tau(P)$ and $\sigma(P)$ holds under certain circumstances. The tangent-linear source transformation $\tau(P)$ fulfills the closure property for a noncritical input code P written in *SPL*. To achieve the closure property of $\sigma(P)$, the input code P has to be noncritical and in addition it has to fulfill the exclusive read property. The exclusive read property was defined in (2.66). It requires that during the parallel execution of P it never happens that two different threads reads from the same memory location concurrently in the context of a right-hand side evaluation of an assignment. In shared memory parallel programming it is a general pattern to let threads read from a shared memory location. These codes where such patterns are used must be adjusted by hand such that each thread uses memory locations exclusively, or the source transformation tool must handle these codes adequately. However, the exclusive read property for the input code P is necessary because otherwise the concurrent read access of two threads in P becomes a concurrent store operation during the execution of the adjoint model $\sigma(P)$.

In order to enable the source transformation tool to recognize the exclusive read analysis, we introduce a static program analysis in Chapter 3. Afterwards in Chapter 4, we make the transition from a general parallel region to an OpenMP parallel region. We show how the most important constructs of OpenMP can be transformed into the tangent-linear or the adjoint model.

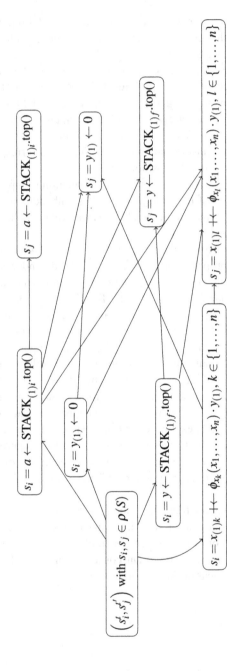

Figure 2.6: This tree shows the possibilities for the case where both s_i' and s_j' are from the reverse section. We show only these cases where the left-hand side reference of s_i is probably read or written in statement s_j.

3 Exclusive Read Analysis

The exclusive read property was introduced in (2.66) on page 119. If the parallel region P fulfills the exclusive read property then it is ensured that it never happens that two threads read the same memory location concurrently in the context of an assignment evaluation. Just to be clear, this situation is not a race condition and the parallel execution of P is correct if two threads read the same memory location. The only problem with such a code is that the concurrent read access in P becomes a concurrent store operation in the adjoint model $\sigma(P)$.

The constraint that the input code for our adjoint transformation has to fulfill the exclusive read property can be handled in different ways. To neglect the pragmas to avoid the parallel execution solves the problem through a sequential execution but this is most likely the worst solution. Another option is to put the responsibility into the hands of the user of the source transformation tool such that the user has to provide an option to the tool whether the exclusive read property holds or not. Depending on the given option the transformation tool implements different versions of the derivative code. The last possibility is that one could assume conservatively that all input codes possibly violate the exclusive read property. In order to resolve possible race conditions, the source transformation tool generates derivative code that uses synchronization or additional memory to avoid the race condition. Both, synchronization and additional memory, imply overhead that should be avoided whenever possible.

This motivates a static program analysis that provides as much information as needed to avoid synchronization constructs. With this information at hand, the source transformation tool only augments these adjoint assignments with a synchronization construct where the static program analysis cannot ensure that the original assignment from P fulfills the exclusive read property. The dataflow analysis that we introduce in this chapter is called *exclusive read analysis* (ERA). The target language of the ERA is *SPL* that we defined in Section 2.2. The main part of ERA is basically spoken an integer interval analysis.

Beginning at the point where the parallel directive is placed, the analysis propagates an approximation of the possible value range of all integer variables through all the possible computation paths. This is done by solving an equation system iteratively. The equations are called transfer functions and represent the dataflow of the code. The iterative process finishes with a fixed point that hopefully provides enough information to decide the exclusive read property. For example, when a statement assigns a constant value to an integer, we propagate this information further through the transfer functions. The interval of the value range of each integer variable changes depending on what assignments each variable encounters. Unfortunately, the theorem of Rice [68] shows that we can only expect an approximation of the actual value range at compile time.

For simplification purposes, we assume that the number of threads $p \in \mathbb{N}$ divides the number of data elements $n \in \mathbb{N}$. This implies that $n = k \cdot p$ for a $k \in \mathbb{N}$ and therefore the inequality $n \geq p$ is valid.

A detailed introduction into static program analysis is beyond the scope of this thesis. Therefore, we refer to textbooks with a detailed description of dataflow analyses [59, 43]. A more common introduction into the domain of static program analyses used in nowadays compiler can be found in [4, 79, 56, 31, 1, 42, 80].

As an example, let us consider a private integer variable, say i. With the information about the value range of i for each thread, the source transformation tool is able to decide whether or not a read access to a floating-point memory location, say x[i], is exclusive or not. This is the case if during runtime each thread sees its own range of values in variable i. For example, thread 0 uses values from zero to four during the runtime whereby thread 1 encounters the values five to nine in variable i. In this particular case, the memory access to x[i] fulfills the exclusive read property. In case that both threads use the whole value range from zero to nine for variable i, the memory access to x[i] is not exclusive since both threads read all the ten elements of the array x. The difficulty is to decide at compile time what value range the threads will see at runtime.

Example 23. The following example shows an implementation of a computation of $n \cdot \frac{n}{5}$ polynomial values. The n^2 coefficients of the polynomials are stored in a two dimensional array A which means that the highest degree of the polynomials is four. The points where the polynomials should be evaluated are stored in the one dimensional array x.

A possible application where such a code may occur is the Galerkin method in the context of the finite element method [17, 8]. The finite element method is nowadays very popular for finding a numerical solution for elliptic and parabolic partial differential equations. Simply speaking, an approximation of the solution is obtained by computing the value of a linear combination of a set of polynomial basis functions.

As introduced in Chapter 1, we use the SPMD model to divide the polynomials among the group of threads. The data decomposition is done by computing a lower bound and an upper bound. This is shown in Listing 3.1, line 14 to line

18. To minimize the effect of false sharing[1], the sum of the polynomial values is first computed in a private variable and afterwards at the end of the parallel region copied into the shared array thread_result. For readability reasons, the computation of the polynomial value is coded as one would probably calculate it on a sheet of paper. In general holds that a parallel evaluation of a polynomial is possible in $\log_2 n$ steps where n is the polynomial degree [46, 65].

```
1   #pragma omp parallel
2   {
3       int  tid←0;
4       int  p←0;
5       int  c←0;
6       int  lb←0;
7       int  ub←0;
8       int  i←0;
9       int  j←0;
10      int  k←0;
11      double  y;
12      double  local_sum←0.;
13
14      tid←omp_get_thread_num();
15      p←omp_get_num_threads();
16      c←n/p;
17      lb←tid*c;
18      ub←(tid+1)*c−1;
19      i←lb;
20      thread_result_tid←0.;
21      while(i ≤ ub) {
22          y←0.;
23          j←0;
24          k←n−5;
25          while(j ≤ k) {
26              y ← A_{i*n+j+4}*(x_i*x_i*x_i*x_i) + A_{i*n+j+3}*(x_i*x_i*x_i)
27                  + A_{i*n+j+2}*(x_i*x_i) + A_{i*n+j+1}*(x_i) + A_{i*n+j};
28              local_sum+←y;
29              j←j+1;
30          }
31          i←i+1;
32      }
33      thread_result_tid+←local_sum;
```

---

[1]This effect often occurs in the context of cache coherence, see [66, p. 141] or [15, p. 155].

34 }

Listing 3.1: The array A contains the coefficients of $n \cdot \frac{n}{5}$ polynomial functions of degree four. A can be considered as a two dimensional array with n rows where each row contains n values. These values represent the coefficients of the polynomial functions. Each polynomial function that has its coefficients in row i is evaluated at point x_i. The n rows of A are divided into chunks where each thread gets a chunk of rows. The polynomial function values of each chunk are evaluated concurrently. Each thread puts the sum of the polynomial values into the private variable local_sum. At the end of the parallel region the value of the local sum is put into a component of the shared array thread_result.

Our focus in this example code are the lines 26 and 27. The different elements inside the arrays A and x are accessed by different offsets. The offsets themselves are composed of integer expressions. Our goal in this chapter is to establish a static program analysis that decides whether or not these offsets are used only by one thread of probably by multiple threads.

For example, variable i takes only values between the lower bound lb and the upper bound ub. As shown in Listing 3.1, these boundaries depend on the individual thread identifier number provided by omp_get_thread_num() and the size of the chunk c that each thread should process. The chunk size depends on the overall number of elements n that should be processed and the number of threads that is supplied by omp_get_num_threads(). The upper bound of thread t ub^t has the offset value that is one below the lower bound offset value of thread $(t+1)$, namely lb^{t+1}. Therefore, when we use t as a thread ID, p as the number of concurrent threads, and n as the number of data elements, then we can write the range of values that i takes in thread t as:

$$t \cdot \frac{n}{p} \leq i \leq (t+1) \cdot \frac{n}{p} - 1 \tag{3.1}$$

When we multiply with n we get:

$$t \cdot \frac{n}{p} \cdot n \leq i \cdot n \leq (t+1) \cdot \frac{n}{p} \cdot n - n$$

By respecting that variable j takes values from zero to (n-5), we obtain the values of the expression $i \cdot n + j$ for thread t:

$$t \cdot \frac{n}{p} \cdot n \leq i \cdot n + j \leq (t+1) \cdot \frac{n}{p} \cdot n - 5 \tag{3.2}$$

The expressions in (3.2) depend on the unique thread identification number t, the number of threads p, the size of the data n, and a linear term. In case that we propagate the information about the possible value range of all integer variables in P along the dataflow, we probably may perform an interval analysis.

In order to decide the exclusive read property, we examine whether or not the intervals of the threads t and $t + 1$ overlap. For example, let us consider inequality (3.2). We check that the value of the upper bound of thread t is smaller than the value of the lower bound of thread $t + 1$:

$$(t+1) \cdot \frac{\mathsf{n}}{p} \cdot \mathsf{n} - 5 < (t+1) \cdot \frac{\mathsf{n}}{p} \cdot \mathsf{n} \qquad (3.3)$$

$$-5 < 0$$

This inequality is valid for all values of t, n, and p. Hence, the memory reference A_{i*n+j} in line 26 fulfills the exclusive read property. The fact that memory reference x_i is read exclusively can be recognized from inequality (3.1). $\quad\triangle$

The following sections contain the base ingredients that are necessary to define the dataflow analysis called ERA. Section 3.1 displays how the control flow of a given *SPL* code is formalized. Our analysis and its formalism bases on the interval analysis presented in the textbook [59]. Therefore, we briefly present the idea of this integer interval analysis in Section 3.2. Since we need to track the value propagation from, for example, omp_get_thread_num() to the definition of the lower and upper boundaries for each thread, we need an extension of this interval analysis. Therefore, in Section 3.3 we define some order-theoretic foundations for directed acyclic graphs (DAG). DAGs are mainly used in our implementation of the source transformation tool. The value range of integer variables in our *SPL* code is going to be represented as an interval where the lower and the upper boundary are formed by a DAG. Hence, Section 3.4 clarifies how we define a partial order of intervals of DAGs. All these sections serve as introduction for the dataflow analysis that is defined in Section 3.5. The result of this dataflow analysis is an approximation of the value range of all integer variables in the given *SPL* code. This approximation is necessary to decide whether or not a memory reference that is addressed by a base address and an offset expression, is accessed only by one thread or maybe by multiple threads at the same time. This check is made similar as shown in Example 23 and is explained in Section 3.6.

3.1 Control Flow in *SPL* code

The control flow of a parallel region P can be represented with help of the *initial* and *final labels*. With these two definitions, we can define the *flow relation*.

Definition 48 (Initial and final labels). Every statement in *SPL* has a single entry which is called *initial label*. The mapping $init : SPL \to \mathbb{N}$ maps a given *SPL* statement to its *initial label*. Suppose P has the shape

$$\#\text{pragma omp parallel} \quad \{ \quad D \quad S \quad \}$$

then *init* is defined as follows:

$$
\begin{array}{ll}
init(P) := init(D) & \text{if } D \neq \emptyset \\
init(P) := init(S) & \text{if } D = \emptyset \\
init(D) := init(d_1) & \text{if } D = (d_1; d_2; \ldots; d_q), q \in \mathbb{N} \\
init(S) := init(s_1) & \text{if } S = (s_1; s_2; \ldots; s_q), q \in \mathbb{N}
\end{array}
$$

$$
init(l : d) := l \qquad \text{if } d =
\begin{cases}
\text{unsigned int } a \\
\text{unsigned int } a \leftarrow N \\
\text{int } a \\
\text{int } a \leftarrow Z \\
\text{double } x \\
\text{double } x \leftarrow R
\end{cases}
$$

$$
init(l : s) := l \qquad \text{if } s =
\begin{cases}
\text{STACK}_*.\text{push}(a) \\
\text{STACK}_*.\text{pop}() \\
a \leftarrow e \\
y \leftarrow \phi(x_1, \ldots, x_n) \\
y \leftarrow\!\!\!\leftarrow \phi(x_1, \ldots, x_n) \\
\text{if } (b) \{ S \} \\
\text{while } (b) \{ S \} \\
\#\text{pragma omp barrier}
\end{cases}
$$

The mapping *final* : $SPL \to \mathscr{P}(\mathbb{N})$ provides the set of exit labels.

$$
\begin{array}{ll}
final(P) := final(S) & \\
final(D) := d_q & \text{if } D = (d_1; d_2; \ldots; d_q), q \in \mathbb{N} \\
final(S) := s_q & \text{if } S = (s_1; s_2; \ldots; s_q), q \in \mathbb{N}
\end{array}
$$

$$\mathit{final}(l:d) := l \qquad\qquad \text{if } d = \begin{cases} \text{unsigned int } a \\ \text{unsigned int } a \leftarrow N \\ \text{int } a \\ \text{int } a \leftarrow Z \\ \text{double } x \\ \text{double } x \leftarrow R \end{cases}$$

$$\mathit{final}(l:s) := l \qquad\qquad \text{if } s = \begin{cases} \text{STACK}_*.\text{push}(a) \\ \text{STACK}_*.\text{pop}() \\ a \leftarrow e \\ y \leftarrow \phi(x_1,\ldots,x_n) \\ y +\!\!\leftarrow \phi(x_1,\ldots,x_n) \\ \#\text{pragma omp barrier} \end{cases}$$

$$\mathit{final}\,(\,l : \text{if } (b) \ \{ \ S \ \} \ ) := \{l\} \cup \mathit{final}(S)$$
$$\mathit{final}\,(\,l : \text{while } (b) \ \{ \ S \ \} \ ) := \{l\}$$

The final label in a branch statement is either the branch statement itself in case that the test expression b is not valid or if b is evaluated to true, the final label is the final label of the sub block S. The loop statement allows in our *SPL* language only the test expression b as exit. The break statement in C/C++ would make this definition more complex. $\square$

Definition 49 (Flow relation). The relation $\mathit{flow} \subset \mathbb{N} \times \mathbb{N}$ returns for a given *SPL* statement its control flow as a set of pairs of labels. The empty set as flow is defined for the following statements

$$\mathit{flow}(s) := \emptyset \qquad\qquad \text{if } s = \begin{cases} \text{STACK}_*.\text{push}(a) \\ \text{STACK}_*.\text{pop}() \\ a \leftarrow e \\ y \leftarrow \phi(x_1,\ldots,x_n) \\ y +\!\!\leftarrow \phi(x_1,\ldots,x_n) \\ \#\text{pragma omp barrier} \end{cases}$$

In case of a conditional or a loop statement, we define:

$$flow(l : \text{if } (b) \ \{ \ S \ \} \ ) := \ flow(S) \cup \{ (l, init(S)) \}$$
$$flow(l : \text{while } (b) \ \{ \ S \ \} \ ) := \ flow(S) \cup \{ (l, init(S)) \}$$
$$\cup \{ (l', l) \mid l' \in final(S) \}$$

In case of a sequence of definitions $D = (d_1; \ldots; d_q)$ or a sequence of statements $S = (s_1; \ldots; s_q)$ we define:

$$flow(D) := flow(d_1) \cup flow(d_2) \cup \ldots \cup flow(d_q)$$
$$\cup \{ (l, init(d_{i+1})) \mid l \in final(d_i) \} \text{ with } i \in \{1, \ldots, q-1\}$$
$$flow(S) := flow(s_1) \cup flow(s_2) \cup \ldots \cup flow(s_q)$$
$$\cup \{ (l, init(s_{i+1})) \mid l \in final(s_i) \} \text{ with } i \in \{1, \ldots, q-1\}$$

The overall flow of parallel region P is given by

$$flow(P) := flow(D) \cup flow(S)$$

This means that the flow of the parallel region consists of the join of the flow sets of the sequence of declarations and the sequence of statements. □

Example 24. In this example we show the application of Definition 48 and of Definition 49. Suppose that the OpenMP code of Listing 3.2 is given. The corresponding control flow graph is shown in Figure 3.2.

The code contains a loop that is executed by all threads as long as the shared variable t has a value below 100. t is increased only by the master thread (ID 0) while all the other threads wait at the OpenMP barrier construct.

```
1 double *x;
2 double *y;
3 double t←0.;
4 /* ... */
5 #pragma omp parallel
6 {
7    int tid←0;
```

```
8    tid←omp_get_thread_num();
9    y_tid←0.;
10   while(t<100.) {
11       y_tid←y_tid+t*sin(x_tid*x_tid);
12       #pragma omp barrier
13       if(tid = 0) {
14           t←t+1.;
15       }
16       #pragma omp barrier
17   }
18 }
```

Listing 3.2: Example code that is represented as a control flow graph in Figure 3.2. The barriers prevent the threads from computing the value of y_{tid} during the update of the master thread.

Let us consider the node 59 in Figure 3.2. It represents the branch statement in line 13 of Listing 3.2.

$$init(59 : \text{if (tid=0)} \{ t \leftarrow t+1.; \}) = \{59\}$$
$$final(59 : \text{if (tid=0)} \{ t \leftarrow t+1.; \}) = \{57, 59\}$$

The *init* set of node 59 contains only the label 59 because the control flow of the program first tests the boolean expression before it decides where to continue the execution. The *final* set of node 59 is $\{57, 59\}$. The label 59 corresponds to the case that the boolean expression is false and the runtime execution continues with the code following the conditional statement, here the next code statement is the barrier construct. In case that the boolean expression is true, the execution continues with the code labeled with 57 which is also the last statement inside the body of the conditional statement. Finally, the *flow* set of the conditional statement is given by

$$flow(59 : \text{if (tid=0)} \{ t \leftarrow t+1.; \}) = flow(57 : t \leftarrow t+1.) \cup \{(59, 57)\}$$
$$= \{(59, 57)\}$$

which corresponds with the edge going from node 59 to 57 in Figure 3.2. The edge going from node 59 to node 60 is evoked by the *flow* relation of the sequence of statements that represents the body of the loop in line 10.   △

3.2 Integer Interval Analysis

The theory of partially ordered sets and complete lattices play an important role in program analysis. However, we want to drop the presentation of the foundations of this theory and refer to [59, appx. A] for details. We define the complete lattice $(Interval, \sqsubseteq)$ of intervals over $\mathbb{Z}$ corresponding to the definition in [59, p. 222]:

$$Interval = \{\perp\} \cup \{[z_1, z_2] \mid z_1 \leq z_2, z_1 \in \mathbb{Z} \cup \{-\infty\}, z_2 \in \mathbb{Z} \cup \{\infty\}\}$$

We extend the common ordering $\leq$ on $\mathbb{Z}$ to an ordering on $\mathbb{Z}' = \mathbb{Z} \cup \{-\infty, \infty\}$ by setting $-\infty \leq z$, $z \leq \infty$ for all $z \in \mathbb{Z}$, and $-\infty \leq \infty$. The element $\perp$ is called bottom and denotes the least element in $Interval$, which is represented by the empty interval. The interval $[z_1, z_2]$ is from z_1 to z_2 including the end points if they are in $\mathbb{Z}$. The relation $[y_1, y_2] \sqsubseteq [z_1, z_2]$ corresponds to

$$\{y \mid y \in [y_1, y_2]\} \subseteq \{z \mid z \in [z_1, z_2]\}.$$

The empty interval $\perp$ is a subset of all intervals $I \in Interval$. Figure 3.1 clarifies the idea behind the partial ordering $\sqsubseteq$ on $Interval$. The figure can be read from the bottom to the top with a "contained in" relation. The $\perp$ element is contained in all intervals. The intervals with only one integer number are contained in intervals with two integer numbers and so forth. The greatest interval in $Interval$ is $[-\infty, \infty]$ that contains all integer numbers together with $-\infty$ and ∞. This interval is called the top element and is denoted by the symbol $\top$. We define the following order on the sets of intervals.

$$[y_1, y_2] \sqsubseteq [z_1, z_2] : \iff y_1 \geq z_1 \text{ and } y_2 \leq z_2 \tag{3.4}$$

The *least upper bound* is indicated by $\bigsqcup$ and the *greatest lower bound* is denoted by $\bigsqcap$. In case that every subset of $Interval$ has a least upper bound, we can conclude that $(Interval, \sqsubseteq)$ is a complete lattice, see [59, app. A]. Suppose J is a subset of $Interval$. We define the least upper bound of J as

$$\bigsqcup J = \begin{cases} \perp & \text{if } J = \perp \text{ or } J = \{\perp\} \\ [Z_1, Z_2] & \text{otherwise} \end{cases}$$

where

$$Z_1 := \bigsqcap \{n_1 \mid [n_1, n_2] \in J\} \tag{3.5}$$

$$Z_2 := \bigsqcup \{n_2 \mid [n_1, n_2] \in J\}. \tag{3.6}$$

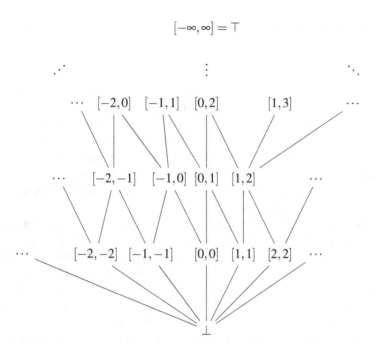

$$[-\infty, \infty] = \top$$

Figure 3.1: The complete lattice *Interval*. The empty interval $\bot$ is a subset of all intervals in *Interval*. The interval $[0,0]$ consists of the element zero and $[1,1]$ is comprised of the value one. Both, $[0,0]$ and $[1,1]$, are subsets of the interval $[0,1]$ that covers the values zero and one. Thus, an interval is contained in all intervals that lies on a path from this interval to the interval $[-\infty, \infty]$.

Please note that $\sqcap$ in (3.5) is the greatest lower bound of $\mathbb{Z} \cup \{-\infty\}$ and $\sqcup$ in (3.6) is the least upper bound of $\mathbb{Z} \cup \{\infty\}$. Each interval in J is surrounded by the interval $\sqcup J$. The definitions (3.5) and (3.6) ensure that $\sqcup J$ is indeed the least upper bound of J. Since J was chosen arbitrary each subset J of *Interval* has the least upper bound $\sqcup J$. Hence, $(Interval, \sqsubseteq)$ is a complete lattice.

This concludes the section about the base idea for our static program analysis. The next section introduces our approach for the exclusive read analysis where we extend the complete lattice $(Interval, \sqsubseteq)$ in such a way that the complete lattice does not consist of intervals of integer numbers but of intervals of DAGs. An analysis as shown in the current section would also be possible but this would lead to a relative coarse grained information. For example, the lower bound lb in Listing 3.1 is computed by

```
tid←omp_get_thread_num();
p←omp_get_num_threads();
c←n/p;
lb←tid*c;
```

In case that we use the analysis from this section we cannot trace back the information that lb depends on the thread identifier number, the number of threads, and the number of elements to process. In addition, intervals that result from expressions with several operations such as a multiplication tend to be an overestimation such that they reach the top element $\top = [0,\infty]$. This happens since the analysis has to be conservative and in case that the interval cannot be determined exactly the analysis information becomes the top element. The information of the top element is obviously not sufficient for our purpose.

For the exclusive read analysis, we need an information whether or not an integer variable has overlapping value ranges in different threads. Therefore, we have to go a step further and to track the value propagation of a computation with help of DAGs. These DAGs are used as lower and upper boundaries for intervals. After a fixed point computation, we yield intervals of DAGs that can be compared to achieve the information if certain intervals are used individually by threads or whether the occurring values are the same for different threads. The first step for extending the current analysis is to show that the set of possible directed acyclic graphs *DAG* and a relation $\sqsubseteq$ form a partial order.

3.3 Directed Acyclic Graphs and Partial Orders

Since we want to compare expressions as shown in inequality (3.2), we have to implement an internal representation of these expressions. Our source transformation tool makes extensive use of directed acyclic graphs (DAG). Therefore, we choose this data structure as representation for expressions. The disadvantage of this method is that the representation is not unique. Nevertheless, we choose this representation since it fits well into the implementation of our source transformation tool. Probably, this will be changed in a later implementation of the proposed static program analysis.

The set of possible directed acyclic graphs is referred to as *DAG*. We define a DAG such that it contain nodes from the set

$$V = \{a, Z, +, -, *, / \mid a \in INT_P, Z \in \mathbb{Z} \cup \{-\infty, \infty\}\}.$$

An edge from $v_1 \in V$ to $v_2 \in V$ represents a dataflow dependence from node v_1 to v_2. Nodes that represent an operation such as $+, -, *$ or $/$ have two incoming

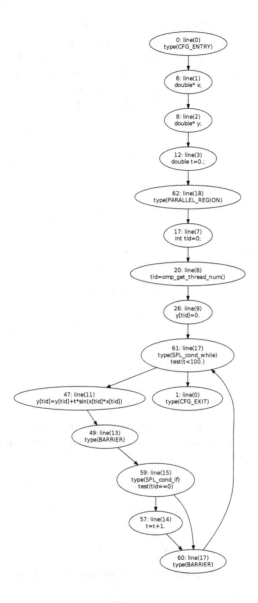

Figure 3.2: The corresponding control flow graph to Listing 3.2.

edges representing the dependence on the two operands.

What we want to achieve is a representation that is capable of displaying the information what value range a certain integer variable has. For example, let us consider inequality (3.1) where the value range of variable i is shown. The boundaries consist of expressions of integer variables. To express this inequality as an interval of DAGs, we note $[d_1, d_2]$ where the lower bound of the interval is defined by

$$d_1 :=$$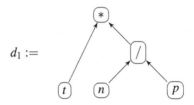

and the upper bound is

$$d_2 :=$$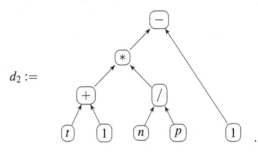

For simplicity reasons, we use the expression that the DAG represents to describe its appearance instead of showing the DAG as graph. For example, we simply write the above graphs as

$$d_1 := t * (n/p) \tag{3.7}$$

and

$$d_2 := ((t+1) * (n/p)) - 1. \tag{3.8}$$

In case that $d_l \in DAG$ only consist of one node and this node represents an integer number Z then we write

$$d_l = Z.$$

One of the essential operations on DAGs is the equality operation. We define two DAGs, d_l and d_r, as equal if and only if they consist of the exact same number of nodes and all the nodes have the same content and the same set of predecessors and successors. We express the equality by writing

$$d_l = d_r.$$

Please note that this definition does not cover cases where we have algebraic equality due to commutativity. In the simple implementation of the analysis that we want to introduce here, the expressions $(t \cdot \frac{n}{p})$ and $(\frac{n}{p} \cdot t)$ are not equal, as shown in Figure 3.3.

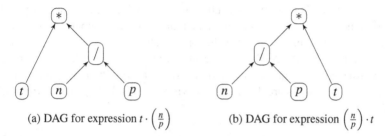

(a) DAG for expression $t \cdot \left(\frac{n}{p}\right)$ (b) DAG for expression $\left(\frac{n}{p}\right) \cdot t$

Figure 3.3: Both expressions are algebraically equal due to the commutativity of the multiplication of two natural numbers. However, we define these DAGs as not equal since they have a different structure.

The relation $\sqsubseteq: DAG \times DAG \to \{true, false\}$ is defined for $d_a, d_b \in DAG$ as

$$d_a \sqsubseteq d_b :=$$

$$
\begin{cases}
true & \text{if } d_a = d_b \\
true & \text{if } d_a = 0 \text{ and } d_b \in \mathbb{N}_0 \\
true & \text{if } d_a = -\infty \in \mathbb{Z} \cup \{-\infty\} \\
true & \text{if } d_b = \infty \in \mathbb{Z} \cup \{\infty\} \\
Z_a \leq Z_b & \text{if } d_a = Z_a, d_b = Z_b, \text{ where } Z_a, Z_b \in \mathbb{Z} \\
true & \text{if } d_b = d_a + d, \text{ where } d = Z \in \mathbb{Z},\ Z \geq 0 \\
true & \text{if } d_a = d_1,\ d_b = d_2, \text{ see } (3.7),(3.8) \\
true & \text{if } d_a = ((((t+1)*(n/p))-1)*n)+(n-1), \\
      & \text{and } d_b = ((t+1)*(n/p))*n \\
false & \text{otherwise}
\end{cases}
\tag{3.9}
$$

where the need for the certain expressions becomes more clear at the end of this chapter. The relation provides true in case that both DAGs are equal. When d_a

represents the smallest element in the domain which is $-\infty$, or when d_b represents the biggest element ∞ then the relation $d_a \sqsubseteq d_b$ is true. Another special case is that d_a represents zero and the other DAG represents some value out of the natural number range with zero. In this case zero is for sure the lowest boundary and the relation provides also the value true. If both DAGs represent an integer numbers, say Z_a and Z_b, we use the relation $Z_a \leq Z_b$ to determine the result where $\leq \subseteq \mathbb{Z} \times \mathbb{Z}$ is the general ordering of integer numbers.

It often occurs the situation where an integer variable is incremented and this variable is used as an offset for accessing some data. This means for our relation that we have two DAGs from consecutive iterations with a similar structure. The only difference between them is an increment of an integer number, let us say Z. If d_b is d_a plus Z and Z has a positive value or the value is zero then $d_a \sqsubseteq d_b$ is true.

The remaining two cases where the result is true are special cases which reflect the often used SPMD data decomposition code patterns. The first special case is the one where the DAGs are defined as shown in (3.7) and (3.8). The corresponding inequality

$$t \cdot \frac{n}{p} \leq \left((t+1) \cdot \frac{n}{p} \right) - 1 \tag{3.10}$$

can be rewritten as

$$t \cdot \frac{n}{p} \leq t \cdot \frac{n}{p} + \frac{n}{p} - 1$$

what leads to

$$1 \leq \frac{n}{p} \quad . \tag{3.11}$$

As a reminder, n represents here the number of data elements that should be processed by the group of threads. p is the number of threads that process the n elements. Therefore, n and p are positive natural numbers and we obtain that the inequality (3.10) is true for $n \geq p$. We assume for our program analysis that the number of data sets is always bigger than the number of threads. In practice this is almost always the case but has to be tested by an actual implementation.

The second special case reflects the following inequality.

$$\left(\left((t+1) \cdot \frac{n}{p} \right) - 1 \right) \cdot n + n - 1 \leq (t+1) \cdot \frac{n}{p} \cdot n$$

$$(t+1) \cdot \frac{n}{p} \cdot n - n + n - 1 \leq (t+1) \cdot \frac{n}{p} \cdot n \tag{3.12}$$

$$-1 \leq 0$$

Therefore, the inequality is always true, no matter what the actual values of n, p, and t are. In order to show that $(DAG, \sqsubseteq)$ is a partial order, we have to show that the relation $\sqsubseteq \subseteq DAG \times DAG$ from (3.9) is reflexive, transitive, and anti-symmetric.

Lemma 50. $(DAG, \sqsubseteq)$ *is a partial order where DAG is the domain of possible DAGs and the relation* $\sqsubseteq \subseteq DAG \times DAG$ *is an ordering relation. This means for* $d_1, d_2, d_3 \in DAG$ *it holds:*

1. *Reflexivity:* $d_1 \sqsubseteq d_1$

2. *Transitivity:* $d_1 \sqsubseteq d_2 \wedge d_2 \sqsubseteq d_3 \implies d_1 \sqsubseteq d_3$

3. *Anti-Symmetry:* $d_1 \sqsubseteq d_2 \wedge d_2 \sqsubseteq d_1 \implies d_1 = d_2$

Proof. Let $d_1, d_2, d_3 \in DAG$ be out of the set of possible *DAGs*.

1. Reflexivity: $d_1 \sqsubseteq d_1$ is true since this situation is tested by the first case in (3.9)

2. Transitivity: Let us assume that

$$d_1 \sqsubseteq d_2 \wedge d_2 \sqsubseteq d_3$$

holds. Since $d_1 \sqsubseteq d_2$ is true, we consider the different cases from (3.9) where the relation returns the value true.

a) $d_1 = d_2$: With $d_2 \sqsubseteq d_3$ we conclude $d_1 = d_2 \sqsubseteq d_3$.

b) $d_1 = -\infty$: In this case $d_1 \sqsubseteq d_3$ is true independent of the value in d_3.

c) $d_2 = \infty$: From $d_1 \sqsubseteq d_2$ and $d_2 \sqsubseteq d_3$ we conclude that $d_3 = \infty$ and therefore $d_1 \sqsubseteq d_3$.

d) $d_1 = Z_1$ and $d_2 = Z_2$ with $Z_1, Z_2 \in \mathbb{Z}$ and $Z_1 \leq Z_2$: From $d_2 \sqsubseteq d_3$ we infer $d_1 \sqsubseteq d_3$ with help of the relation $\leq \subseteq \mathbb{Z} \times \mathbb{Z}$.

e) Suppose $d_1 \sqsubseteq d_1 + d$ is true where $d = Z \in \mathbb{Z}$. Therefore, we get $d_2 = d_1 + d \sqsubseteq d_3$ and with $Z \geq 0$ we conclude $d_1 \sqsubseteq d_3$.

f) $d_1 = t \cdot \frac{n}{p}$, $d_2 = ((t+1) \cdot \frac{n}{p}) - 1$: We know from (3.10) that with $n \geq p$ it holds

$$d_1 = t \cdot \frac{n}{p} \sqsubseteq ((t+1) \cdot \frac{n}{p}) - 1 = d_2.$$

With $d_2 \sqsubseteq d_3$ we conclude $d_1 \sqsubseteq d_3$.

g) $d_1 = ((((t+1) \cdot \frac{n}{p}) - 1) \cdot n) + (n-1)$ and $d_2 = ((t+1) \cdot \frac{n}{p}) \cdot n$: We showed in (3.12) that $d_1 \sqsubseteq d_2$ is always valid. Therefore, together with $d_2 \sqsubseteq d_3$, we conclude $d_1 \sqsubseteq d_3$.

3. Anti-Symmetry:

a) $d_1 = d_2$: In this case nothing is to show.

b) $d_1 = -\infty$: From $-\infty \sqsubseteq d_2$ and $d_2 \sqsubseteq -\infty$ we conclude $d_2 = -\infty$.

c) $d_2 = \infty$: With $d_1 \sqsubseteq \infty$ and $\infty \sqsubseteq d_1$ we achieve $d_1 = \infty$.

d) $d_1 = Z_1$ and $d_2 = Z_2$, where $Z_1, Z_2 \in \mathbb{Z}$: With $Z_1 \leq Z_2$ and $Z_2 \leq Z_1$ we conclude $Z_1 = Z_2$.

e) $d_2 = d_1 + d$ where $d = Z \in \mathbb{Z}$: Since $d_1 \sqsubseteq d_1 + Z \sqsubseteq d_1$, and $Z \geq 0$ the only possible value for Z is 0. Thus, it holds $d = 0$ and $d_1 = d_2$.

f) $d_1 = t \cdot \frac{n}{p}$ and $d_2 = ((t+1) \cdot \frac{n}{p}) - 1$: From (3.11) we know that this leads to the inequalities $p \leq n$ and $p \geq n$. Hence, it holds $n = p$ and therefore $d_1 = d_2$.

g)

$$d_1 = \left(\left(\left((t+1) \cdot \frac{n}{p}\right) - 1\right) \cdot n\right) + (n-1)$$

and

$$d_2 = \left((t+1) \cdot \frac{n}{p}\right) \cdot n :$$

In this case the relation $d_1 \sqsubseteq d_2$ is true but $d_2 \sqsubseteq d_1$ provides false due to (3.9).

$\square$

The greatest lower bound (GLB) of a subset of DAGs is denoted by $\bigsqcap$ and we define it recursively. This means that we trace back the GLB of a subset of DAGs to the problem of a pair of DAGs.

$$\bigsqcap\{d_1, d_2, d_3, \ldots, d_n\} = \bigsqcap\left\{\bigsqcap\{d_1, d_2, d_3, \ldots, d_{n-1}\}, d_n\right\}$$

$$\vdots$$

$$= \bigsqcap\left\{\bigsqcap\left\{\bigsqcap\left\{\ldots\bigsqcap\{d_1, d_2\}\ldots\right\}, d_{n-1}\right\}, d_n\right\}$$

Suppose that d_1 and d_2 are out of *DAG* and suppose that $n \geq p$, then the GLB of a set of two DAGs is defined as:

$$\sqcap\{d_1,d_2\} :=$$

$$
\begin{cases}
d_1 & \text{if } d_1 = d_2 \\
d_1 & \text{if } d_1 = 0 \text{ and } d_2 \in \mathbb{N}_0 \\
d_1 & \text{if } d_1 = -\infty \in \mathbb{Z} \cup \{-\infty\} \\
d_2 & \text{if } d_2 = -\infty \in \mathbb{Z} \cup \{-\infty\} \\
d_1 & \text{if } d_2 = \infty \in \mathbb{Z} \cup \{\infty\} \\
d_2 & \text{if } d_1 = \infty \in \mathbb{Z} \cup \{\infty\} \\
\sqcap\{Z_1,Z_2\} & \text{if } d_1 = Z_1, d_2 = Z_2, \text{ where } Z_1, Z_2 \in \mathbb{Z} \\
d_1 & \text{if } d_1, d_2 \text{ given as shown in } (3.7), (3.8) \\
d_1 & \text{if } d_1 = ((((t+1)*(n/p))-1)*n)+(n-1), \\
& \text{and } d_2 = ((t+1)*(n/p))*n \\
d_1 & \text{if } d_1 = d_3 + d_4, d_2 = d_3 + d_5, \text{ with} \\
& d_3, d_4, d_5 \in DAG, d_4 = Z_4, d_5 = Z_5, \text{ and} \\
& Z_4, Z_5 \in \mathbb{Z}, Z_5 - Z_4 \geq 0 \\
-\infty & \text{otherwise}
\end{cases}
\tag{3.13}
$$

The first case provides d_1 as result when both DAGs are equal. In case that one DAG represents $-\infty$ we return this DAG since $-\infty \leq z$ for all integer numbers $z \in \mathbb{Z} \cup \{-\infty\}$. This holds analogously for the case where one DAG represents ∞. If both DAGs represent integer numbers Z_1 and Z_2, we use the relation $\leq \subseteq \mathbb{Z} \times \mathbb{Z}$ to determine the GLB. Afterwards, we handle the two special cases that we already introduced in (3.10) and (3.12). During the fixed point iteration there is often a comparison of two DAGs necessary where the difference between them is only an addition or a subtraction of an integer number. Therefore, this case is also covered in (3.13).

The fact that we define the GLB only for a set of two DAGs can be exploited in the definition of the lowest upper bound (LUB). The LUB of two DAGs is defined by the result of the GLB of these two DAGs.

$$
\sqcup\{d_1,d_2\} :=
\begin{cases}
d_1 & \text{if } \sqcap\{d_1,d_2\} = d_2 \\
d_2 & \text{if } \sqcap\{d_1,d_2\} = d_1 \\
\infty & \text{otherwise}
\end{cases}
\tag{3.14}
$$

We defined the partial order $(DAG, \sqsubseteq)$ which is indeed a partial order as Lemma 50 showed. In addition, we defined the least upper and the greatest lower bound of two DAGs. The next section defines the partial order that has intervals of DAGs as domain.

3.4 Intervals of Directed Acyclic Graphs

In this section, we want to define a partial order where the domain consist of intervals of DAGs. The interval boundaries are given by DAGs as they are defined in Section 3.3. This can be seen as extension of the approach in Section 3.2 where we defined intervals of integer numbers. The set of intervals containing all possible DAG intervals is referred to as *DAGIntervals* and is defined by

$$DAGIntervals := \{\bot\} \cup \{ [d_1, d_2] \mid d_1 \sqsubseteq d_2, \ d_1, d_2 \in DAG \} \ .$$

The ordering $\sqsubseteq$ on *DAG* is the one defined in (3.9). The empty interval is denoted by $\bot$. Therefore, we initialize at the begin of the dataflow analysis all information with $\bot$. The relation $\sqsubseteq$ of two intervals is defined by the equivalence

$$[d_1, d_2] \sqsubseteq [d_3, d_4] \quad :\Leftrightarrow \quad d_3 \sqsubseteq d_1 \wedge d_2 \sqsubseteq d_4 \tag{3.15}$$

The greatest interval in *DAGIntervals* is $[-\infty, \infty]$ where $-\infty$ and ∞ are DAGs with only one node with the corresponding content. $[-\infty, \infty]$ covers all integer numbers. The expressions represented by *DAG* can all be evaluated to have a value out of $\mathbb{Z} \cup \{-\infty, \infty\}$. This means that $[-\infty, \infty] \in DAGIntervals$ contains all DAGs that can be evaluated to have an integer number as a result. The interval $[-\infty, \infty] \in DAGIntervals$ is called the top element and is denoted by the symbol $\top$.

Lemma 51. *(DAGIntervals, $\sqsubseteq$) is a partial order. This means for*

$$[d_1, d_2], [d_3, d_4], [d_5, d_6] \in DAGIntervals$$

with $d_1, d_2, d_3, d_4, d_5, d_6 \in DAG$ it holds:

- *Reflexivity: $[d_1, d_2] \sqsubseteq [d_1, d_2]$*

- *Transitivity: $[d_1, d_2] \sqsubseteq [d_3, d_4] \wedge [d_3, d_4] \sqsubseteq [d_5, d_6] \implies [d_1, d_2] \sqsubseteq [d_5, d_6]$*

- *Anti-Symmetry: If $[d_1, d_2] \sqsubseteq [d_3, d_4]$ and $[d_3, d_4] \sqsubseteq [d_1, d_2]$ are valid then it holds $[d_1, d_2] = [d_3, d_4]$.*

Proof. Let $[d_1, d_2], [d_3, d_4], [d_5, d_6]$ be out of the domain *DAGIntervals* with $d_1, d_2,$ $d_3, d_4, d_5, d_6 \in DAG.$

- Reflexivity: We obtain $[d_1, d_2] \sqsubseteq [d_1, d_2]$ if and only if $d_1 \sqsubseteq d_1 \wedge d_2 \sqsubseteq d_2$ from (3.15). The reflexivity of DAGs was shown in Lemma 50.

- Transitivity:

$$[d_1, d_2] \sqsubseteq [d_3, d_4] \quad \wedge \quad [d_3, d_4] \sqsubseteq [d_5, d_6]$$
$$\stackrel{(3.15)}{\Longleftrightarrow} \quad d_3 \sqsubseteq d_1 \wedge d_2 \sqsubseteq d_4 \quad \wedge \quad d_5 \sqsubseteq d_3 \wedge d_4 \sqsubseteq d_6$$
$$\Longrightarrow \quad d_5 \sqsubseteq d_3 \sqsubseteq d_1 \quad \wedge \quad d_2 \sqsubseteq d_4 \sqsubseteq d_6$$
$$\Longrightarrow \quad [d_1, d_2] \sqsubseteq [d_5, d_6]$$

- Anti-Symmetry:

$$[d_1, d_2] \sqsubseteq [d_3, d_4] \quad \wedge \quad [d_3, d_4] \sqsubseteq [d_1, d_2]$$
$$\stackrel{(3.15)}{\Longleftrightarrow} \quad d_3 \sqsubseteq d_1 \wedge d_2 \sqsubseteq d_4 \quad \wedge \quad d_1 \sqsubseteq d_3 \wedge d_4 \sqsubseteq d_2$$
$$\Longrightarrow \quad d_1 \sqsubseteq d_3 \sqsubseteq d_1 \quad \wedge \quad d_2 \sqsubseteq d_4 \sqsubseteq d_2$$
$$\Longrightarrow \quad d_1 = d_3 \quad \wedge \quad d_2 = d_4$$
$$\Longrightarrow \quad [d_1, d_2] = [d_3, d_4]$$

This shows that $(DAGIntervals, \sqsubseteq)$ is a partial order. □

The next step is to show that the partial order $(DAGIntervals, \sqsubseteq)$ is a complete lattice. We claim that each subset of $DAGIntervals$ has a least upper bound. With this claim, we can conclude that $(DAGIntervals, \sqsubseteq)$ is a complete lattice [59, app. A]. Suppose that J is a subset of $DAGIntervals$. We define the least upper bound of J in the following way

$$\bigsqcup J = \begin{cases} \bot & \text{if } J = \bot \text{ or } J = \{\bot\} \\ [D_1, D_2] & \text{otherwise} \end{cases}$$

where D_1 and D_2 are defined with help of the GLB and LUB of DAG.

$$D_1 := \bigsqcap \{d_1 \mid [d_1, d_2] \in J\} \tag{3.16}$$
$$D_2 := \bigsqcup \{d_2 \mid [d_1, d_2] \in J\}. \tag{3.17}$$

D_1 is the greatest lower bound of the set of all left-sided boundaries of intervals in J. D_2 is the lowest upper bound of all right-sided boundaries of intervals in J. The definitions (3.16) and (3.17) ensure that all intervals in J are contained in $\bigsqcup J$.

The fact that $\bigsqcup J$ is indeed the least upper bound of J can be understood by assuming that we have another upper boundary $[l_1, l_2]$ with

$$[l_1, l_2] \sqsubseteq \bigsqcup J = [D_1, D_2].$$

This can be rewritten equivalently as

$$D_1 \sqsubseteq l_1 \wedge l_2 \sqsubseteq D_2 \quad.$$

Since $[l_1, l_2]$ is an upper boundary of J it holds

$$[d_1, d_2] \sqsubseteq [l_1, l_2]$$

for all intervals $[d_1, d_2]$ contained in J. Because of (3.15) this is equivalent to

$$l_1 \sqsubseteq d_1 \wedge d_2 \sqsubseteq l_2.$$

D_1 is defined as GLB in (3.16). Thus, it holds $D_1 \sqsubseteq d_1$ for all $[d_1, d_2] \in J$. Together with the above relations we obtain

$$D_1 \sqsubseteq l_1 \sqsubseteq d_1$$

for the left-sided boundaries. We reason analogously for the right-sided boundaries, which brings us to

$$d_2 \sqsubseteq l_2 \sqsubseteq D_2$$

because D_2 is defined as LUB in (3.17). The fact that D_1 and D_2 are boundaries that indeed occur in intervals $[d_1, d_2] \in J$ leaves the only conclusion that

$$l_1 = D_1 \wedge l_2 = D_2 \quad.$$

Therefore, $\bigsqcup J$ is indeed the least upper bound of J and since J was chosen arbitrary we imply that each $J \subseteq DAGIntervals$ has a least upper bound. This shows that $(DAGIntervals, \sqsubseteq)$ is a complete lattice.

With the partial order $(DAGIntervals, \sqsubseteq)$ that we defined in this section, we have the core requirement to define a dataflow analysis with intervals of DAGs. The next sections shows how we propagate the analysis information along the control flow.

3.5 Data Flow Analysis with DAG Intervals

In this section, we introduce transfer functions ϕ_l for transferring the analysis information δ from one code statement $[s]^l$ with the label $l \in LABEL(P)$ to another code statement. δ is a mapping that provides the connection between integer variables of the parallel region and its corresponding value range interval. For evaluating expressions we have to define how intervals of DAGs should be combined when the corresponding variables are part of an expression. For example, a multiplication of two expressions leads necessarily to an operation that combines the analysis information of these two expressions. With these definitions, we can define the evaluation function $val_\delta(e)$ that evaluates the expression e depending on the analysis information δ.

After the definition of the transfer functions ϕ_l we will have to ensure that our static analysis reaches a fixed point. This problem is solved by introducing a widening operator in Section 3.5.1. We will see that this method tends to supply an overestimate what motivates the definition of a narrowing operator. Because of the need of very precise interval information, Section 3.5.2 presents a further improvement of our program analysis by taking the test expressions of loops and conditional branches into account.

Our analysis is a forward problem which means that the dataflow is analyzed from the beginning to the end of the control flow. Therefore, the analysis starts at $init(P)$ and ends at an element of the set $final(P)$. The analysis information at the entry is initialized such that all integer variables that have a shared status are assumed to have a certain well defined value. For example, the size of the data that the threads are going to process can be contained in the integer variable n. The value of n must be set outside the parallel region and all the threads need this variable to calculate the size of the chunk that they have to process. Therefore, variable n is shared among the group of threads and the interval for this variable is defined to $[n, n]$ at the entry of the parallel region.

The possible value range of an integer variable a is represented by intervals out of *DAGIntervals*. The association of a variable $a \in INT_P$ with its value range is done by the mapping $\delta : INT_P \to DAGIntervals$. The complete lattice in our program analysis is therefore $(D, \sqsubseteq)$ where the domain D is defined by

$$D := \{\delta \mid \delta : INT_P \to DAGIntervals\} \tag{3.18}$$

and the relation $\sqsubseteq \subseteq D \times D$ corresponds the following equivalence:

$$\delta_1 \sqsubseteq \delta_2 \quad :\Leftrightarrow \quad \delta_1(a) \sqsubseteq \delta_2(a) \quad \forall a \in INT_P$$

We need a set of operations on intervals from *DAGIntervals*. For example, let us consider a right-hand side of an integer assignment with an addition $a_1 + a_2$. The static program analysis has the interval information for a_1 as well as for a_2, but we have to define what the resulting interval of combining these two intervals is. We define the operations $\oplus$, $\ominus$, $\otimes$, and $\oslash$ for combining intervals where the corresponding operations on the variables are $+, -, *$ and $/$. The DAGs d_1, d_2, d_3, and d_4 represent the boundaries of the two intervals that are combined by such an operation. The first operation is the addition of two intervals which is defined as

$$[d_1, d_2] \oplus [d_3, d_4] := \begin{cases} \bot & \text{if } [d_1, d_2] = \bot \text{ or } [d_3, d_4] = \bot \\ [d_1 + d_3, d_2 + d_4] & \text{otherwise} \end{cases}$$

In case of a subtraction, the operation $\ominus$ is defined as follows.

$$[d_1, d_2] \ominus [d_3, d_4] := \begin{cases} \bot & \text{if } [d_1, d_2] = \bot \text{ or } [d_3, d_4] = \bot \\ [d_1 - d_4, d_2 - d_3] & \text{otherwise} \end{cases}$$

The multiplication $\otimes$ and the division operation $\oslash$ are defined with help of the GLB and the LUB.

$$[d_1, d_2] \otimes [d_3, d_4] := \begin{cases} \bot & \text{if } [d_1, d_2] = \bot \text{ or } [d_3, d_4] = \bot \\ [D_5, D_6] & \text{otherwise} \end{cases}$$

where

$$D_5 = \bigsqcap \{d_1 * d_3, d_1 * d_4, d_2 * d_3, d_2 * d_4\},$$

and

$$D_6 = \bigsqcup \{d_1 * d_3, d_1 * d_4, d_2 * d_3, d_2 * d_4\}.$$

Please note that for better readability reasons, we use the more common notation of a set than the pairwise notation that we use in the definition of the GLB and LUB in (3.13) and (3.14). The remaining definition is the one for an expression where a division is involved.

$$[d_1, d_2] \oslash [d_3, d_4] := \begin{cases} \bot & \text{if } [d_1, d_2] = \bot \text{ or } [d_3, d_4] = \bot \\ [D_5, D_6] & \text{otherwise} \end{cases}$$

where

$$D_5 = \bigsqcap \{d_1/d_3, d_1/d_4, d_2/d_3, d_2/d_4\},$$

and

$$D_6 = \bigsqcup \{d_1/d_3, d_1/d_4, d_2/d_3, d_2/d_4\}.$$

With the above definitions we can define a mapping *val* that takes variables, constants, or expressions as input and provides the corresponding interval from

DAGIntervals as output. The simplest case is when *val* is applied to a constant $Z \in \mathbb{Z}$ because the resulting interval is $[Z,Z]$.

$$val_\delta(Z) := [Z,Z], \text{ where } Z \in \mathbb{Z}$$

The current value range of a variable $a \in INT_P$ can be determined by $\delta(a)$.

$$val_\delta(a) := \delta(a), \text{ where } a \in INT_P$$

When an operator is applied to two expressions e_1 and e_2, *val* is recursively applied to the operands and the two results are combined by the corresponding operation for intervals.

$$val_\delta(e_1 + e_2) := val_\delta(e_1) \oplus val_\delta(e_2)$$
$$val_\delta(e_1 - e_2) := val_\delta(e_1) \ominus val_\delta(e_2)$$
$$val_\delta(e_1 * e_2) := val_\delta(e_1) \otimes val_\delta(e_2)$$
$$val_\delta(e_1 / e_2) := val_\delta(e_1) \oslash val_\delta(e_2)$$

In case that a right-hand side expression corresponds to an OpenMP runtime function, *val* returns a specific interval. The function omp_get_num_threads() returns the number of threads in the current parallel execution. Therefore, a call to this function provides the interval $[p,p]$. The value p represents the number of threads executing the current parallel region. In case that *val* gets the OpenMP function omp_get_thread_num() as input, it returns the interval $[t,t]$ where t represents the unique thread identification number that all threads possess.

$$val_\delta(\text{omp\_get\_thread\_num}()) := [t,t]$$
$$val_\delta(\text{omp\_get\_num\_threads}()) := [p,p]$$

In the coming examples our implementation uses the specific intervals

[THREAD_ID, THREAD_ID]

and

[NUM_OF_THREADS, NUM_OF_THREADS]

to represent these runtime related values.

The remaining possible right-hand side of an integer assignment is the stack operation top() that pops the current top value from the integer stack STACK$_i$ or from the control flow stack STACK$_c$. Since we are not interested in following the value propagation into the stack data structure and back, we return the top element $\top$. This means that the variable can possess all values from $[-\infty, \infty]$.

$$val_\delta(\text{STACK}_{(i|c)}.\text{top}()) := \top$$

Eventually, we have the ingredients at hand to define the transfer functions

$$\{\phi_l \mid l \in LABEL(P)\}$$

where l is a label associated uniquely with a certain statement in P. The transfer function ϕ_l takes the dataflow information δ that is valid before the evaluation of the statement $[s]^l$ labeled with l and provides the dataflow information as output that is valid after the evaluation of $[s]^l$.

$$\phi_l(\delta) := \begin{cases} \delta & \text{if } [s]^l = \text{STACK}_{(i|c|f)}.\text{push}(a) \\ \delta & \text{if } [s]^l = \text{STACK}_{(i|c|f)}.\text{pop}() \\ \delta & \text{if } [s]^l = y \leftarrow \phi(x_1,\ldots,x_n) \\ \delta & \text{if } [s]^l = y \mathrel{+\!\!\leftarrow} \phi(x_1,\ldots,x_n) \\ \delta[a \to \bot] & \text{if } [s]^l = \text{unsigned int } a \\ \delta[a \to [N,N]] & \text{if } [s]^l = \text{unsigned int } a \leftarrow N \\ \delta[a \to \bot] & \text{if } [s]^l = \text{int } a \\ \delta[a \to [Z,Z]] & \text{if } [s]^l = \text{int } a \leftarrow Z \\ \delta[a \to val_\delta(e)] & \text{if } [s]^l = a \leftarrow e \\ \delta & \text{if } [s]^l = \text{if } (b) \ \{ \ S \ \} \\ \delta & \text{if } [s]^l = \text{while } (b) \ \{ \ S \ \} \end{cases} \tag{3.19}$$

As stated in (3.19), the dataflow information δ is only altered for declarations and assignments of integers, in all other cases δ keeps the same as it was before the statement. If the control flow encounters an integer declaration (int a) then the value of the variable a is undefined what is represented by the $\bot$ element. In case that the declaration of a is combined with an initialization with the value Z (int $a \leftarrow Z$), we change the information of a to be the interval $[Z,Z]$. An integer assignment $a \leftarrow e$ influences δ in such a way that it resets the information for a to the interval that is given by $val_\delta(e)$.

To convince the reader that our static program analysis reaches a fixed point, we present the theorem of Tarski and Knaster [45, 76]. This includes the introduction of further terminology. During the fixed point iteration the approximation of the intervals could possibly take the following *chain* of intervals:

$$\emptyset \sqsubseteq [2,2] \sqsubseteq [1,3] \sqsubseteq [0,4] \sqsubseteq [0,5] \sqsubseteq [0,6] \sqsubseteq [0,7] \sqsubseteq \cdots \tag{3.20}$$

A subset S of the partially ordered set *DAGIntervals* is called chain [59, app. A] if for all intervals $[l_1, u_1], [l_2, u_2] \in S$ pertains that

$$[l_1, u_1] \sqsubseteq [l_2, u_2] \text{ or } [l_2, u_2] \sqsubseteq [l_1, u_1].$$

A sequence $(l_i, u_i)_{i \in \mathbb{N}}$ of elements from the partial order $(DAGIntervals, \sqsubseteq)$ is called an *ascending chain* if $[l_i, u_i] \sqsubseteq [l_{i+1}, u_{i+1}]$ for each $i \in \mathbb{N}$. The termination of the fixed point iteration is guaranteed by the *Ascending Chain Condition* (ACC). A partial order $(DAGIntervals, \sqsubseteq)$ satisfies the ACC if and only if each ascending chain

$$[l_1, u_1] \sqsubseteq [l_2, u_2] \sqsubseteq \ldots$$

eventually stabilizes. This is there exists an $n \in \mathbb{N}$ such that

$$\ldots \sqsubseteq [l_{n-1}, u_{n-1}] \sqsubseteq [l_n, u_n] = [l_{n+1}, u_{n+1}] = \ldots$$

Unfortunately, the complete lattice property and the ACC property are unrelated. Therefore, we have to ensure that $(DAGIntervals, \sqsubseteq)$ satisfies the ACC.

In order to avoid a fixed point iteration that alternates between increasing and decreasing the approximation intervals, we must ensure that the transfer functions are *monotonic*. This means that

$$[l_1, u_1] \sqsubseteq [l_2, u_2] \implies \phi([l_1, u_1]) \sqsubseteq \phi([l_2, u_2])$$

for all $[l_1, u_1], [l_2, u_2] \in (DAGIntervals, \sqsubseteq)$. The definition of ϕ in (3.19) is the identity mapping, the constant values $\bot$ or $[N, N]$, or the interval is determined by $val_\delta(e)$. In case that $val_\delta(e)$ uses $\oplus$ and $\ominus$, the monotonicity is given because of the usage of the linear operations $+$ and $-$ to obtain the resulting interval. On the other hand, the operations $\otimes$ and $\oslash$ use the monotonicity of $\sqcup$ and $\sqcap$. This indicates that the transfer function ϕ is monotonic.

The *join operator* $\sqcup : DAGIntervals \times DAGIntervals \to DAGIntervals$ for joining two different data flow informations is defined as

$$\bot \sqcup d := d \sqcup \bot := d \quad , d \in DAGIntervals$$

$$[d_1, d_2] \sqcup [d_3, d_4] := [d_5, d_6]$$

$$\text{with } d_1, d_2, d_3, d_4, d_5, d_6 \in DAG \text{ and}$$

$$d_5 := \bigsqcap \{d_1, d_3\}$$

$$d_6 := \bigsqcup \{d_2, d_4\}.$$

To guarantee that our interval analysis find a solution we have to ensure that the requirements of the following theorem are fulfilled.

Theorem 52. (Fixpoint Theorem by Tarski and Knaster [45, 76]

Let $(D, \sqsubseteq)$ be a complete lattice satisfying ACC and $\phi : D \to D$ monotonic. Then

$$fix(\phi) := \bigsqcup \left\{ \phi^k(\bot) \mid k \in \mathbb{N} \right\}$$

is the least fixed point of ϕ where $\phi(d) = d$ and $\phi^{k+1}(d) := \phi \circ \phi^k(d)$

Proof. See [59, app. A.4]   □

At this point we could implement the interval analysis but we would quickly determine that there are codes where the static program analysis would not terminate. The analysis iterates over and over again and increases the value range of the corresponding intervals until the internal representation reaches its limits. An example chain for this behavior is shown in (3.20). A counting loop is a typical example code where a fixed point search at this point would fail since the interval that represents the counting variable would be increased in every iteration of the fixed point search. The missing requirement of Theorem 52 is the ACC. Hence, the next section introduces a so called widening operator.

3.5.1 Widening and Narrowing Operators

The complete lattice $(DAGIntervals, \sqsubseteq)$ does not fulfill the ACC and therefore we cannot be sure to reach a fixed point. To ensure that the iterative sequence for approximating the least upper bound eventually stabilizes, we use a *widening operator* ∇ [59, p. 226] instead of a join operator $\sqcup$ for combining different dataflows.

Let $(D, \sqsubseteq)$ be a complete lattice. $\nabla : D \times D \to D$ is called a *widening operator* if

1. for all $d_1, d_2 \in D$ hold $d_1 \sqcup d_2 \sqsubseteq d_1 \nabla d_2$, and

2. for all ascending chains $d_0 \sqsubseteq d_1 \sqsubseteq \dots$, the ascending chain $d_0^\nabla \sqsubseteq d_1^\nabla \sqsubseteq \dots$ eventually stabilizes where $d_0^\nabla := d_0$ and $d_{i+1}^\nabla := d_i^\nabla \nabla d_{i+1}$ for each $i \in \mathbb{N}$.

The sequence $(d_i^\nabla)_{i \in \mathbb{N}}$ is an ascending chain as

$$d_i^\nabla \sqsubseteq d_i^\nabla \sqcup d_{i+1} \sqsubseteq d_i^\nabla \nabla d_{i+1} = d_{i+1}^\nabla .$$

In contrast to $\sqcup$, the widening operator ∇ does not need to be commutative, associative, monotonic, nor absorptive $(d \nabla d = d)$. Applied to our interval analysis, we define the following widening operator

$$\nabla : DAGIntervals \times DAGIntervals \to DAGIntervals$$

with

$$\bot \triangledown d := d \triangledown \bot := d \quad , d \in DAGIntervals$$
$$[d_1, d_2] \triangledown [d_3, d_4] := [d_5, d_6]$$

with $d_1, d_2, d_3, d_4, d_5, d_6 \in DAG$ and

$$d_5 := \begin{cases} d_1 & \text{if } d_1 \sqsubseteq d_3 \\ -\infty & \text{otherwise} \end{cases} \tag{3.21}$$

$$d_6 := \begin{cases} d_2 & \text{if } d_4 \sqsubseteq d_2 \\ \infty & \text{otherwise} \end{cases}$$

With this widening operator, the ascending chain from (3.20) is transformed into

$$\emptyset \sqsubseteq [2,2] \sqsubseteq [-\infty, \infty] \sqsubseteq [-\infty, \infty] \dots$$

because

$$d_0^\triangledown = \emptyset$$
$$d_1^\triangledown = \emptyset \triangledown [2,2] = [2,2]$$
$$d_2^\triangledown = [2,2] \triangledown [1,3] = [-\infty, \infty]$$
$$d_3^\triangledown = [-\infty, \infty] \triangledown [0,4] = [-\infty, \infty].$$

Algorithm 3 describes a possible implementation for computing a fixed point approximation. Please note that this algorithm is far from being optimal and the intention was just to implement a very simple fixed point computation. A better algorithm is the worklist algorithm [59, p. 365] which does not iterate over all the labels but instead it tracks where new analysis information could arise depending on the control flow.

Example 25. Listing 3.3 shows the implementation of a data decomposition that we often use in our example codes. A data set of size n is divided into chunks of size c where c is n divided by the number of threads p. We assume for simplicity reasons that the data size is divisible by the number of threads. Each chunk consist of a certain index range [lb, ub] and each thread is uniquely associated with certain chunks. The lower bound of this interval is lb, the upper bound by ub.

The thread identifier that is returned by the function omp_get_thread_num() is put into the variable tid. The code that precedes the loop in line 11 is a typical pattern in a parallel region that uses the SPMD model. This code is often implicitly given as, for example, in the case for the OpenMP parallel for directive. Regardless

---

Algorithm 3 Calculation of the possible value ranges of the integer variables occurring in parallel region P. The fixed point is calculated by using the widening operator from (3.21).

Require: D (see (3.18)), $LABEL(P)$, $init(P)$ and $flow(P)$

Ensure: $fix^\nabla(\phi_L)$, where $fix^\nabla(\phi_L) \sqsupseteq fix(\phi_L)$ and $L \subseteq LABEL(P)$

  for all $l \in L$ **do**

    if $l \in init(P)$ **then**

      $AI \leftarrow \{\delta[v \rightarrow [v,v]] \mid v \in SHARED_P \cap INT_P\}$

    else

      $AI \leftarrow \bot_D$

    end if

  end for

  $cont \leftarrow true$

  while $cont$ **do**

    $cont \leftarrow false$

    for all $l \in L$ **do**

      $AI_{new} \leftarrow \bot$

      for all $(l',l) \in flow(P)$ **do**

        $AI_{new} \leftarrow AI_{new} \sqcup \phi_{l'}(AI_{l'})$

      end for

      if $AI_{new} \sqsubseteq \phi_{l'}(AI_{l'})$ and $AI_{new} \neq \phi_{l'}(AI_{l'})$ **then**

        $AI_l \leftarrow AI_l \nabla AI_{new}$

        $cont \leftarrow true$

      end if

    end for

  end while

---

of whether through implicit or explicit code, if we consider a SPMD program then the data is somehow decomposed into chunks such that the threads have more or less the same portion of work.

The lower bound lb of the data elements that should be processed by the current thread tid is given by the value of tid*c. The thread with the identifier number tid+1 starts its work at the element with index (tid+1)*c, and so forth. Therefore, the last component that the thread tid has to process is the one with the offset (tid+1)*c-1. This offset value is stored in the variable ub.

```
1 #pragma omp parallel
2 {
3    int tid =0; int p=0; int c=0;
```

```
 4    int lb=0; int ub=0; int i=0;
 5    tid=omp_get_thread_num();
 6    p=omp_get_num_threads();
 7    c=n/p;
 8    lb=tid*c;
 9    ub=(tid+1)*c-1;
10    i=lb;
11    while(i<=ub) {
12      // Thread processes data at position i.
13      // ...
14      i=i+1;
15    }
16  }
```

Listing 3.3: Example code for applying of Algorithm 3.

Let us consider the dataflow information of the integer assignment in line 14. After one iteration, we yield the following dataflow information:

```
n   : [ n, n ]
tid : [ THREAD_ID, THREAD_ID ]
p   : [ NUM_OF_THREADS, NUM_OF_THREADS ]
c   : [ n/NUM_OF_THREADS, n/NUM_OF_THREADS ]
lb  : [ THREAD_ID*(n/NUM_OF_THREADS), THREAD_ID*(n/NUM_OF_THREADS) ]
ub  : [ ((THREAD_ID+1)*(n/NUM_OF_THREADS))-1,
        ((THREAD_ID+1)*(n/NUM_OF_THREADS))-1 ]
i   : [ THREAD_ID*(n/NUM_OF_THREADS), THREAD_ID*(n/NUM_OF_THREADS) ]
```

The shared variable n is assumed to be defined from the code outside the parallel region and is therefore given through the interval $[n, n]$. The special value THREAD_ID is used by the program analysis to represent the unique thread identifier number that is supplied by the runtime function omp_get_thread_num(). The number of threads is coded through the specific value NUM_OF_THREADS and is at runtime provided by omp_get_num_threads(). At this point, variable i has the same value as variable lb since the incrementation statement in line 14 has no effect yet. The effect of the incrementation in line 14 is visible from the second iteration forward. Let us assume that only variable i is changed inside the loop. Hence, the only interval that is changed through the next fixed point iterations is the one for variable i. In case that we use the common join operation for intervals rather than using the widening operator, we get

```
i: [ THREAD_ID*(n/NUM_OF_THREADS), THREAD_ID*(n/NUM_OF_THREADS)+1 ]
```

as result for variable i after the second iteration. The third iteration yields

```
i: [ THREAD_ID*(n/NUM_OF_THREADS), THREAD_ID*(n/NUM_OF_THREADS)+2 ]
```

and so forth. Since the fixed point iteration does not consider the condition of the loop test, the iteration would never terminate[2].

In the following, we consider what happens if we use the widening operator from (3.21). After applying the transfer function ϕ, we also get

```
i: [ THREAD_ID*(n/NUM_OF_THREADS), THREAD_ID*(n/NUM_OF_THREADS)+1 ]
```

as result when the computation path that contains the incrementation statement comes the first time into play. But we do not join the results but rather use the widening operator on

```
i: [ THREAD_ID*(n/NUM_OF_THREADS), THREAD_ID*(n/NUM_OF_THREADS) ]
```

and

```
i: [ THREAD_ID*(n/NUM_OF_THREADS), THREAD_ID*(n/NUM_OF_THREADS)+1 ]
```

and therefore we obtain

```
i: [ THREAD_ID*(n/NUM_OF_THREADS), PLUS_INFINITY ]
```

as result of the second iteration. The upper boundary is set to ∞ since

```
THREAD_ID*(n/NUM_OF_THREADS)+1 ⋢ THREAD_ID*(n/NUM_OF_THREADS) .
```

This result stays the same in the next fixed point iteration which means that we have reached a fixed point.

But as you can imagine this result is far from being sufficient to provide the needed information to allow a test whether or not a thread accesses data exclusively. In fact, the above result implies that the intervals for i overlap for different thread identifier since they all have the upper bound ∞. Since the intervals have overlapping value ranges, the conclusion is that the threads do not read exclusively.

Δ

---

[2]Or the implementation of the analysis would crash or provide unpredicted results due to an overflow of the internally used value range.

As Example 25 indicates, we need a better approximation for the fixed point than the analysis provides at this point. For this situation, the authors of [59, ch. 4.2.2] suggest a so called *narrowing operator*. Let us denote the fixed point solution obtained by widening with $fix^\nabla(\phi_L)$ where $L \subseteq LABEL(P)$. Then, the idea behind narrowing is to calculate another sequence

$$(\phi_L^k(fix^\nabla(\phi_L)))_{k=1,2,\ldots}$$

starting from $fix^\nabla(\phi_L)$. The widening operator fulfills the property that

$$[l_1, u_1] \sqcup [l_2, u_2] \sqsubseteq [l_1, u_1] \nabla [l_2, u_2] .$$

Thus, for the fixed point approximation we obtain

$$fix^\nabla(\phi_L) \sqsupseteq fix(\phi_L) .$$

Since the transfer function ϕ_L is monotonic, the sequence $(\phi_L^k)_{k=1,2,\ldots}$ is also monotonic. The sequence $(\phi_L^k(fix^\nabla(\phi_L)))_{k=1,2,\ldots}$ is a descending chain and it holds

$$\phi_L^k(fix^\nabla(\phi_L)) \sqsupseteq \phi_L^k(fix(\phi_L)) \sqsupseteq fix(\phi_L)$$

Unfortunately, we have no reason to believe that the descending chain eventually stabilizes. Nevertheless, the common strategy is to stop at an arbitrary point of the sequence. Algorithm 4 shows the algorithm that we have implemented. This means that we first compute a fixed point approximation with Algorithm 3 and then we apply Algorithm 4 to improve this solution.

Example 26. *(Narrowing)* Let us consider what the impact of narrowing is. In order to illustrate this, we apply Algorithm 3 and Algorithm 4 to Listing 3.4.

```
1 j=lb ;
2 while(i<=ub) {
3   // Thread processes data at position i.
4   // ...
5   j=ub;
6   // ...
7   i=i+1;
8 }
```

Listing 3.4: Example code for presenting the effect of the narrowing operator.

We show a portion of code without the whole parallel region because it contains only slight changes comparing to Listing 3.3. The only difference is that we use

---

Algorithm 4 After a fixed point $fix^\nabla(\phi_L)$ has been found by Algorithm 3, we try to improve this approximation by using narrowing. The maximal number of narrowing steps is here defined to be 10.

Require: $fix^\nabla(\phi_L)$ from Algorithm 3 where $L \subseteq LABEL(P)$

Ensure: $\phi_L^k(fix(\phi_L))$ with $\phi_L^k(fix^\nabla(\phi_L)) \sqsupseteq \phi_L^k(fix(\phi_L)) \sqsupseteq fix(\phi_L)$

    cont ← *true*
    counter ← 1
    while cont *or* counter ≤ 10 **do**
        counter ← counter + 1
        cont ← *false*
        for all $l \in L$ **do**
            for all $(l',l) \in flow(P)$ **do**
                if $AI_l \neq \phi_{l'}(AI_{l'})$ **then**
                    $AI_l \leftarrow AI_l \sqcup \phi_{l'}(AI_{l'})$
                    cont ← *true*
                end if
            end for
        end for
    end while

---

here another integer variable j that is the left-hand side of two assignments in line 1 and line 5. To get a better context between the results and the code, we display the control flow graph with the internal information of our compiler in Figure 3.4. In the following, we show firstly the results that the widening algorithm provides (Algorithm 3) where we only consider variable j because its value range is more precise after narrowing.

```
Node 83 (SPL_cond_while):
j: [THREAD_ID*(n/NUM_OF_THREADS), PLUS_INFINITY]

Node 76 (SPL_int_assign):
j: [THREAD_ID*(n/NUM_OF_THREADS), PLUS_INFINITY]

Node 1 (CFG_EXIT):
j: [THREAD_ID*(n/NUM_OF_THREADS), PLUS_INFINITY]
```

These results serve as input for the narrowing algorithm (Algorithm 4) which provides the following improved result:

```
Node 83 (SPL_cond_while):
```

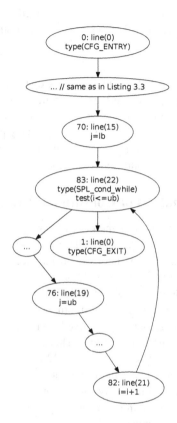

Figure 3.4: Extract from the control flow graph of the code shown in Listing 3.4.

```
j: [THREAD_ID*(n/NUM_OF_THREADS),((THREAD_ID+1)*
                                  (n/NUM_OF_THREADS))-1]

Node 76 (SPL_int_assign):
j: [THREAD_ID*(n/NUM_OF_THREADS),((THREAD_ID+1)*
                                  (n/NUM_OF_THREADS))-1]

Node 1 (CFG_EXIT):
j: [THREAD_ID*(n/NUM_OF_THREADS),((THREAD_ID+1)*
                                  (n/NUM_OF_THREADS))-1]
```

The upper bound of the value range of j can be reduced from plus infinity because our relation $\sqcup$ from (3.14) can decide the maximum of both possible values lb and ub. $\triangle$

We recognize in Example 26 that the narrowing operator improves the widening result in this example. However, further experiments with other example codes reveals that the results supplied by the narrowing operator are still too inaccurate. For example, for the code from Listing 3.1 the narrowing process does not change the widening result at all. Therefore, we developed another dataflow technique which takes the conditions of conditional branches into account.

3.5.2 Data Flow Analysis of Conditional Branches

In Figure 3.5, we display a part of the control flow graph from the code shown in Listing 3.1. The reader may recognize the labels with conditions at the edges going from a conditional branch statement to its successors. For example, the control flow goes from node 227 to node 92 only if the condition i≤ub is fulfilled. Otherwise, the control flow continues at node 232 because the condition i>ub is true. According to the floating-point assignment in node 211 we achieve with the widening and narrowing operators a fixed point approximation which states that variable j has the value range $[0,\infty]$. This is because each time when the control flow encounters the statement of node 220, the value of j is incremented. Without the knowledge how many iterations the loop performs, the analysis has to assume that there are infinite many and therefore there is no upper limit in the value range for j. We may improve this approximate estimate by taking the conditions of conditional branches and of loops into account. For example, this means we exploit the knowledge that at the entry of the loop body (node 211) the condition j≤k is true. The idea behind this is to adjust the static analysis information appropriately with respect to the conditions that are true in a particular control flow path. In [73], the author describes a method where the language supported by the compiler is extended by an assert-statement. This additional statement serves as a filter where this filter only allows this information to pass that fits the valid assertion. We want to implement this approach without extending the language *SPL* but rather by adjusting the transfer function ϕ_l for the cases that $[s]^l$ is a conditional statement.

We adjust the transfer function ϕ_l for the case that the transfered analysis information is intended for the statement that is executed when the condition is valid. Let us assume that the statement $[s]^l$ with label l is a conditional statement. This is $[s]^l$ has the form if (b) $\{S\}$ or while (b) $\{S\}$. Further, we assume that the label l' identifies the first statement in the sequence of statements S. Therefore, $[s]^{l'} = init(S)$ is the statement that is executed first after validating the branch's condition b as true. We define different adjustments of the dataflow δ depending on the shape of the condition b and depending on the value ranges of contained variables.

Suppose that a_1 and a_2 are integer variables and d_1, d_2, d_3, and d_4 are DAGs, then we define

$$
\phi_{(l,l')}(\delta) := \begin{cases}
\delta[a_1 \to [d_1,d_4]] & \text{if } b = (a_1 \leq a_2),\ \delta(a_1) = [d_1,d_2], \\
& \quad \delta(a_2) = [d_3,d_4],\ \text{and} \\
& \quad d_4 \sqsubset d_2,\ d_1 \sqsubseteq d_4 \\
\delta[a_1 \to [d_1,d_4-1]] & \text{if } b = (a_1 < a_2),\ \delta(a_1) = [d_1,d_2], \\
& \quad \delta(a_2) = [d_3,d_4],\ \text{and} \\
& \quad d_4 \sqsubset d_2,\ d_1 \sqsubseteq d_4 - 1 \\
\delta[a_1 \to [d_1,d_4]] & \text{if } b = (a_1 \leq a_2),\ \delta(a_1) = [d_1,\infty], \\
& \quad \delta(a_2) = [d_3,d_4],\ \text{and} \\
& \quad d_4 \neq \infty,\ d_1 \sqsubseteq d_4 \\
\delta[a_1 \to [d_1,d_4-1]] & \text{if } b = (a_1 < a_2),\ \delta(a_1) = [d_1,\infty], \\
& \quad \delta(a_2) = [d_3,d_4],\ \text{and} \\
& \quad d_4 \neq \infty,\ d_1 \sqsubseteq d_4 - 1
\end{cases}
\tag{3.22}
$$

where $[s]^l \in \{\ \text{if } (b)\ \{\ S\ \}\ , \text{while } (b)\ \{\ S\ \}\ \}$, $l' = init(S)$, and $d_1, d_2, d_3, d_4 \in DAG$, $a_1, a_2 \in INT_P$.

We use here the relation $a \sqsubset b$ without definition since it only serves as an abbreviation for $a \sqsubseteq b \wedge a \neq b$. As this definition shows, we only adjust the upper boundary of the interval. This means, for example, in case that we have a loop that decrements a counting variable, the resulting value range would still be minus infinity. We want only to illustrate the idea of this analysis and let further improvements as the adjustment of the lower boundary to the people who implement this.

Definition (3.22) consists of four different cases. The first and second case are very similar and the only difference is the relation of the test expression. The relation in the first case is a lower or equal relation ($\leq$) and in the second case it is a lower relation ($<$). This also holds for the cases three and four which is the reason why we only explain case one and three.

The first case adjusts the information δ for variable a_1 to $[d_1,d_4]$ in case that the test expression is $a_1 \leq a_2$, the current value ranges of a_1 is $[d_1,d_2]$, the value

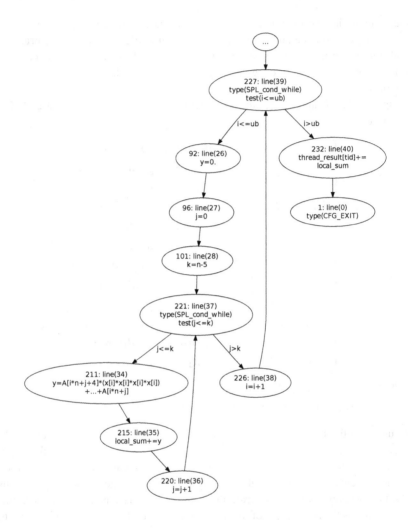

Figure 3.5: Extract from the control flow graph of the code shown in Listing 3.1. The edge from node 227 to node 92 is labeled with $i \leq$ ub which means that the control flow only takes this path if this condition is true. If the condition is not true then the control flow takes the path from node 227 to 232 (i>ub).

range for a_2 is $[d_3, d_4]$, and the relation $d_4 \sqsubset d_2$ is valid. This means that the upper boundary of the value range of variable a_2 can be carried over to the value range of variable a_1. The fact that $d_1 \sqsubseteq d_4$ should be true is based on the fact that the adjusted interval must fit the requirement that the lower boundary is lower or equal than the upper boundary. The third case is similar and covers the situation where variable a_1 has infinity as upper boundary and the variable a_2 has the upper boundary d_4 that is different from infinity. In this case we can be sure that the value range of a_1 can be changed to d_4.

We saw in Example 26 that we gain better analysis information by applying narrowing. However, this does not hold for Listing 3.1 and the next example presents the results that we obtain when we use the adjusted transfer function $\phi_{(l,l')}$ from (3.22).

Example 27. *(Exclusive-read analysis of Listing 3.1)* For the sake of clarity, we only present the results that differ from the ones that we obtained by using widening and narrowing. As we recognize in the below output, the upper boundary of the intervals are not infinity. For example, let us consider node 211. We only list the two variables i and j because the intervals of these two variables can be improved by applying (3.22). The upper boundary of variable j is not longer infinity but rather n-5. This better approximation can be made because we know that the condition j≤k is valid at node 211 and since k has the upper bound n-5, we refine the upper bound of j to n-5.

```
Node 92 (SPL_float_assign):
i: [THREAD_ID*(n/NUM_OF_THREADS),((THREAD_ID+1)*
                         (n/NUM_OF_THREADS))-1]
list of assertions:
  1. 'i<=ub'
  2. 'j>k'


Node 96 (SPL_int_assign):
i: [THREAD_ID*(n/NUM_OF_THREADS),((THREAD_ID+1)*
                         (n/NUM_OF_THREADS))-1]
list of assertions:
  1. 'i<=ub'
  2. 'j>k'


Node 101 (SPL_int_assign):
i: [THREAD_ID*(n/NUM_OF_THREADS),((THREAD_ID+1)*
                         (n/NUM_OF_THREADS))-1]
list of assertions:
  1. 'i<=ub'


Node 221 (SPL_cond_while):
```

```
i: [THREAD_ID*(n/NUM_OF_THREADS),((THREAD_ID+1)*
                                  (n/NUM_OF_THREADS))-1]
list of assertions:
   1. 'i<=ub'

Node 211 (SPL_float_assign):
i: [THREAD_ID*(n/NUM_OF_THREADS),((THREAD_ID+1)*
                                  (n/NUM_OF_THREADS))-1]
j: [ 0, n-5 ]
list of assertions:
   1. 'i<=ub'
   2. 'j<=k'

Node 215 (SPL_float_plus_assign):
i: [THREAD_ID*(n/NUM_OF_THREADS),((THREAD_ID+1)*
                                  (n/NUM_OF_THREADS))-1]
j: [ 0, n-5 ].
list of assertions:
   1. 'i<=ub'
   2. 'j<=k'

Node 220 (SPL_int_assign):
i: [THREAD_ID*(n/NUM_OF_THREADS),((THREAD_ID+1)*
                                  (n/NUM_OF_THREADS))-1]
j: [ 0, n-5 ].
list of assertions:
   1. 'i<=ub'
   2. 'j<=k'

Node 226 (SPL_int_assign):
i: [THREAD_ID*(n/NUM_OF_THREADS),((THREAD_ID+1)
                                  *(n/NUM_OF_THREADS))-1]
list of assertions:
   1. 'i<=ub'
   2. 'j>k'
```

For all the intervals above holds that they only depends at most on three different input values:

- the size of the data (here given by the shared variable n),

- the unique thread identifier THREAD_ID,

- and the special value NUM_OF_THREADS that represents the number of threads that execute the current parallel region.

This shows that we may obtain the actual value range intervals by taking the conditional test expressions into account. $\triangle$

Eventually, we achieve the fine-grained information that we need for analyzing whether or not a thread reads a memory location exclusively. The next section explains how we perform the implication that the corresponding intervals not overlap when we consider different threads.

3.6 Towards the Exclusive Read Property

The previous sections showed that a static interval analysis can achieve a very precise information about the possible value range of a particular integer variable occurring in the parallel region P. In case that P contains a floating-point assignment that uses a memory access of the shape $b + o$ where b is a base address and o being an integer expression, the analysis has to examine whether it is possible that the value of o at runtime could occur in more than one thread. For example, the C array double x[10] has 10 elements, the base address b is given by &x, the offset values o can be between 0 and 9.

Let us assume that the parallel region P contains a memory reference x[i]. We want to use two threads where each thread uses a private integer variable i as an offset to access components of x. The exclusive read analysis has to decide if it is possible that the value ranges of i with respect to different threads can overlap. Since i is a private variable, each i has its own scope inside a certain thread. In case that i may has the same value in thread 0 and thread 1, let us say five, both threads potentially read from the memory location x[5]. This leads to the fact that this memory access is not exclusive. On the other hand, if thread 0 uses the value range from zero to four in variable i and thread 1 uses the values five to nine for i, the corresponding memory access x[i] is read exclusively.

In the following, we explain how the exclusive read analysis uses the results from interval analysis to imply that a certain memory access made inside of a floating-point assignment is exclusive or not. As an example, we take the results provided by Example 27. In particular, the floating-point assignment from Listing 3.1, line 26 serves as an example statement.

$$y \leftarrow A_{i*n+j+4} * \left(x_i * x_i * x_i * x_i \right) + A_{i*n+j+3} * \left(x_i * x_i * x_i \right)$$
$$+ A_{i*n+j+2} * \left(x_i * x_i \right) + A_{i*n+j+1} * \left(x_i \right) + A_{i*n+j} \, ;$$

This statement is internally identified by our implementation with node 211. The result for node 211 that we achieved in Example 27 is:

```
Node 211 (SPL_float_assign):
i: [THREAD_ID*(n/NUM_OF_THREADS),((THREAD_ID+1)*
                                   (n/NUM_OF_THREADS))-1]
j: [ 0, n-5 ]
list of assertions:
  1. 'i<=ub'
  2. 'j<=k'
```

Let us consider whether or not the memory access by x[i] is exclusive or not. In order to compare the corresponding intervals for the threads t and $t + 1$, we replace THREAD_ID by t and $t + 1$. The internal name NUM_OF_THREADS is replaced by p what represents the number of threads. The total number of data elements processed by the group of threads is here n. Variable i used in thread t has a value range of

$$\left[t \cdot \frac{n}{p}, (t+1) \cdot \frac{n}{p} - 1 \right].$$

The thread with identifier $t + 1$ uses the values

$$\left[(t+1) \cdot \frac{n}{p}, (t+2) \cdot \frac{n}{p} - 1 \right]$$

for the variable i. This holds for all threads with the identification numbers between 0 and $p - 1$. In order to ensure that thread t does not use offset values which are also used by thread $t + 1$, we compare the upper bound of the value range of thread t with the lower bound of thread $t + 1$. Formally,

$$(t+1) \cdot \frac{n}{p} - 1 \sqsubset (t+1) \cdot \frac{n}{p}$$

must be valid. This is obviously true since the two expressions only differ by a subtraction of one. However, as often in the program analysis domain this is not easy to decide for the compiler tool. Our implementation uses DAGs as internal representation and therefore the implementation of the program analysis must be able to compare these certain kinds of DAGs. For this reason, we defined in (3.13) certain shapes of DAGs that can be compared. These shapes are implemented in our tool and sophisticated enough to decide the exclusive read property for the examples that we present in this work. For a productive implementation this set of possible shapes probably must be extended.

The case we discussed above is the easiest shape that occurs, namely an array reference [i] where only one variable is used. In case that variable j is used as an offset, the value range is $[0, n - 5]$. This value range is the same for all threads since it does not depend on the unique thread identification number THREAD_ID. This fact alone is sufficient to imply that a memory reference x[j] cannot fulfill the exclusive read property.

Another shape from the above example is [i*n+j] what is typically used when a two dimensional array is referenced. As shown in the example, we use this expression together with an addition of a constant value to reference multiple elements during one iteration. In fact, during one iteration of the inner loop of our example code, the elements referenced by A[i*n+j], A[i*n+j+1], ..., A[i*n+j+4] are read. The two dimensional array is A which consist of n rows and n columns.

To decide whether or not the memory location referenced by A[i*n+j+3] is read exclusively, we have to take three different intervals into account. We focus on the valid intervals in node 211 for the variables i, j. The value range of the shared variable n is $[n, n]$. With these intervals at hand, we can imply that thread t encounters the following value range with respect to the expression i*n+j+3.

$$\left[t \cdot \frac{n}{p} \cdot n + 0 + 3, \left((t+1) \cdot \frac{n}{p} - 1 \right) \cdot n + (n-5) + 3 \right]$$

Analogously, thread $t + 1$ uses the following interval to reference the memory location A[i*n+j+3].

$$\left[\left((t+1) \cdot \frac{n}{p} \right) \cdot n + 0 + 3, \left((t+2) \cdot \frac{n}{p} - 1 \right) \cdot n + (n-5) + 3 \right]$$

As before in the simple reference case, we compare the upper bound of thread t with the lower bound of thread $t + 1$ to ensure that these intervals does not overlap. Formally, the relation

$$\left((t+1) \cdot \frac{n}{p} - 1 \right) \cdot n + (n-5) + 3 \quad \sqsubseteq \quad \left((t+1) \cdot \frac{n}{p} \right) \cdot n + 0 + 3 \qquad (3.23)$$

must be valid to ensure that thread t is not reading any components that are also accessed by thread $t + 1$. We rewrite this in order to use the implementation of the greatest lower bound.

$$\prod \left\{ \left((t+1) \cdot \frac{n}{p} - 1 \right) \cdot n + (n-5) + 3, \left((t+1) \cdot \frac{n}{p} \right) \cdot n + 0 + 3 \right\} \qquad (3.24)$$

Since our tool is only capable of comparing a certain kind of DAGs, we have to rewrite the given DAGs in (3.24) to achieve a shape that is known to our definition of the greatest lower bound. For this reason, we implemented the distributivity rule to enable our tool to rewrite the expression

$$\left((t+1) \cdot \frac{n}{p} - 1 \right) \cdot n + (n-5) + 3$$

into

$$\left((t+1) \cdot \frac{n}{p} \cdot n \right) - n + n + (-5 + 3) \ .$$

Subsequently, a normalization procedure looks for adjacent nodes in the DAGs that can be simplified by simple algebraic rules. For example, the expression $(-n + n)$ is recognized to be zero and is erased from the *DAG*. Furthermore, the expression where two constants are adjacent is simplified to one constant when the connecting operation is a subtraction or an addition. In our example the expression $(-5 + 3)$ is replaced by the constant -2, and the expression $(0 + 3)$ is replaced by 3. Eventually, the reformulation leads to

$$\prod \left\{ \left((t+1) \cdot \frac{n}{p} \cdot n \right) - 2, \left((t+1) \cdot \frac{n}{p} \cdot n \right) + 3 \right\}$$

where we achieve two DAGs where a subtree is exactly the same and both DAGs only differ by a constant term. This can be decided by the implementation of the definition of the greatest lower bound in (3.13) and the program analysis can imply that the memory reference through A[i*n+j+3] is exclusive for each thread.

The illustrated method from above to decide the exclusive read property is described in Algorithm 5. It provides a set of labels *ExclusiveRead* where each contained label l is associated with a memory reference that fulfills the exclusive read property. As a starting point, the algorithm takes the result $\phi_L^k(fix(\phi_L))$ that we obtain by applying Algorithm 4 and the adjustments for taking the conditional branches into account.

The approach in Algorithm 5 is as follows. Each floating-point assignment $[s]^l$ in P is examined. All the right-hand side references x occurring in $[s]^l$ are tested. There are two distinct shapes of memory references possible. The first kind reads a memory location addressed by a base address b plus an offset value o. The second kind is that a memory location is associated with a scalar variable x.

The fairly easy case to decide is the one where we consider a scalar variable. In this case the exclusive read property depends on the sharing status of x according to the OpenMP memory model[3]. Only when x is a private variable, the associated memory location is read exclusively by a thread.

---

[3]For information see Section 1.3 on page 48

Algorithm 5 This algorithm provides a set of labels where the associated statements are verified to contain only read accesses that are exclusive for all threads.

Require: $\phi_L^k(\mathit{fix}(\phi_L))$ from Algorithm 4 where $L \subseteq \mathit{LABEL}(P)$
Ensure: $\mathit{ExclusiveRead} \subseteq \mathit{LABEL}(P)$.

   $\mathit{ExclusiveRead} \leftarrow \emptyset$
   for all $[s]^l \in P$ **do**
     if $[s]^l = y \leftarrow \phi(x_1,\ldots,x_n)$ or $[s]^l = y +\!\leftarrow \phi(x_1,\ldots,x_n)$ **then**
       for all $\&x \in \mathit{RHSREF}\big([s]^l\big)$ **do**
         if $\&x = b + o$ **then**
           $[l_t, u_t] \leftarrow \mathit{val}_\delta(o)$
           if $\mathtt{THREAD\_ID} \in l_t \wedge \mathtt{THREAD\_ID} \in u_t$ **then**
             if $b \in \mathit{SHARED}_P$ **then**
               $[l_{t+1}, u_{t+1}] \leftarrow \mathit{val}_\delta(o)[t/(t+1)]$
               if $u_t \sqsubset l_{t+1}$ **then**
                 $\mathit{ExclusiveRead} \leftarrow \mathit{ExclusiveRead} \cup \{\&x\}$
               else
                 $\mathit{ExclusiveRead} \leftarrow \mathit{ExclusiveRead} \setminus \{\&v \mid \&v = b + o'\}$
               end if
             else
               $\mathit{ExclusiveRead} \leftarrow \mathit{ExclusiveRead} \cup \{\&x\}$
             end if
           end if
         else
           if $x \in \mathit{PRIVATE}_P$ **then**
             $\mathit{ExclusiveRead} \leftarrow \mathit{ExclusiveRead} \cup \{\&x\}$
           end if
         end if
       end for
     end if
   end for

On the other hand, is the memory location $\&x$ addressed by a base address b plus an offset o. The first check is whether or not the base address b is shared. If b is private, the memory access is exclusive for each thread.

In case that b is shared, it must be verified that offset o has a different value range for each thread. For this important case we need the results from the interval analysis. The interval of the offset expression o with respect to thread t is

$$[l_t, u_t] = val_\delta(o).$$

Obviously, in order to enable the thread to exclusively use a certain value range, the thread identifier THREAD_ID must be contained in the lower boundary l_t as well as in the upper boundary u_t. The value range of offset o with respect to thread $t+1$ is

$$[l_{t+1}, u_{t+1}] = val_\delta(o)[t/(t+1)],$$

where $val_\delta(o)[t/(t+1)]$ means the interval provided by $val_\delta(o)$ and each occurrence of t is replaced by the expression $(t+1)$, what is indicated by $[t/(t+1)]$. To ensure that the interval of thread t and thread $t+1$ does not overlap the following relation must be true:

$$u_t \sqsubset l_{t+1}$$

Please note that the relation $\sqsubset$ again serves as an abbreviation for

$$u_t \sqsubseteq l_{t+1} \quad \lor \quad u_t \neq l_{t+1}.$$

In case that the memory reference $\&x$ is read exclusively, it is inserted into the set *ExclusiveRead*:

$$ExclusiveRead \leftarrow ExclusiveRead \cup \{\&x\}$$

This can be seen as the typical *gen* step in a dataflow analysis. In contrast to this the *kill* step must be performed if a memory reference is not exclusively read:

$$ExclusiveRead \leftarrow ExclusiveRead \setminus \{\&v \mid \&v = b + o'\}$$

For example, suppose that the analysis recognizes x[i] as exclusively read. Later in the dataflow analysis process, it is recognized that the memory reference x[j] is not exclusively read. This means j has potentially values that are also used as offset in another thread and even worse the value j may potentially has the same value as i. Therefore, we have to remove these memory references in *ExclusiveRead* where the same base address b is used in case that a memory reference is exposed to be not exclusively read.

Example 28. *(Continuation of the exclusive-read analysis of Listing 3.1)* By applying Algorithm 5, we get the exclusive read status of all the right-hand side floating-point references.

```
Read access in node 211 (line 26) A[((i*n)+j)+4]: exclusive read: yes
Read access in node 211 (line 26) x[i]         : exclusive read: yes
Read access in node 211 (line 26) x[i]         : exclusive read: yes
Read access in node 211 (line 26) x[i]         : exclusive read: yes
Read access in node 211 (line 26) x[i]         : exclusive read: yes
Read access in node 211 (line 26) A[((i*n)+j)+3]: exclusive read: yes
Read access in node 211 (line 26) x[i]         : exclusive read: yes
Read access in node 211 (line 26) x[i]         : exclusive read: yes
Read access in node 211 (line 26) x[i]         : exclusive read: yes
Read access in node 211 (line 26) A[((i*n)+j)+2]: exclusive read: yes
Read access in node 211 (line 26) x[i]         : exclusive read: yes
Read access in node 211 (line 26) x[i]         : exclusive read: yes
Read access in node 211 (line 26) A[((i*n)+j)+1]: exclusive read: yes
Read access in node 211 (line 26) x[i]         : exclusive read: yes
Read access in node 211 (line 26) A[(i*n)+j]   : exclusive read: yes
Read access in node 211 (line 28) y            : exclusive read: yes
Read access in node 211 (line 33) local_sum    : exclusive read: yes
```

All the memory references fulfill the exclusive read property which means that the associated adjoint assignments do not need synchronization. In Section 5.3 we will compare the runtime results for the resulting adjoint code with synchronization and the one without synchronization. Δ

Example 29. *(Kill effect of the ERA)* This example illustrates the situation where we have to remove the results from set *ExclusiveRead* when it turns out that a certain memory reference is not exclusively read. For example, take the code from Listing 3.1 and change one offset of an arbitrary x[i] reference to x[j]. This leads to the following output from our implementation.

```
Read access in node 211 (line 26) A[((i*n)+j)+4]: exclusive read: yes
Read access in node 211 (line 26) x[i]         : exclusive read: no
Read access in node 211 (line 26) x[i]         : exclusive read: no
Read access in node 211 (line 26) x[i]         : exclusive read: no
Read access in node 211 (line 26) x[i]         : exclusive read: no
Read access in node 211 (line 26) A[((i*n)+j)+3]: exclusive read: yes
Read access in node 211 (line 26) x[i]         : exclusive read: no
Read access in node 211 (line 26) x[i]         : exclusive read: no
Read access in node 211 (line 26) x[i]         : exclusive read: no
Read access in node 211 (line 26) A[((i*n)+j)+2]: exclusive read: yes
Read access in node 211 (line 26) x[i]         : exclusive read: no
Read access in node 211 (line 26) x[i]         : exclusive read: no
```

```
Read access in node 211 (line 26) A[((i*n)+j)+1]: exclusive read: yes
Read access in node 211 (line 26) x[i]         : exclusive read: no
Read access in node 211 (line 26) A[(i*n)+j]   : exclusive read: yes
Read access in node 211 (line 28) y            : exclusive read: yes
Read access in node 211 (line 33) local_sum    : exclusive read: yes
```

First, the memory references x[i] are recognized as exclusive but x[j] is recognized as not exclusively read since the offset j has the value range $[0 : n - 5]$ and this value range holds for all threads. Therefore, we have to assume that the offsets i and j potentially have the same values and x[i] therefore is not exclusive anymore. Hence, we adjust the results such that all memory references where x is the base pointer are removed from the set *ExclusiveRead*. $\triangle$

This concludes this chapter where we have seen that it is possible to examine at compile time whether or not memory references are accessed exclusively at runtime by one thread. With this knowledge, the source transformation tool can decide which adjoint assignment needs synchronization and which one can be executed freely.

3.7 Summary

The basic idea of the exclusive read analysis (ERA) is that the very general code pattern of the data decomposition in a parallel region can be analyzed in such a way that the result is a fairly precise approximation of possible value ranges of integer variables. These value ranges are crucial when the task is to decide if a certain array reference is exclusively read by only one thread. For example, if there is a floating-point reference x[i], then we have to examine what value range the offset variable i has in order to decide whether or not x[i] is read exclusively or if it is possible that more than one thread accesses this memory location. In this chapter, we considered a static analysis of expressions that are typical for referencing one or two dimensional arrays.

The typical code pattern of a data decomposition for parallel computations detects the number of threads p, it divides the number of data elements n by p and determines the size of the chunk that each thread should process. To perform this, the lower and upper bound is set in dependence of the thread identifier number t. Subsequently, the lower and upper boundary of each chunk define where a certain thread accesses the data. In order to decide if a certain memory reference is exclusively read or not, the interval analysis must cover the value propagation from the statements where the values t and p are determined until the assignments where certain memory references are used. To simplify the static program analysis, we assumed that

- the number of threads p divides the number of data elements n, and

- n is bigger or equal to p.

Despite of these assumptions, we had to perform some effort to achieve good results. One can imagine that the cases where the above assumptions do not hold lead to even more effort to gain practical results. But this is beyond the scope of this work. Our intend was only to introduce a method that serves as a proof of concept.

In the case that the above assumptions do not hold or in case that the array references inside an original code are more complex than the one we covered here, it is conceivable that a mixture between a static program analysis and dynamic information can be used. With help of the dynamic information obtained at runtime and the value range intervals that we achieved through a static analysis, an interval analysis can be performed efficiently. However, in this chapter we introduced a pure static analysis that tries to achieve as much information as possible during compile time. The hybrid approach of a static analysis together with dynamic information is a possible part of further research.

We presented an integer interval analysis from [59] that we took as a starting point. Since this analysis often provides not very precise intervals, we extended this approach by using DAGs as components of the intervals. The advantage is that at a certain point in the code, the DAGs can reflect the whole expression of computation steps which were performed to reach this point. This allows a very precise analysis.

To convince the reader that this approach can be taken as basis for a static program analysis, we had to go through some order-theoretic foundations. We have shown that the set of possible DAGs and the relation $\sqsubseteq: DAG \times DAG$ form a partial order. Afterwards, we clarified that intervals of DAGs (*DAGIntervals*) also form a partial order together with the relation

$$\sqsubseteq: DAGIntervals \times DAGIntervals.$$

We developed the interval analysis where we first introduced a widening operator to ensure that the analysis reaches a fixed point, and then we presented a narrowing operator which provides better results in some cases. An important improvement of the approximation for the fixed point was realized through an adjustment of the transfer function that takes conditional test expressions of branch and loop statements into account.

The actual ERA takes the results from the interval analysis as input and decides for each right-hand side reference &v of a floating-point assignment in the parallel region P whether &v is exclusively read or not. The concluding examples illustrated that the ERA in fact can provide fruitful information that can be exploited by the AD source transformation tool. Only these adjoint assignments that really need synchronization are augmented, for example, with an OpenMP atomic construct.

One could think of ways to strengthen the comparison skills of the program analysis by using methods typically lying the domain of computer algebra systems. A work that focuses on solving inequalities can be found in [38]. A common way to compare inequalities would enable the ERA to compare arbitrary expressions that are used as an offset to reference an array. Our approach was only to support the comparison of certain shapes of expressions. It is not part of this work to cover all the possible cases of occurring expressions. However, it would be worth of thinking about using an interface to an open source computer algebra system such as Axiom[4] or Maxima[5] to examine the possibilities to retrieve information about inequalities. It should be possible to gain useful information since the inequalities we encounter in the context of the data decomposition are not complex.

Another domain where it could be interesting to explore if there are connections that can help us with the ERA are theorem prover. In recent years there are more and more projects where a so called Satisfiability modulo theories (SMT) solver[6] is implemented. Two examples are the MathSAT5[7] and the Z3[8] solver. One example application that builds on the Z3 solver is the Verifier for Concurrent C (VCC[9]). The VCC uses assertions in the preamble of C functions to proof certain properties of the code [22]. This approach of using assert statements at certain points in the parallel region P may improve the ERA results significantly similar to the approach where we took the conditional branches into account.

---

[4] http://www.axiom-developer.org/
[5] http://maxima.sourceforge.net/
[6] Overview of SMT solvers: http://smtlib.cs.uiowa.edu/solvers.html
[7] http://mathsat.fbk.eu/
[8] http://z3.codeplex.com/
[9] http://research.microsoft.com/en-us/projects/vcc/

4 Source Transformation of OpenMP Constructs

In Chapter 2, we used the context-free grammar $\mathscr{G}_{SPL} = (V, \Sigma, R, P)$, defined in Definition 40, for *SPL* codes. In this chapter, we extend this grammar in order to recognize OpenMP constructs inside of the parallel region P. The non-terminal symbol that is used in the production rules for these constructs is c. The set of terminal symbols Σ and the start symbol P remain the same as they were in $\mathscr{G}_{SPL}$. The production rules R_{OMP} is at the moment equal to R but is extended by rules during the coming sections. The new context-free grammar is therefore defined as

$$\mathscr{G}_{OMP} = (V \cup \{c\}, \Sigma, R_{OMP}, P).$$

The language produced by $\mathscr{G}_{OMP}$ is called *SPLOMP*. In the following sections, we extend $\mathscr{G}_{OMP}$ by rules for the different compiler directives defined in OpenMP 3.1. We introduced the OpenMP 3.1 standard and its constructs in Section 1.3.

Since we often speak about sequences of statements we denote for the sake of clarity that S_0 is the sequence inside the scope of P. Hence, P has the following shape

$$\text{\#pragma omp parallel} \{ D \quad S_0 \}$$

where D is a sequence of definitions and S_0 is a sequence of statements.

4.1 Stack Implementation for the Adjoint Code

For executing the adjoint model, we need a data structure that implements the behavior of a stack with the typical methods push(), pop(), for inserting and removing elements, and top() for testing the element on top of the stack. This interface conforms with the template class stack defined in the C++ *Standard Template Library* (STL). An approach where the implementation uses these STL stacks would ensure best possible cross-platform interoperability. However, this may not be the most efficient way to implement these stacks and its advantage is that the size of the stack grows dynamically. dcc uses static arrays together with an integer counter to implement these kind of stacks efficiently and at the same time with a maximum of portability. The downside of this approach is that the developer has to define a certain size for the static arrays and in case that this size is not sufficient to host all the values the execution of the adjoint code ends up with a stack overflow. Thus, a balance between a static fast solution and a dynamic flexible solution must be chosen. We present an example for both approaches and let the decision to choose the method what fits best to the software engineer who implements the derivative code.

Suppose you want to use a class STACK that is somehow defined as indicated in Listing 4.1. As pointed out in the definition of the *SPL* language, we need three different stacks, two value stacks for integer and floating-point values (STACK$_{(1)i}$, and STACK$_{(1)f}$), and one stack to store the labels of the control flow (STACK$_{(1)c}$). To declare a globally defined class as thread local, OpenMP knows the threadprivate directive. This means that the definition is placed in the global scope but each thread uses its own stack. This is necessary when the original parallel region uses calls to subroutines and the stacks therefore have to record values from multiple levels of scopes. Thus, the different scopes have to write all to the same stack which is possible by defining the stack in the global scope.

```
class STACK { ... /* This is left to the user of AD */};
STACK STACK(1)c ;
STACK STACK(1)i ;
STACK STACK(1)f ;
#pragma omp threadprivate (STACK(1)c , STACK(1)i , STACK(1)f )
```

Listing 4.1: The global section provides some definition of a C++ class, here named STACK. Since these stacks has to be defined as thread local, they are contained in a threadprivate directive.

In case that the parallel region P does not contain subroutine calls it would probably be more wise to place the definition of the stacks STACK$_{(i|f|c)}$ into the scope of the parallel region. This has the advantage that the memory for the data is placed on the stack that is provided from the operating system (OS). Hence, the data has a certain locality which can make an huge difference when the target machine has a *non-uniform memory access* (NUMA) architecture in opposite to the *symmetric multiprocessing* (SMP) architecture. In case of a SMP machine, all processor cores are connected to a single shared main memory, whereby the NUMA architecture splits the main memory into parts such that each processor core has a local memory and the latency of an access to this local memory is much lower than the latency that is necessary to access the memory from distinct processors. For example, thread t is associated by the OS to core C_0. Thus, we are well advised to place the stacks for the adjoint code into the memory that is associated with C_0 to ensure a fast access time.

```
#pragma omp parallel
{
    unsigned int  STACK(1)c[SZ];
    unsigned int  counter1=0;
    int           STACK(1)i[SZ];
    unsigned int  counter2=0;
    double        STACK(1)f[SZ];
    unsigned int  counter3=0;
    ...
    /* a1_STACKc_push(34) is then implemented as */
    STACK(1)c[counter1] = 34;
    counter1 = counter1 + 1;
    ...
}
```

Listing 4.2: On a NUMA architecture the adjoint stacks should be defined inside the scope of the parallel region. The whole code of the concurrent execution must have access to these arrays. Thus, the scope of the parallel region must not be left, for example, through a subroutine call.

A possible implementation for a NUMA architecture is to estimate the proper size that is necessary to host all the data that arises during the augmented forward section. Let us say SZ is a C/C++ macro that defines this value. The arrays that represent the stacks $STACK_{(1)c}$, $STACK_{(1)i}$, and $STACK_{(1)f}$, are defined inside the scope of the parallel region as indicated in Listing 4.2. The current position in the array where the next element can be placed is shown by the integer variable counter. This variable is incremented each time when a value is pushed onto the stack. During the reverse section, this counter is decremented until it reaches again the value zero at the end of the reverse section.

4.2 *SPLOMP*[1] - Synchronization Constructs

This section describes the extension of our basic language *SPL* by the synchronization constructs that are provided by OpenMP 3.1. The resulting language is called *SPLOMP*[1]. The first subsection starts with the barrier construct. Subsequently, we continue with the master, critical, and atomic constructs. This brings us to a point where we are able to show the closure of *SPLOMP*[1] in Section 4.2.5. There, adjoint assignments are synchronized and the only restriction to the original parallel region is that it has to be noncritical.

We saw in Chapter 2 the importance of providing a parallel region without any critical references. Unfortunately, there are often situations where it is necessary to allow a write access to a shared resource. This access must be synchronized in order to prevent a race condition. For example, suppose P can be structured into two sequences of statements $S_1 = (s_1; s_2; \ldots; s_{i-1})$ and $S_i = (s_i; s_{i+1}; \ldots; s_q)$.

```
#pragma omp parallel
{
    S₁
    Sᵢ
}
```

Listing 4.3: Example structure of a code that needs synchronization.

Suppose S_1 can be executed by each thread independently because there are no data dependencies between the codes executed by different threads. Afterwards, the results of the computation in S_1 must be joined into one or few values which is realized through some statements contained in S_i. Thus, there is a data dependence between S_1 and S_i. It must be ensured that the join does not start before all threads have finished the computation in S_1. Otherwise, a thread could use a value for the join that is not yet present because the corresponding thread has not finished its computation.

We consider two arbitrary threads t and t' where $t \neq t'$. Let us assume that $s_k \in S_1$ and $s_l \in S_i$ are critical, which leads to the fact that $(s_k^t, s_l^{t'})$ may not have the same behavior as $(s_l^{t'}, s_k^t)$. For such cases, parallel programming APIs usually provide a synchronization mechanism that restricts the set of possible interleavings. The OpenMP standard provides, for example, the barrier construct for these situations. For our example, this means that by placing a barrier between S_1 and S_i, the interleavings where the pair $(s_l^{t'}, s_k^t)$ occurs are no longer possible. The pair $(s_k^t, s_l^{t'})$ are subject to no restrictions.

Example 30. In the following listing there are two consecutive worksharing loops. The first assigns values to the vectors x and y, the second loop uses these vectors to compute a value for the matrix A. We have to define a barrier before the second loop to ensure that each thread has finished the first loop, otherwise the computation in the second loop may use values that have not been defined by the first loop.

```
#pragma omp parallel
{
    int i;
    int j;
```

```
#pragma omp for nowait
for( i←0; i<N; i++) {
    x_i ← f( i ,x );
    y_i ← g( i ,y );
}
#pragma omp barrier
#pragma omp for
for( i←0; i<N; i++)
    for( j←0; j<N; j++)
        A_{ij} ← h( i ,j ,x ,y );
}
```

The nowait clause in the first loop construct suppresses the implicit barrier. Without this clause, we would not need to define the explicit barrier because there is an implicit barrier by default at the end of each worksharing construct. $\triangle$

4.2.1 Synchronization with Barriers

This section describes the extension of our basic language *SPL* by the barrier construct. This construct will play an important role in all worksharing constructs. The worksharing construct is defined for a certain sequence of statements S and OpenMP determines that an implicit barrier follows S to join the results. We will see that this implicit barrier becomes an explicit barrier in the reverse section of the adjoint code.

```
#pragma omp parallel
{
    S_1
    #pragma omp barrier
    S_i
}
```

Listing 4.4: Example structure of an OpenMP parallel region with a barrier inside.

Suppose that the parallel region $P \in SPLOMP$[1] is structured as shown in Listing 4.4. There are two consecutive sequences of statements $S_1 = (s_1; s_2; \ldots; s_{i-1})$ and $S_i = (s_i; s_{i+1}; \ldots; s_q)$, with a barrier between them. The barrier between S_1 and S_i means for the set of possible interleavings that the interleavings where a statement from S_i is executed before a statement from S_1 are not possible. This is formally expressed by

$$\forall I \in \mathscr{I}(P,q,p): \quad s_k^t \prec s_l^{t'}, \tag{4.1}$$

where $k \in \{1,\ldots,i-1\}$, $l \in \{i,\ldots,q\}$, $t \neq t'$. For example, suppose that there is a critical reference $\&v$ with

$$\&v \in LHSREF\left(s_k^t\right) \wedge \&v \in REF\left(s_l^{t'}\right).$$

The execution leads likely to different results because it depends on the order of s_k^t and $s_l^{t'}$ in the interleaving that represents the concurrent execution. We stated earlier that this is called a race condition because the thread that writes last determines the value in $\&v$. This non-deterministic behavior is avoided by restricting the set of possible interleavings $\mathscr{I}(P,q,p)$ to such interleavings where $s_k^t \prec s_l^{t'}$ holds. This can be achieved by defining a barrier between s_k and s_l. In other words the developer who puts a barrier between s_k and s_l defines that only the order $(s_k^t, s_l^{t'})$ may occur in an interleaving, not the other way around.

We extend the set of production rules R_{OMP} by the rule for a barrier construct where the barrier construct is classified to be a straight-line code statement.

$$R_{OMP} = R_{OMP} \cup \left\{ \quad c \quad : \quad \#\text{pragma omp barrier} \quad \right\}$$

The fact that the barrier construct is defined as straight-line code is a question of the abstraction level. On our level of abstraction the barrier construct can be seen as a straight-line code statement because it does not change the control flow of a certain thread. The barrier only leads to the situation that the thread stops its execution until all threads have reached the barrier. Then the thread continues with the statement that follows the barrier. The control flow is not changed. On a lower level, the implementation of the barrier construct probably changes the control flow because the method for waiting is usually implemented through a loop.

The tangent-linear as well as the adjoint transformation rules map the barrier construct by the identity mapping:

$$\tau(\#\text{pragma omp barrier}) := \#\text{pragma omp barrier} \tag{4.2}$$

$$\sigma(\#\text{pragma omp barrier}) := \#\text{pragma omp barrier} \tag{4.3}$$

$$\rho(\#\text{pragma omp barrier}) := \#\text{pragma omp barrier} \tag{4.4}$$

As mentioned in OpenMP 3.1 citation 26 (page 45), we have to guarantee that either each thread that executes the parallel region encounters the barrier or none at all. Without any assumption about the original code, this is undecidable.[1] We assume that this property holds for the parallel region $P \in SPLOMP^1$ that serves as input for our source transformation. However, we have to show that the source transformation of P does not invalidate this property.

Proposition 53. *Suppose that an OpenMP parallel region P contains a barrier that is encountered by all threads or none at all. Then this also holds for the barrier contained in $\tau(P)$ and $\sigma(P)$.*

Proof. The tangent-linear code $\tau(P)$ as well as the forward section $\sigma(S_0)$ do not change the control flow in comparison to P. Hence, the property that each thread encounters a given barrier in P is inherited to $\tau(P)$ and $\sigma(S_0)$.

The reverse section $\rho(S_0)$ reverses the control flow of P. We show that this reversal does not disallow any thread from reaching the barrier that is emitted by ρ. Assume that there is a barrier construct inside of a sequence S contained in P. As in Chapter 2, we distinguish between the cases that S is straight-line code (*SLC*) or not. If a control flow statement (*CFSTMT*) is contained in S, we split the sequence until we achieve just *SLC* sequences or control flow statements.

1. Case $S \in SLC$:
   Assume that the barrier in P succeeds statement $s_i \in S = (s_1; \ldots; s_q)$. We apply (2.39) to S and afterwards we apply (4.3) to the barrier. This yields

$$\text{STACK}_{(1)c}.\text{push}(1);$$
$$\sigma(s_1);$$
$$\vdots$$
$$\sigma(s_i); \hspace{6cm} (4.5)$$
$$\#\text{pragma omp barrier}$$
$$\vdots$$
$$\sigma(s_q); \quad ,$$

   where $LABEL(s_1) = 1$ is the label of the first statement in S. According to the assumptions, the barrier in S is reached by each thread or by none at all. In addition, the control flow in P is the same as in $\sigma(S)$. This means that all threads reach the barrier in (4.5) or none at all. Hence, either each thread has the label 1 on their $\text{STACK}_{(1)c}$ or none at all. When we apply ρ first

---

[1]This can be reduced to the halting problem. Imagine a barrier after the sequence S. The question whether or not a thread reaches the barrier is equivalent to the question whether or not S terminates.

to S by rule (2.49) and then to the barrier by applying rule (4.4), we obtain the code shown in (4.6). This shows that either each thread encounters the barrier contained in (4.6) or no thread at all, depending on the content of stack $STACK_{(1)c}$.

Please note the special cases where the barrier is the first or the last statement in sequence S. In case that the barrier is the first statement in $\sigma(S)$, then the barrier is the last statement in $\rho(S)$. This means that each thread first must execute the adjoint code of S before encountering the barrier, which is exactly what one would expect when reversing the dataflow. In case that the barrier precedes $\sigma(s_1)$, the barrier in the reverse section follows after $\rho(s_1)$.

$$
\begin{aligned}
&\text{if } \left(STACK_{(1)c}.\text{top}() = 1 \right) \{ \\
&\qquad STACK_{(1)c}.\text{pop}(); \\
&\qquad \rho(s_q); \\
&\qquad \vdots \\
&\qquad \#\text{pragma omp barrier} \\
&\qquad \rho(s_i); \\
&\qquad \vdots \\
&\qquad \rho(s_1); \\
&\}
\end{aligned}
\qquad (4.6)
$$

2. Case $S \notin SLC$:
   Split S into S_1, s_i, S_{i+1} with

   $$S_1 = (s_1; \ldots; s_{i-1}), \ S_{i+1} = (s_{i+1}; \ldots; s_q) \in SLC, \text{ and } s_i \in CFSTMT.$$

   The previous case holds if the barrier is contained in S_1 or S_{i+1}. Thus, let us assume that the barrier succeeds s_i. By applying rule (2.38) to S and (4.3) to the barrier, we get

   $$
   \begin{aligned}
   &\sigma\left((s_1; \ldots; s_{i-1}) \right); \\
   &\sigma(s_i); \\
   &\sigma\left((\#\text{pragma omp barrier}; s_{i+1}; \ldots; s_q) \right);
   \end{aligned}
   \qquad (4.7)
   $$

and the reverse section comprises after applying rule (2.48) and (4.4):

$$\rho\left((s_1;\ldots;s_{i-1})\right);$$

$$\rho(s_i);\tag{4.8}$$

$$\rho\left((\#\text{pragma omp barrier};s_{i+1};\ldots;s_q)\right);$$

Since $(\#\text{pragma omp barrier};s_{i+1};\ldots;s_q) \in SLC$ holds, we can apply the previous case to this sequence.

This shows the claim.   □

The following proposition clarifies that a barrier is in fact necessary in the result of the source transformation. Since a barrier synchronization in parallel programming is always a very expensive construct in terms of runtime efficiency, it should be ensured that no barrier is produced where none is needed.

Proposition 54. *Suppose $P \in SPLOMP^1$ is given as in Listing 4.4. The parallel region P consists of a sequence S_0. S_0, in turn, consists of a sequence S_1, afterwards a barrier, and then a concluding sequence S_i.*

The barrier construct introduced in the source transformations $\tau(P)$, $\sigma(S_0)$, and $\rho(S_0)$ cannot be avoided. For a data dependence between $s_1 \in S_1$ and $s_i \in S_i$, we have to ensure that

$$\forall I \in \mathscr{I}(\tau(P),q,p): \quad \tau(s_1^t) \prec \tau\left(s_i^{t'}\right)\tag{4.9}$$

$$\forall I \in \mathscr{I}(\sigma(S_0),q,p): \quad \sigma(s_1^t) \prec \sigma\left(s_i^{t'}\right)\tag{4.10}$$

$$\forall I \in \mathscr{I}(\rho(S_0),q,p): \quad \rho(s_i^t) \prec \rho\left(s_1^{t'}\right)\tag{4.11}$$

where $t \neq t'$.

Proof. In case that there is no data dependence between statements from sequence S_1 to statement $s_i \in S_i$, we could move the barrier to the point in front of statement s_{i+1}. If this would again not lead to a data dependence from S_1 to S_i, we could move the barrier further, until we get a data dependence or until we get the end of the sequence. Hence, let us assume that there is a data dependence between statements from sequence S_1 to statement $s_i \in S_i$. Without loss of generality we assume that the statement from S_1 is s_1. Otherwise, we only need to change the statement index. The statements are assumed to be assignments, and s_1 is executed by thread t, whereby thread t' executes s_i:

$$s_1^t = y_1^t \leftarrow \phi(x_{1,1}^t, \ldots, x_{1,n}^t)$$
$$s_i^{t'} = y_i^{t'} \leftarrow \phi(x_{i,1}^{t'}, \ldots, x_{i,n}^{t'})$$

with $\&y_1^t = \&x_{i_k}^{t'}, t \neq t', k \in \{1, \ldots, n\}$. This means that a memory location is used by two different threads t and t' and thread t uses this location as left-hand side reference and t' uses it as right-hand side reference. We have the situation that thread t computes a value that is read by thread t'. Let us consider the derivative code of s_1 and s_i.

Thread t executes $\tau(s_1)$:   Thread t' executes $\tau(s_i)$:

$$y_1^{(1)t} \leftarrow \sum_{k=1}^{n} \phi_{x_{1_k}}(x_{1_1}^t, \ldots, x_{1_n}^t) \cdot x_{1_k}^{(1)t}; \qquad y_i^{(1)t'} \leftarrow \sum_{k=1}^{n} \phi_{x_{i_k}}(x_{i_1}^{t'}, \ldots, x_{i_n}^{t'}) \cdot x_{i_k}^{(1)t'};$$

$$y_1^t \leftarrow \phi(x_{1_1}^t, \ldots, x_{1_n}^t); \qquad\qquad y_i^{t'} \leftarrow \phi(x_{i_1}^{t'}, \ldots, x_{i_n}^{t'}); \qquad (4.12)$$

1. Tangent-linear source transformation by $\tau(s_1)$ and $\tau(s_i)$:
   In the above code, we show the result for $\tau(s_1)$ on the left side and the result of applying $\tau(s_i)$ on the right side. The left side is executed by thread t and the right side by thread t' with $t \neq t'$. The equality $\&y_1^t = \&x_{i_k}^{t'}$ from the original code P also holds for $\tau(P)$. Therefore, this equality leads to a race condition in (4.12). This race condition must be synchronized by a barrier between the two assignments in (4.12). In addition, the assignments for computing the derivative values also need synchronization because of

$$\&y_1^t = \&x_{i_k}^{t'} \overset{Lem.42}{\Longrightarrow} \&y_1^{(1)t} = \&x_{i_k}^{(1)t'}.$$

   The data dependence is inherited to the tangent-linear code and therefore we cannot omit the barrier construct when transforming in tangent-linear mode.

2. Adjoint source transformation by σ and ρ:
   Suppose that thread t executes the code that arise from applying $\sigma(s_1)$ and

$\rho(s_1)$.

Forward section:

$\text{STACK}_{(1)f}.\text{push}(y_1^t);$

$$y_1^t \leftarrow \phi^1(x_{1,1}^t,\ldots,x_{1,n}^t); \tag{4.13}$$

$\vdots$

Reverse section:

$\vdots$

$y_1^t \leftarrow \text{STACK}_{(1)f}.\text{top}();$

$\text{STACK}_{(1)f}.\text{pop}();$

$$x_{(1)1,1}^t \mathrel{+\!\!+\!\!\leftarrow} \phi_{x_{1,1}^t}(x_{1,1}^t,\ldots,x_{1,n}^t)\cdot y_{(1)1}^t; \tag{4.14}$$

$\vdots$

$$x_{(1)1,n}^t \mathrel{+\!\!+\!\!\leftarrow} \phi_{x_{1,n}^t}(x_{1,1}^t,\ldots,x_{1,n}^t)\cdot y_{(1)1}^t;$$

$$y_{(1)1}^t \leftarrow 0; \tag{4.15}$$

Another thread t' processes the code obtained by applying $\sigma(s_i)$ and $\rho(s_i)$.

Forward section:

$\text{STACK}_{(1)f}.\text{push}(y_i^{t'});$

$$y_i^{t'} \leftarrow \phi(x_{i,1}^{t'},\ldots,x_{i,n}^{t'}); \tag{4.16}$$

$\vdots$

Reverse section:

$\vdots$

$y_i^{t'} \leftarrow \text{STACK}_{(1)f}.\text{top}();$

$\text{STACK}_{(1)f}.\text{pop}();$

$$x_{(1)i,1}^{t'} \mathrel{+\!\!+\!\!\leftarrow} \phi_{x_{i,1}^{t'}}(x_{i,1}^{t'},\ldots,x_{i,n}^{t'})\cdot y_{(1)i}^{t'}; \tag{4.17}$$

$\vdots$

$$x^{t'}_{(1)i,k} \mathrel{+}\!\!\leftarrow \phi_{x'_{i,k}} (x'_{i,1}, \ldots, x'_{i,n}) \cdot y^{t'}_{(1)i}; \tag{4.18}$$

$$\vdots$$

$$x^{t'}_{(1)i,n} \mathrel{+}\!\!\leftarrow \phi_{x'_{i,n}} (x'_{i,1}, \ldots, x'_{i,n}) \cdot y^{t'}_{(1)i}; \tag{4.19}$$

$$y^{t'}_{(1)i} \leftarrow 0; \tag{4.20}$$

Similar as in the tangent-linear case, the equality $\&y^t_1 = \&x^{t'}_{i,k}$ holds. This implies that we need a barrier in front of (4.16). Otherwise, thread t' may read reference $\&x^{t'}_{i,k}$ in (4.16) before thread t writes this reference in (4.13). The computation of the partial derivatives in the code lines from (4.17) to (4.19) may also read a wrong value if we omit this barrier.

During the reverse section we have another data dependence to take account of, namely the one induced by the implication

$$\&y^t_1 = \&x^{t'}_{i,k} \overset{Lem.42}{\Longrightarrow} \&y^t_{(1)1} = \&x^t_{(1)i,k}.$$

Hence, we have a data dependence from (4.18) to all the assignments from (4.14) to (4.15). To synchronize this race condition we need a barrier construct following $\rho(s_i)$.

The barrier in P must be placed before s_i what leads to a barrier after $\rho(s_i)$ in $\rho(S)$. We showed that a data dependence in the original code P leads to multiple data dependencies in the adjoint code of P. Therefore, we cannot omit the barrier in the reverse section.

The above shows that in case of a true dependence between S_1 and S_i, we cannot omit the barrier construct. A *true dependence* [66, p. 98] between the statements s_1 and s_i means that s_1 writes a memory reference that is concurrently read through statement s_i. An *output dependency* means that two assignments write to the same reference. In case of

$$\&y^t_1 = \&y^{t'}_i \overset{Lem.42}{\Longrightarrow} \&y^t_{(1)1} = \&y^{t'}_{(1)i},$$

we obtain a race condition between (4.20) and (4.15) in the reverse section.

Summarizing, we showed that a barrier construct in P that does not have its pendant in the source transformation result, leads to a race condition during the derivative code execution. $\square$

4.2.2 Synchronization per master Construct

The next construct for our language *SPLOMP*[1] is the master construct. This construct was introduced by OpenMP 3.1 citation 1.3 (page 43) and an example of what this construct means for the possible set of interleavings was shown in Example 16 on page 72. To briefly recap, the master construct defines a structured block that is only executed by the master thread which has the identification number 0. The parallel region P that we consider in this section is structured as follows:

```
#pragma omp parallel
{
    S₁
    #pragma omp master
    {
        Sᵢ
    }
    Sⱼ
}
```

Listing 4.5: Example structure of an OpenMP parallel region with a master construct.

S_1, S_i, S_j are sequences of statements with

$$S_1 = (s_1; s_2; \ldots; s_{i-1}) \quad ,$$
$$S_i = (s_i; s_{i+1}; \ldots; s_{j-1}) \quad , \text{ and}$$
$$S_j = (s_j; s_{j+1}; \ldots; s_q).$$

The statements from sequence S_i are only executed by the master thread, which implies that statements from S_i in all possible interleavings have always the thread identifier zero:

$$\forall s^t_{\{i,\ldots,j-1\}} \in I \in \mathscr{I}(P,q,p): \quad t = 0.$$

Since the master construct does not define an implicit barrier, there is no order restriction inside of $\mathscr{I}(P,q,p)$. The production rule for the master construct is

$$R_{OMP} = R_{OMP} \cup \left\{ \begin{array}{cl} c & : \quad \texttt{\#pragma omp master} \\ & \quad \{S\} \end{array} \right\}$$

where the master construct counts as control flow statement and is therefore an element of *CFSTMT*. Given a master construct statement as input, the tangent-linear model and the forward section are defined recursively.

$$\tau \left(\begin{array}{l} \texttt{\#pragma omp master} \\ \{S\} \end{array} \right) := \begin{array}{l} \texttt{\#pragma omp master} \\ \{\tau(S)\} \end{array} \qquad (4.21)$$

$$\sigma \left(\begin{array}{l} \texttt{\#pragma omp master} \\ \{S\} \end{array} \right) := \begin{array}{l} \texttt{\#pragma omp master} \\ \{\sigma(S)\} \end{array} \qquad (4.22)$$

$$\stackrel{(2.39)}{=} \left\{ \begin{array}{l} \texttt{\#pragma omp master} \\ \{ \\ \qquad \text{STACK}_{(1)c}.\text{push}(l); \\ \qquad \vdots \\ \} \end{array} \right.$$

We indicate above that label l is the label of the first statement in S. This label is pushed onto the stack $\text{STACK}_{(1)c}$ that belongs to the master thread. All the other threads do not execute this push statement. Hence, only the master thread encounters this element during the reverse section which means that we do not need the master construct during the reverse section to restrict the execution of $\rho(S_i)$ to the master thread.

$$\rho \left(\begin{array}{l} \texttt{\#pragma omp master} \\ \{S\} \end{array} \right) := \rho(S) \qquad (4.23)$$

4.2.3 Synchronization per critical Construct

The critical construct ensures for an associated subsequence of statements S' that only one thread at a time executes S'. The production rule for the critical construct is

$$R_{OMP} = R_{OMP} \cup \left\{ \begin{array}{ll} c & : \quad \texttt{\#pragma omp critical} \\ & \quad \{S\} \end{array} \right\}$$

where the new element is contained *CFSTMT*. Despite the fact that the critical construct is associated with a subsequence of statements, we consider only the assignment

$$s : y \leftarrow y \cdot x$$

as contained in the subsequence. This statement is sufficient to explain the approach of our source transformation.

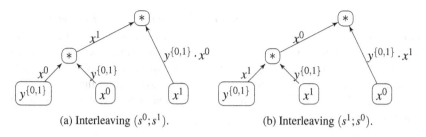

(a) Interleaving $(s^0; s^1)$.   (b) Interleaving $(s^1; s^0)$.

Figure 4.1: DAG of the parallel execution of $s : y \leftarrow y * x$.

The order in which the threads enter the critical region in P is arbitrary. This means for the computation which takes place inside the critical region that it must be independent of the order in which the threads enter the region. When we assume that we use two threads to execute the statement s, we obtain the two possible interleavings $(s^0; s^1)$ and $(s^1; s^0)$. We display the computation of both interleavings by a DAG which is shown in Figure 4.1. No matter which interleaving we consider, the result in the topmost node is the same since the multiplication is commutative.

$$y^{\{0,1\}} \cdot x^0 \cdot x^1 = y^{\{0,1\}} \cdot x^1 \cdot x^0$$

We indicate the affinity of each variable by a superscript index. x^0 belongs to thread zero, x^1 to thread one, and $y^{\{0,1\}}$ belongs to all threads which means that this variable is shared. The result obtained by the computation shown in Figure 4.1 is $y^{\{0,1\}} \cdot x^0 \cdot x^1$. The three derivative values of this result with respect to the input values are as follows:

$$\frac{\partial \left(y^{\{0,1\}} \cdot x^0 \cdot x^1 \right)}{\partial y^{\{0,1\}}} = x^0 \cdot x^1 \tag{4.24}$$

$$\frac{\partial \left(y^{\{0,1\}} \cdot x^0 \cdot x^1 \right)}{\partial x_0} = x^1 \cdot y^{\{0,1\}} \tag{4.25}$$

$$\frac{\partial \left(y^{\{0,1\}} \cdot x^0 \cdot x^1 \right)}{\partial x_1} = y^{\{0,1\}} \cdot x^0 \tag{4.26}$$

No matter whether we use the tangent-linear or the adjoint model, these derivative results should be computed. To examine this, we consider the possible interleavings of the derivative codes concerning the example from above. The tangent-linear model of s is obtained by applying the rule (2.17):

$$\tau(s) = \begin{cases} s_1 = & y^{(1)} & \leftarrow x \cdot y^{(1)} + y \cdot x^{(1)}; \\ s_2 = & y & \leftarrow y \cdot x; \end{cases}$$

The statement s_1 computes the tangent-linear component $y^{(1)}$ and s_2 computes the value for y. If we execute $\tau(s)$ with two threads then we obtain according to Lemma 36 six interleavings. These interleavings are shown in the first column of Table 4.1. The second column shows the expression that is supplied in the variable $y^{(1)}$ after the execution of the corresponding interleaving from the first column. In case that all three derivative values are computed correctly, the third column contains a 'yes'.

Interleaving	$y^{(1)}$	Correct
$s_1^0; s_2^0; s_1^1; s_2^1;$	$x^1 \cdot (x^0 \cdot y^{(1)} + y \cdot x^{(1)0}) + (y \cdot x^0) \cdot x^{(1)1}$	yes
$s_1^0; s_1^1; s_2^0; s_2^1;$	$x^1 \cdot (x^0 \cdot y^{(1)} + y \cdot x^{(1)0}) + y \cdot x^{(1)1}$	no
$s_1^1; s_1^0; s_2^0; s_2^1;$	$x^0 \cdot (x^1 \cdot y^{(1)} + y \cdot x^{(1)1}) + y \cdot x^{(1)0}$	no
$s_1^0; s_1^1; s_2^1; s_2^0;$	$x^1 \cdot (x^0 \cdot y^{(1)} + y \cdot x^{(1)0}) + y \cdot x^{(1)1}$	no
$s_1^1; s_1^0; s_2^1; s_2^0;$	$x^0 \cdot (x^1 \cdot y^{(1)} + y \cdot x^{(1)1}) + y \cdot x^{(1)0}$	no
$s_1^1; s_2^1; s_1^0; s_2^0;$	$x^0 \cdot (x^1 \cdot y^{(1)} + y \cdot x^{(1)1}) + (y \cdot x^1) \cdot x^{(1)0}$	yes

Table 4.1: Parallel computation of the derivative of $y^{\{0,1\}} \cdot x^0 \cdot x^1$ in tangent-linear mode. The only two interleavings that result in the correct derivative values for all three input variables are the first and the last interleaving. For better readability, we omit the set $\{0,1\}$ as superscript for variable y.

Let us illustrate this by taking the first two lines in Table 4.1 as an example. The interleaving $(s_1^0; s_2^0; s_1^1; s_2^1)$ leads to the expression

$$x^1 \cdot (x^0 \cdot y^{(1)} + y \cdot x^{(1)0}) + (y \cdot x^0) \cdot x^{(1)1}$$

that defines the value for the variable $y^{(1)}$ after the tangent-linear code has been executed. To verify that the derivative values are correct, we initialize the tangent-linear vector

$$\begin{pmatrix} y^{(1)\{0,1\}} \\ x^{(1)0} \\ x^{(1)1} \end{pmatrix} \text{ with } \begin{pmatrix} 1 \\ 0 \\ 0 \end{pmatrix}, \begin{pmatrix} 0 \\ 1 \\ 0 \end{pmatrix}, \text{ and } \begin{pmatrix} 0 \\ 0 \\ 1 \end{pmatrix}.$$

The results are the correct derivative values shown in (4.24), (4.25), and (4.26) depending on which Cartesian basis vector was used. Analogously, we obtain the value y for the derivative with respect to x^1 when we use the expression

$$x^1 \cdot (x^0 \cdot y^{(1)} + y \cdot x^{(1)0}) + y \cdot x^{(1)1}$$

which represents the value that is computed by the interleaving $(s_1^0; s_1^1; s_2^0; s_2^1)$. This is wrong because the correct value is shown in (4.26). The verification of the remaining interleavings reveals that only the first and the last interleaving provide all three derivative values correctly.

The interleavings $s_1^0; s_2^0; s_1^1; s_2^1;$ and $s_1^1; s_2^1; s_1^0; s_2^0;$ have in common that the statements s_1 and s_2 are executed right after another. This shows that we have to define the tangent-linear model of s inside one critical region. Hence, we define the tangent-linear transformation of a critical region by

$$\tau \left(\begin{array}{l} \text{\#pragma omp critical} \\ \{S\} \end{array} \right) := \left\{ \begin{array}{l} \text{\#pragma omp critical} \\ \{\tau(S)\} \end{array} \right. \qquad . \qquad (4.27)$$

The adjoint source transformation requires more effort and is not as straightforward as the tangent-linear transformation is. As we will explain below, we need to trace the order of the threads in which they enter the critical region to ensure the correct derivative computation. This is implemented by using a counter value that is put onto the stack $\text{STACK}_{(1)c}$ which is a departure of the common usage of this stack.

For the explanation, we assume the content of the critical region to be again the statement s. The forward section of s is obtained by applying (2.40):

$$\sigma(s) = \left\{ \begin{array}{ll} s_1 = & \text{STACK}_{(1)f}.\text{push}(y); \\ s_2 = & y \leftarrow y \cdot x; \end{array} \right. \qquad .$$

Table 4.2 displays the different interleavings of the forward section and their impact on the floating-point stacks $\text{STACK}_f{}^0$ and $\text{STACK}_f{}^1$ which are defined as thread local. The interleavings where statement s_1 is executed by both threads before the first execution of statement s_2 takes place, result in the same stack content y in both threads. This is wrong because this leads to a wrong value recovery during the reverse section. Therefore, only the first and the last interleaving are correct in the sense that they provide different values in the stacks. Similar to the tangent-linear case, this shows that we have to ensure that $\sigma(s)$ is executed as a compound. This is achieved by putting the forward section of s into a critical region.

$$\rho(s) = \left\{ \begin{array}{lll} s_1 = & y & \leftarrow \text{STACK}_f.\text{top}(); \text{STACK}.\text{pop}(); \\ s_2 = & x_{(1)} & +\!\!\leftarrow y \cdot y_{(1)}; \\ s_3 = & y_{(1)} & \leftarrow x \cdot y_{(1)}; \end{array} \right. \qquad (4.28)$$

The code for the reverse section of s is emitted by (2.50) and contains actually four statements. However, we only consider three statements as shown in (4.28) to

Interleaving	STACK_f^0	STACK_f^1	Correct
$s_1^0; s_2^0; s_1^1; s_2^1;$	y	$y * x[0]$	yes
$s_1^0; s_1^1; s_2^0; s_2^1;$	y	y	no
$s_1^1; s_1^0; s_2^0; s_2^1;$	y	y	no
$s_1^0; s_1^1; s_2^1; s_2^0;$	y	y	no
$s_1^1; s_1^0; s_2^1; s_2^0;$	y	y	no
$s_1^1; s_2^1; s_1^0; s_2^0;$	$y * x[1]$	y	yes

Table 4.2: Parallel evaluation of $\sigma(s)$. The first and the last interleaving result in the correct stack content.

keep the number of combinations for the interleavings down. We stick together the first two statements, namely the recovery of value y and the pop-statement. This is possible because the pop-statement only has a thread local effect. Please note that the statement s_3 is not an incremental assignment but a real assignment. This is not what the source transformation provides but it is semantically equivalent as long as the transformation tool uses temporary variables for intermediate values as shown in Section 1.2. The reason for the difference of this adjoint assignment is the fact that variable y occurs on both sides of the original assignment s.

To examine the adjoint statements of s, let us assume that the forward section has already been executed. The interleaving that reflects the execution of the forward section is $s_1^0; s_2^0; s_1^1; s_2^1$ (first line in Table 4.2). This means that first thread zero executes $\sigma(s)$ and then thread one. Figure 4.1a illustrates this computation. This results in the following values for the floating-point stacks and for variable $y^{\{0,1\}}$:

STACK_f^0	STACK_f^1	Result in $y^{\{0,1\}}$
$y^{\{0,1\}}$	$y^{\{0,1\}} \cdot x^0$	$y^{\{0,1\}} \cdot x^0 \cdot x^1$

With this program state in mind we can examine the effect of the different interleavings of the adjoint statements (4.28). According to Lemma 36 we have

$$\frac{(3 \cdot 2)!}{3!^2} = 20$$

different interleavings for the three statements, executed by two threads. To make the result more readable, we only present the result for the different derivative values and not the expressions that result from the different interleavings. Comparing the values shown in the third, fourth, and fifth column of Table 4.3 with the correct

No.	Interleaving	$x^0_{(1)}$	$x^1_{(1)}$	$y^{\{0,1\}}_{(1)}$	Correct
1	$s_1^0;s_2^0;s_3^0;s_1^1;s_2^1;s_3^1;$	y	$y \cdot x^0 \cdot x^0$	$x^0 \cdot x^1$	false
2	$s_1^0;s_2^0;s_1^1;s_3^0;s_2^1;s_3^1;$	y	$y \cdot x^0 \cdot x^0$	$x^0 \cdot x^1$	false
3	$s_1^0;s_1^1;s_2^0;s_3^0;s_2^1;s_3^1;$	$y \cdot x^0$	$y \cdot x^0 \cdot x^0$	$x^0 \cdot x^1$	false
4	$s_1^1;s_1^0;s_2^0;s_3^0;s_2^1;s_3^1;$	y	$y \cdot x^0$	$x^0 \cdot x^1$	false
5	$s_1^0;s_2^0;s_1^1;s_2^1;s_3^0;s_3^1;$	y	$y \cdot x^0$	$x^0 \cdot x^1$	false
6	$s_1^0;s_1^1;s_2^0;s_2^1;s_3^0;s_3^1;$	$y \cdot x^0$	$y \cdot x^0$	$x^0 \cdot x^1$	false
7	$s_1^1;s_2^0;s_2^1;s_3^0;s_1^0;s_3^1;$	y	y	$x^0 \cdot x^1$	false
8	$s_1^0;s_1^1;s_2^1;s_2^0;s_3^0;s_3^1;$	$y \cdot x^0$	$y \cdot x^0$	$x^0 \cdot x^1$	false
9	$s_1^1;s_1^0;s_2^1;s_2^0;s_3^0;s_3^1;$	y	y	$x^0 \cdot x^1$	false
10	$s_1^1;s_2^1;s_1^0;s_2^0;s_3^0;s_3^1;$	y	$y \cdot x^0$	$x^0 \cdot x^1$	false
11	$s_1^0;s_2^0;s_1^1;s_2^1;s_3^1;s_3^0;$	y	$y \cdot x^0$	$x^0 \cdot x^1$	false
12	$s_1^0;s_1^1;s_2^0;s_2^1;s_3^1;s_3^0;$	$y \cdot x^0$	$y \cdot x^0$	$x^0 \cdot x^1$	false
13	$s_1^1;s_1^0;s_2^0;s_2^1;s_3^1;s_3^0;$	y	y	$x^0 \cdot x^1$	false
14	$s_1^0;s_1^1;s_2^1;s_2^0;s_3^1;s_3^0;$	$y \cdot x^0$	$y \cdot x^0$	$x^0 \cdot x^1$	false
15	$s_1^1;s_1^0;s_2^1;s_2^0;s_3^1;s_3^0;$	y	y	$x^0 \cdot x^1$	false
16	$s_1^1;s_2^1;s_1^0;s_2^0;s_3^1;s_3^0;$	y	$y \cdot x^0$	$x^0 \cdot x^1$	false
17	$s_1^0;s_1^1;s_2^1;s_3^1;s_2^0;s_3^0;$	$y \cdot x^0 \cdot x^1$	$y \cdot x^0$	$x^0 \cdot x^1$	false
18	$s_1^1;s_1^0;s_2^1;s_3^1;s_2^0;s_3^0;$	$y \cdot x^1$	y	$x^0 \cdot x^1$	false
19	$s_1^1;s_2^1;s_1^0;s_3^1;s_2^0;s_3^0;$	$y \cdot x^1$	$y \cdot x^0$	$x^0 \cdot x^1$	true
20	$s_1^1;s_2^1;s_3^1;s_1^0;s_2^0;s_3^0;$	$y \cdot x^1$	$y \cdot x^0$	$x^0 \cdot x^1$	true

Table 4.3: The three statements from (4.28) can be interleaved in 20 different ways. The results of these interleavings according to the adjoint variables are presented. The last column shows whether or not the computed values are correct.

derivative values shown in (4.25), (4.26), and (4.24), we recognize that only the two last interleavings provide the correct result.

The correct interleaving in line 20 reveals two things. Firstly, the three statements s_1, s_2, and s_3 have to be executed as a compound without any context switch between threads. This is because $s_1^1;s_2^1;s_3^1$ and $s_1^0;s_2^0;s_3^0$ have all the same thread ID. Secondly, the thread scheduling order is important because otherwise the interleaving in the first line of Table 4.3 would have provided the correct result as well.

The remaining question is why the interleaving in line 19 also provides the correct result. The only difference between the interleavings in line 19 and 20 is that the statements s_1^0 and s_3^1 have switched their position. The statement s_1 restores the value of y from the floating-point stack and s_3 computes the adjoint value $y_{(1)}$. Therefore, these two statements do not have any data dependencies and can be executed simultaneously.

We achieved the correct derivative results, for example, if the forward section was executed first by thread zero, then by thread one. In the reverse section, it is the other way around, the first thread must be thread one, the second thread is thread zero. This means that the order in which the threads enter the critical region in the forward section determines the order in which the reverse section has to execute the adjoint statements. In the following we explain how this is implemented through the adjoint source transformation.

We tried to keep this implementation as simple as possible and achievable with the possibilities that the $SPLOMP$[1] language provides. For that reason, we use a counter variable called a_l where the values of a_l are of the same basic type as the labels from $LABEL(P)$ are. The index l associates this counter variable with the critical region in P which has this label l. This prevents name clashes in the derivative code. The values of a_l are put onto the control flow stack $STACK_{(1)c}$ despite the fact that usually the content consists only of labels.

$$
\sigma \left(\begin{array}{l} l: \quad \text{\#pragma omp critical} \\ \quad\quad \{S\} \end{array} \right) := \left\{ \begin{array}{l} STACK_{(1)c}.\text{push}(l) \\ \text{\#pragma omp critical} \\ \{ \\ \quad \sigma(S) \\ \quad STACK_{(1)c}.\text{push}(a_l) \\ \quad a_l \leftarrow a_l + 1 \\ \} \\ STACK_{(1)c}.\text{push}(l) \end{array} \right.
\qquad (4.29)
$$

Obviously, we have to ensure that the value range of a_l is not interfering with the label values on the control flow stack. Otherwise, a misinterpretation of the values lead to the wrong reversal of the control flow.

To explain transformation (4.29), we consider a critical region inside of P that is labeled with l. Subsequence S is associated to this critical region and is labeled with $l-1$. According to our implementation, this fits the practice because we use a bottom-up parser generated by the tool bison[2]. In case of a top-down parser one has to attend the initialization value for a_l. Please note that the choice of the counter initialization depends on the values of the labels that can arise during the

---

[2]http://www.gnu.org/software/bison/

execution of (4.29). It must be ensured that all the labels occurring in (4.29) are different to the values occurring as counter values during runtime.

We choose $l + 1$ as initial value for the counter a_l, because l is the label of the critical region and in our case the associated subsequence S contains only labels lower than l. This means the branch statements that are emitted by $\sigma(S)$ in (4.29) all contain a test expression that checks for a label value lower than l. Therefore, these label values do not interfere with the values of a_l.

The first thread that enters the critical region that is shown in (4.29), pushes the value $l + 1$ onto $STACK_{(1)c}$, increments the value of a_l and leaves the critical region. The second thread pushes $l + 2$ to $STACK_{(1)c}$, increments a_l and leaves the critical region, and so on. In case that only two threads enter the critical region, the corresponding local control flow stacks have the appearance as shown in Table 4.4.

1st thread	...	l	$l-1$	$l+1$	l	...
2nd thread	...	l	$l-1$	$l+2$	l	...

Table 4.4: This shows the critical region amount that is contained in the $STACK_c$ from the two threads. The order in which the elements were put onto the stack is from left to right. The first l marks that the thread encountered the critical region. The label $l-1$ comes from the execution of $\sigma(S)$ because S is assumed to has the label $l-1$. The third position represents the place where the value of the counter variable a_l is placed. The first thread puts a $l+1$ there, the second thread a $l+2$, and so forth. The concluding label l marks that the thread has left the critical region.

The label l is pushed twice onto $STACK_{(1)c}$, firstly before the thread enters the critical region and secondly after the thread has left the critical region. These two labels l enclose the region associated with the critical region. The label $l-1$ follows what indicates that the execution of the subsequence $\sigma(S)$ has started. We assume in our simple example that only one label is put onto the control flow stack during the execution of $\sigma(S)$, in practice there are likely more. The third position in Table 4.4 represents the position where the values of a_l are placed. Therefore, there are the values $l+1$ for the first thread and $l+2$ for the second thread. This element must be recognized as a counting number instead of a label identifier. When the threads leave the critical region they put another l to $STACK_{(1)c}$ to indicate that this is the end of the critical region. This final label is important during the execution of the reverse section.

When the execution of the reverse section encounters the final l which is the rightmost l in Table 4.4 the consumption of all the values from a_l starts. This is shown in (4.30). Firstly, we test if the current top of the control flow stack contains label l. If this is the case, we pop this label and start a loop which iterates until the current top of the stack is again label l. This means that the loop iterates as long as the leftmost label l in Table 4.4 has not been reached. In addition, this clearly shows why we put the label l twice onto the stack.

After the two threads in our example have popped the final l off their local stack $STACK_c$, the stacks have the following appearance:

1st thread	...	l	$l-1$	$l+1$
2nd thread	...	l	$l-1$	$l+2$

The whole loop body is a critical region, such that it is ensured that at each point in time only one thread executes the loop body. The other threads are probably waiting for being allowed to enter this critical region. Once a thread is allowed to enter the critical region, it tests the current top of $STACK_c$, which represents a counter value not a label identifier.

$$\rho \left(\begin{array}{l} l: \quad \text{\#pragma omp critical} \\ \quad \{S\} \end{array} \right)$$

$$:= \left\{ \begin{array}{l} \text{if } \left(STACK_{(1)c}.top() = l \right) \{ \\ \quad STACK_{(1)c}.pop() \\ \quad \text{while } \left(STACK_{(1)c}.top() \neq l \right) \{ \\ \qquad \text{\#pragma omp critical} \\ \qquad \{ \\ \qquad\quad \text{if } \left(STACK_{(1)c}.top() = a_l - 1 \right) \{ \\ \qquad\qquad STACK_{(1)c}.pop() \\ \qquad\qquad a_l \leftarrow a_l - 1 \\ \qquad\qquad \text{while } \left(STACK_{(1)c}.top() \neq l \right) \{ \\ \qquad\qquad\quad \rho(S) \\ \qquad\qquad \} \\ \qquad\quad \} \\ \qquad \} \\ \quad \} \\ \quad STACK_{(1)c}.pop() \\ \} \end{array} \right. \qquad (4.30)$$

The last thread that entered the critical region during the forward section puts the value $a_l - 1$ onto $\text{STACK}_{(1)c}$. Now during the reverse section, this thread has to be the first that executes the corresponding adjoint statements in $\sigma(S)$. Speaking with our example, after the two threads left the critical region, the current value of a_l is $l + 3$. Therefore, the second thread enters the branch body in (4.30) because it has the value $l + 2$ on top of $\text{STACK}_{(1)c}$. This thread removes the value $l + 2$ from its stack and decreases the counting variable a_l. This decrease has to be atomic since all the threads are accessing this reference. But since we are inside a critical region, the exclusive access is ensured. The stacks look now as follows:

1st thread	...	l	$l-1$	$l+1$
2nd thread	...	l	$l-1$	

The value of a_l is now $l + 2$, but the second thread has not left the critical region yet. Hence, we can be sure that the first thread is still waiting for allowance to enter the critical region. Meanwhile, the second thread executes the code from $\rho(S)$. In our example this is only the element $l - 1$. After $l - 1$ is consumed by the innermost loop the stacks contain:

1st thread	...	l	$l-1$	$l+1$
2nd thread	...	l		

l displays the end of the part where $\sigma(S)$ had put labels onto $\text{STACK}_{(1)c}$ and the thread can leave the critical region. Afterwards, it removes the label l from its control flow stack:

1st thread	...	l	$l-1$	$l+1$
2nd thread	...			

As soon as the second thread leaves the critical region, the first thread is allowed to enter this region. The following test expression checks if the value on top of $\text{STACK}_{(1)c}$ is $l + 1$ which true. It consumes the label $l - 1$ by executing $\rho(S)$ and removes the label l after it has left the critical region. This finishes the reverse section part.

Please note that there is a data dependence between (4.29) and (4.30). We must prevent that one thread can enter the critical region in the forward section, while another thread is already consuming these counting numbers during the reverse section. This situation could lead to ambiguous values. Therefore, we have to place a barrier construct before the reverse section to make sure that each thread has finished its forward section before any thread enters the reverse section.

Proposition 55. *Definition* (4.30) *implies the need of a barrier between the forward and the reverse section.*

Proof. Assume that there is no barrier between the forward and the reverse section. In addition, thread zero and thread one are about to enter the critical region in the reverse section. A third thread with ID two is about to enter the critical region in the forward section. The current value of a_l is assumed to be c and the control flow stacks of the threads are as follows:

0	...	l	...	$c-1$	l
1	...	l	...	$c-2$	l
2	...	l			

Thread zero enters the critical region after removing the label identifier l from STACK$_c$. The top of the stack of thread zero shows $c-1$ and therefore it continues the execution with $\rho(S)$. After thread zero has left the critical region the value of a_l is $c-1$.

0	...	l			
1	...	l	...	$c-2$	l
2	...	l			

Assume that thread two now enters the critical region defined in the forward section, see (4.29). Thread two puts the value $c-1$ onto STACK$_c$, increases a_l to c, leaves the forward section and continues its execution until it reaches the code from (4.30).

0	...	l			
1	...	l	...	$c-2$	l
2	...	l	...	$c-1$	l

At this point, both threads attempt to enter the critical region in the reverse section. Thread two is being allowed to execute $\rho(S)$ because its value on top of its stack is $c-1$ and the current value of a_l is c. This is wrong because thread one has to come first. This shows that a wrong reversal of the control flow is possible when we omit a barrier between the forward and the reverse section. □

Another question may arise during the reading of the above. Would it be possible to put a barrier after the critical region in (4.30)? This would have the advantage that while all the threads are competing with each other to enter the critical region, the wrong candidates would be held from competing another time because they are waiting at the barrier. As a general answer we must deny this question. The problem is that the barrier has to be encountered by all threads in the group or by none at all. This is in general not given since we do not know which threads enter the critical region in P and which one not. In case that it is ensured that all threads run through the critical region in P we could place a barrier after the critical region in (4.30).

4.2.4 Synchronization per atomic Construct

The critical construct is a very heavy weighted construct in terms of runtime overhead. The synchronization performed by the runtime system to ensure that at each point in time only one thread is executing the code inside the critical region can effect dramatically the runtime performance of the program. A critical construct should therefore only be chosen when there is no other choice. A potential better alternative is the atomic construct which is defined for certain update operations of a memory location.

In shared memory programming there are often situations where multiple threads have to update a memory location by a value. The memory location is read and written by multiple threads simultaneously and all the threads try to update the value of this memory location. In this case the commutativity of an addition or a multiplication can be exploited in a way that the order in which the threads add or multiply their value to the memory location is not of importance. We saw this already in the previous section where we considered the transformation of a critical region.

In case that we have to ensure that only one assignment is executed atomically and not a whole sequence of statements, the OpenMP standard knows the atomic construct. It is always better to use an atomic construct instead of a critical construct because the atomic construct is often implemented by hardware on a lower level[3].

Despite the fact that the OpenMP 3.1 standard supports a couple of different atomic statements, we define an atomic statement to have the following form:

$$R_{OMP} = R_{OMP} \cup \left\{ \begin{array}{ll} c & : \quad \texttt{\#pragma omp atomic} \\ & \quad y \mathrel{+\!\!\leftarrow} \phi(x_1, \ldots, x_n) \end{array} \right\}$$

The control flow of the executing thread is not changed, therefore we classify this construct as straight-line code. This means for the partitioning into maximal sequences of *SLC* sequences that we do not end a sequence when we encounter an atomic statement.

To explain the source transformation of an atomic construct, we consider the

---

[3]For example the X86 processor family knows the XCHG or the XADD command where atomicity is ensured, see e.g. [55].

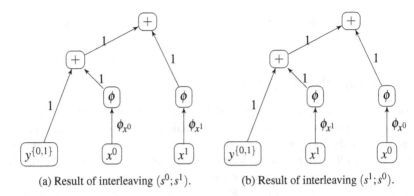

(a) Result of interleaving $(s^0; s^1)$.   (b) Result of interleaving $(s^1; s^0)$.

Figure 4.2: DAG of the execution of statement s defined in (4.31).

following statement s as an example:

$$s = \left\{ \begin{array}{l} \text{\#pragma omp atomic} \\ y \mathrel{+}\!\!\leftarrow \phi(x) \end{array} \right. , \tag{4.31}$$

where y is a shared variable and x is a private variable. For simplification reasons, we only take a ϕ that takes one argument as input. Let us consider the execution of s with two threads. This means we have two possible interleavings, $s^0; s^1$ and $s^1; s^0$. The computation of these two interleavings can be illustrated by a DAG which is displayed in Figure 4.2.

The memory reference $\&y$ occurs in s on the right-hand side as well as on the left-hand side. This means that this reference is initially read and then written. The atomic construct of OpenMP defines that between the read of $\&y$ and the store to $\&y$, no task scheduling points are allowed, see OpenMP 3.1 citation 27 (page 46). A task scheduling point is a point where the runtime environment decides to interrupt the execution of thread t and continues by letting thread t' execute its code, where $t \neq t'$. The lack of a task scheduling point between the read and the store is important when $\&y$ is being accessed by more than one thread. Only when there is no task point between the read and store, it is ensured that each interleaving provides the same result. Similar to the critical construct, the atomic construct ensures atomicity of the associated assignment.

Let us assume that there is a thread t which writes to $\&y$ and in addition there is another thread t' that accesses simultaneously $\&y$ per read or per write operation. In other words, it holds

$$\exists (s^t, s^{t'}) \in I \in \mathscr{I}(P, q, p) : \&y^t = \&y^{t'} \text{ or } \&y^t = \&x^{t'}. \tag{4.32}$$

The derivative computation by the tangent-linear model of the incremental assignment s contains $s^{(1)}$ with

$$s^{(1)} = y^{(1)} \mathrel{+\!\!+\!\!\leftarrow} \phi_x(x) \cdot x^{(1)}.$$

From Lemma 42 we know that

$$\&y^t = \&y^{t'} \quad \text{or} \quad \&y^t = \&x^{t'}$$
$$\implies \&y^{(1)t} = \&y^{(1)t'} \quad \text{or} \quad \&y^{(1)t} = \&x^{(1)t'}.$$

When we apply this implication to (4.32), we see that we have another critical reference in the tangent-linear model of s:

$$\exists (s^{(1)t}, s^{(1)t'}) \in I \in \mathscr{I}(\tau(P), q, p) : \&y^{(1)t} = \&y^{(1)t'} \text{ or } \&y^{(1)t} = \&x^{(1)t'}.$$

From this, we conclude that the critical reference of the original assignment is inherited to the tangent-linear assignment $s^{(1)}$. Hence, we have to put at least an atomic construct in front of the two assignments emitted by $\tau(s)$ to ensure that they are executed atomically by the runtime system of OpenMP. The remaining question is whether is is necessary to put both assignments into one critical region. This can be examined by considering the possible interleavings. We define

$$s_1 := \begin{cases} \text{\#pragma omp atomic} \\ y^{(1)} \mathrel{+\!\!+\!\!\leftarrow} \phi_x(x) \cdot x^{(1)}; \end{cases}$$

$$s_2 := \begin{cases} \text{\#pragma omp atomic} \\ y \mathrel{+\!\!+\!\!\leftarrow} \phi(x) \end{cases}$$

and we suppose that s_1 and s_2 are executed by two threads. This leads to six different interleavings that we have to consider. The correct derivative values are

$$\frac{\partial \left(y^{\{0,1\}} + x^0 + x^1 \right)}{\partial y^{\{0,1\}}} = 1 \tag{4.33}$$

$$\frac{\partial \left(y^{\{0,1\}} + x^0 + x^1 \right)}{\partial x_0} = \phi_{x^0} \tag{4.34}$$

$$\frac{\partial \left(y^{\{0,1\}} + x^0 + x^1 \right)}{\partial x_1} = \phi_{x^1} \tag{4.35}$$

The computation of the derivative value by the tangent-linear model is shown in the following table.

Interleaving				$y^{(1)}\{0,1\}$
s_1^0;	s_2^0;	s_1^1;	s_2^1;	$\phi_{x^0}(x^0)\cdot x^{(1)0} + \phi_{x^1}(x^1)\cdot x^{(1)1}$
s_1^0;	s_1^1;	s_2^0;	s_2^1;	$\phi_{x^0}(x^0)\cdot x^{(1)0} + \phi_{x^1}(x^1)\cdot x^{(1)1}$
s_1^1;	s_1^0;	s_2^1;	s_2^0;	$\phi_{x^1}(x^1)\cdot x^{(1)1} + \phi_{x^0}(x^0)\cdot x^{(1)0}$
s_1^0;	s_1^1;	s_2^1;	s_2^0;	$\phi_{x^0}(x^0)\cdot x^{(1)0} + \phi_{x^1}(x^1)\cdot x^{(1)1}$
s_1^1;	s_1^0;	s_2^1;	s_2^0;	$\phi_{x^1}(x^1)\cdot x^{(1)1} + \phi_{x^0}(x^0)\cdot x^{(1)0}$
s_1^1;	s_2^1;	s_1^0;	s_2^0;	$\phi_{x^1}(x^1)\cdot x^{(1)1} + \phi_{x^0}(x^0)\cdot x^{(1)0}$

Most of the edges in Figure 4.2 are labeled with one. This fact leads to the effect that we see in the above table. All the expressions are the same, no matter what interleaving we are considering. This shows that we do not need to put both assignments s_1 and s_2 into one critical region, as it was necessary when transforming the critical construct by the tangent-linear model. Thus, the tangent-linear transformation of an atomic construct is:

$$\tau\left(\begin{array}{l} \text{\#pragma omp atomic} \\ y \mathrel{+\!\!\!+\!\!\!-} \phi(x_1,\ldots,x_n) \end{array}\right)$$

$$:= \left\{\begin{array}{l} \text{\#pragma omp atomic} \\ y^{(1)} \mathrel{+\!\!\!+\!\!\!-} \sum_{k=1}^n \phi_{x_k}(x_1,\ldots,x_n)\cdot x_k^{(1)}; \\ \text{\#pragma omp atomic} \\ y \mathrel{+\!\!\!+\!\!\!-} \phi(x_1,\ldots,x_n) \end{array}\right.$$
        (4.36)

The above creates the impression that the linear effect of the atomic assignment also makes the life easier for the adjoint source transformation. We examine this in the following. Let us consider the transformation of the assignment in (4.31). We obtain by applying the transformation rule (2.41):

$$\sigma\left(y \mathrel{+\!\!\!+\!\!\!-} \phi(x)\right) = \left\{\begin{array}{ll} s_1 = & \text{STACK}_{(1)f}.\text{push}(y); \\ s_2 = & y \mathrel{+\!\!\!+\!\!\!-} \phi(x); \end{array}\right.$$
        (4.37)

The possible interleavings for executing s_1 and s_2 with two threads are shown in Table 4.5 together with the content of the floating-point value stack. We recognize the same effect that we saw earlier in Section 4.2.3. In case that both threads execute their push-statement before one thread executes the statement s_2 then the content of the stacks in both threads comprise the same value. For the example of the critical region this was wrong. Here, we claim that this effect does not matter since the value from the stacks is not necessary for the adjoint statements in $\rho\left(y \mathrel{+\!\!\!+\!\!\!-} \phi(x)\right)$. This becomes clear by considering (4.38). The value that is recovered by the first statement in $\rho\left(y \mathrel{+\!\!\!+\!\!\!-} \phi(x)\right)$ is not used in the adjoint assignment that follows. Compare (4.38) to the situation that we had in (4.28). There

Interleaving	STACK_f^0	STACK_f^1
$s_1^0;s_2^0;s_1^1;s_2^1;$	y	$y+\phi(x^0)$
$s_1^0;s_1^1;s_2^0;s_2^1;$	y	y
$s_1^1;s_1^0;s_2^0;s_2^1;$	y	y
$s_1^0;s_1^1;s_2^1;s_2^0;$	y	y
$s_1^1;s_1^0;s_2^1;s_2^0;$	y	y
$s_1^1;s_2^1;s_1^0;s_2^0;$	$y+\phi(x^1)$	y s

Table 4.5: The possible interleavings of (4.37). y is the value before any thread executes s_1. The interleavings where the statements $s_1^{0,1}$ are executed before s_2^0 or s_2^1 is executed result in the situation that the stack content of both threads is the same.

was a direct data dependence of the recovery statement with y on the left-hand side and the following adjoint assignment with y on the right-hand side. This data dependence does not exist here.

$$\rho\left(y \mathrel{+\!\!+\!\!\leftarrow} \phi(x)\right) = \begin{cases} y \leftarrow \text{STACK}_{(1)f}.\text{top}(); \\ \text{STACK}_{(1)f}.\text{pop}(); \\ x_{(1)} \mathrel{+\!\!+\!\!\leftarrow} \phi_x(x)\cdot y_{(1)}; \end{cases} \tag{4.38}$$

This brings up the question why one should store all the intermediate values when they are not necessary anyway. The only value we have to store is the value that the variable y has before any thread executes the transformation $\sigma\left(y \mathrel{+\!\!+\!\!\leftarrow} \phi(x)\right)$. For example, we assume that the value of y is c before the first thread executes $\sigma(s)$. This value c is somehow stored but the other intermediate values for variable y are not of interest while the threads execute the forward section.

Once the threads enter the reverse section they encounter at some time the code of $\rho(s)$. The value of y is not important at this time since it does not effect the adjoint information in $\rho(s)$. As soon as the last thread has finished the code $\rho(s)$, we have to ensure that the value of y is again c as it was before executing $\sigma(s)$. Therefore, we restore the value c at some point during the reverse section. Where exactly we perform this is not of importance but after all threads have finished $\rho(s)$ the value must be recovered.

This is implemented with the help of two auxiliary variables. The first auxiliary variable a_l is an integer variable and second auxiliary variable z_l is a floating-point variable that hosts the value that we denoted with c in the above example. The label l is the label of the atomic statement s in the parallel region P. The variable a_l serves as status variable that can take the values zero, one, and two. Since both auxiliary variables have to be shared, the code for storing to them has to be in a critical region. The transformation of an atomic statement for the forward section has the following appearance:

$$
\sigma \left(\begin{array}{l} \text{\#pragma omp atomic} \\ y \mathrel{+\!\!\leftarrow} \phi(x_1,\ldots,x_n) \end{array} \right)
$$
$$
:= \left\{ \begin{array}{l} \text{\#pragma omp critical} \\ \{ \\ \qquad \text{if } (a_l = 0) \; \{ \\ \qquad\qquad a_l \leftarrow 1; \\ \qquad\qquad z_l \leftarrow y; \\ \qquad \} \\ \qquad y \mathrel{+\!\!\leftarrow} \phi(x_1,\ldots,x_n) \\ \} \end{array} \right. \qquad (4.39)
$$

Initially, a_l has the status zero which means that no thread has reached the code in (4.39) yet. As soon as the first thread enters the critical region, the value of y is stored in z_l and the status variable a_l is set to one which prevents that another thread assigns to z_l. From this point forward all threads execute only the assignment $y \mathrel{+\!\!\leftarrow} \phi(x_1,\ldots,x_n)$ in (4.39) and all the intermediate values of y are lost. The code for the reverse section is defined as shown in (4.40).

The first thread that reaches (4.40), finishes the adjoint assignments, and enters the critical region sets the status in a_l to two, which represents the status that the value of y has been restored. Since the current thread has encountered the reverse section, it is ensured that at least one thread has finished the forward section and therefore the status in a_l is one. The new status of a_l is set to two and the variable y is set to its previous value which was stored in z_l.

Before the execution of the adjoint code reaches (4.39) it must ensured that a_l has the status zero. This can be done outside the parallel region or when we want to be sure that a_l is set correctly by our source transformation tool, we can implement the initialization of a_l to zero in a way that it takes place inside a single construct. Another possibility is the initialization inside a master construct and a barrier following the master construct. This barrier is implicit given with single construct and cannot be neglected as we show in the following proposition.

$$\rho \left(\begin{array}{l} \#\text{pragma omp atomic} \\ y \mathrel{+\!\!+\!\!\leftarrow} \phi(x_1,\ldots,x_n) \end{array} \right)$$

$$:= \left\{ \begin{array}{l} x_{(1)1} \mathrel{+\!\!+\!\!\leftarrow} \phi_{x_1}(x_1,\ldots,x_n) \cdot y_{(1)}; \\ \quad \vdots \\ x_{(1)n} \mathrel{+\!\!+\!\!\leftarrow} \phi_{x_n}(x_1,\ldots,x_n) \cdot y_{(1)}; \\ \#\text{pragma omp critical} \\ \{ \\ \quad \text{if } (a_l \neq 2) \ \{ \\ \qquad a_l \leftarrow 2; \\ \qquad y \leftarrow z_l; \\ \quad \} \\ \} \end{array} \right. \tag{4.40}$$

Proposition 56. *When the initial status of a_l is set inside the parallel region, a barrier must be defined after the initialization assignment to a_l and before* (4.39).

Proof. Suppose that $s = a_l \leftarrow 0$ is the assignment that sets the initial status of a_l inside of the parallel region. The subsequence of the critical region of (4.39) is denoted by S. Inside of an interleaving, S can be considered as one statement as S is contained in a critical region. Without a barrier between s and S the following interleaving is possible, where t and t' are two distinct threads.

$$S^t; s^{t'}; S^{t'};$$

This means, the assignment $y \mathrel{+\!\!+\!\!\leftarrow} \phi(x_1,\ldots,x_n)$ in S is executed by thread t before s has been executed by thread t'. The initial value of y is lost. A barrier between s and S ensures $s^t \prec S^{t'}$ and therefore that the status variable a_l is initialized before any thread encounters S. $\qquad \square$

Proposition 57. *The definitions* (4.39) *and* (4.40) *require a barrier between them.*

Proof. The subsequence of the critical region in (4.39) is denoted by S_1, the one in (4.40) is denoted by S_2. Since S_1 and S_2 are inside a critical region, we consider them as one statement that is executed atomically. Therefore, the following interleaving is possible where the critical regions are executed by two threads, t and t'. The dots represent a part of the interleaving that is not of importance here.

$$\ldots; S_1^t; \ldots; S_2^t; S_1^{t'}; \ldots$$

This interleaving has the following effect. The value of y is stored correctly in S_1^t and is restored in S_2^t. The problem is that after the assignment $y \leftarrow z_l$ in S_2^t, the variable y is changed again through the assignment $y +\!\!\leftarrow \phi(x_1, \ldots, x_n)$ in $S_1^{t'}$ because thread t' has not finished the code from (4.39).

A barrier between (4.39) and (4.40) ensures $S_1^t \prec S_2^{t'}$ and prevents that the interleaving above can occur. Instead, we can be sure that all threads have finished their work in S_1. A reasonable point where we can place the barrier is just before the loop that opens the reverse section.   □

The next section shows that we can achieve closure of the adjoint source transformation $\sigma(P)$ no matter the parallel region P fulfills the exclusive read property or not.

4.2.5 Closure of SPLOMP [1]

We showed in Proposition 45 on page 116 that the emitted code of the tangent-linear source transformation $\tau(P)$ is noncritical for a given parallel region $P \in SPL$. The resulting code of applying the adjoint source transformation $\sigma(P)$ is noncritical in case that P fulfills the exclusive read property, as shown in Proposition 47. The need for the exclusive read property can be seen as a big constraint for possible codes P since in shared memory parallel programming there are often situations, where it is desired that a certain memory reference is shared among all threads and its value should be read by all threads. This reference is not critical, as long as all threads only read this memory reference. The problem arises in the reverse section of the adjoint source transformation, where this particular read operation becomes a write operation. This write operation must be synchronized to ensure correctness. In the current section, we show how we can synchronize such write operations with the help of synchronization constructs, which were introduced in Section 4.2.

To recap the problem, consider the assignment $s = y \leftarrow x$ with y being a thread local variable and x is a shared variable. Each thread that encounters this assignment, reads the value of x and writes the value to the local variable y. The reverse section of s is composed of

$$\rho(s) = \begin{cases} s_1 = & x_{(1)} +\!\!\leftarrow y_{(1)}; \\ s_2 = & y_{(1)} \leftarrow 0; \end{cases}$$

Reference $\&x_{(1)}$ and reference $\&y_{(1)}$ are read in assignment s_1 and the sum of both values is assigned to reference $\&x_{(1)}$. Both, the read and the stores of $\&x_{(1)}$ are critical since all threads read and write to this memory location simultaneously. Since we have two critical references in this assignment, the statement does not fulfill the LCR property, which means that we cannot use our interleaving abstraction.

Fortunately, the atomic construct of OpenMP allows us to declare s_1 as to be executed atomically. This means that no thread schedule may happen between the read access of $\&x_{(1)}$ and the succeeding store.

$$
\begin{aligned}
&\text{\#pragma omp atomic} \\
s_1 = \quad & x_{(1)} +\!\!\!\!+\!\!\!\leftarrow y_{(1)}; \\
s_2 = \quad & y_{(1)} \leftarrow 0;
\end{aligned}
$$

The explanation above suggests that we use the atomic construct in case that the original assignment does not fulfill the exclusive read property. The problem is that the source transformation tool does not gain the information of the exclusive read property without performing a static program analysis of the original code. We presented such a static analysis in Chapter 3 called exclusive read analysis.

Without performing the exclusive read analysis, the source transformation tool has two options. On the one hand, it can be conservative in a way that it generates atomic constructs preceding each adjoint assignment. This transformation is shown in (4.41).

$$
\rho\left(y \leftarrow \phi(x_1,\ldots,x_n)\right) :=
$$

$$
\left\{
\begin{aligned}
& y \leftarrow \text{STACK}_{(1)f}.\text{top}(); \\
& \text{STACK}_{(1)f}.\text{pop}(); \\
& \text{\#pragma omp atomic} \\
& x_{(1)1} +\!\!\!\!+\!\!\!\leftarrow \phi_{x_1}(x_1,\ldots,x_n)\cdot y_{(1)}; \\
& \text{\#pragma omp atomic} \\
& x_{(1)2} +\!\!\!\!+\!\!\!\leftarrow \phi_{x_2}(x_1,\ldots,x_n)\cdot y_{(1)}; \\
& \quad\vdots \\
& \text{\#pragma omp atomic} \\
& x_{(1)n} +\!\!\!\!+\!\!\!\leftarrow \phi_{x_n}(x_1,\ldots,x_n)\cdot y_{(1)}; \\
& y_{(1)} \leftarrow 0;
\end{aligned}
\right.
\tag{4.41}
$$

Obviously, this would have a major impact on the performance and therefore the source transformation tool could be implemented in a way that it lays the responsibility into the hands of the user. For example, the tool could provide a custom pragma that can be used by the user to declare assignments when it contains a reference that does not fulfill the exclusive read property. Let us call this pragma #pragma ad exclusive read failure. Then the source transformation augments only assignments with this pragma as in (4.42). All the other assignments are transformed without a preceding atomic directive. Obviously, the correctness of the resulting code depends on the correct placement of the compiler directive

#pragma ad exclusive read failure

The transformation rule for an assignment preceded by such a pragma looks as follows:

$$
\rho \left(\begin{array}{l} \text{\#pragma ad exclusive read failure} \\ y \leftarrow \phi(x_1,\ldots,x_n) \end{array} \right) :=
$$

$$
\left\{ \begin{array}{l} y \leftarrow \text{STACK}_{(1)f}.\text{top}(); \\ \text{STACK}_{(1)f}.\text{pop}(); \\ \text{\#pragma omp atomic} \\ x_{(1)1} \mathrel{+\!\!\!+\!\!\!\leftarrow} \phi_{x_1}(x_1,\ldots,x_n) \cdot y_{(1)}; \\ \text{\#pragma omp atomic} \\ x_{(1)2} \mathrel{+\!\!\!+\!\!\!\leftarrow} \phi_{x_2}(x_1,\ldots,x_n) \cdot y_{(1)}; \\ \vdots \\ \text{\#pragma omp atomic} \\ x_{(1)n} \mathrel{+\!\!\!+\!\!\!\leftarrow} \phi_{x_n}(x_1,\ldots,x_n) \cdot y_{(1)}; \\ y_{(1)} \leftarrow 0; \end{array} \right.
\qquad (4.42)
$$

The above solutions should only be taken into account when the implementation of a static program analysis, as the one presented in Chapter 3, is not an option. With help of such an analysis, the source transformation could obtain information whether a reference is read exclusive by a thread or not. For example, let us assume that reference $\&x_k$ does not fulfill the exclusive read property but all the other references $\&x_{\{1,\ldots,k-1,k+1,\ldots,n\}}$ fulfill this property. With this knowledge, we can define the transformation by ρ as in (4.43).

With the synchronization constructs of OpenMP we have adequate tools at hand to show the closure property from Definition 39 for the source transformations. The only requirement for the source transformation is that P is noncritical.

$$
\rho\left(y \leftarrow \phi(x_1,\ldots,x_n)\right) :=
$$

$$
\left\{
\begin{array}{l}
y \leftarrow \text{STACK}_{(1)f}.\text{top}(); \\[4pt]
\text{STACK}_{(1)f}.\text{pop}(); \\[4pt]
x_{(1)1} \mathrel{+\!\!+\!\leftarrow} \phi_{x_1}(x_1,\ldots,x_n)\cdot y_{(1)}; \\[2pt]
\quad\vdots \\[2pt]
x_{(1)k-1} \mathrel{+\!\!+\!\leftarrow} \phi_{x_{k-1}}(x_1,\ldots,x_n)\cdot y_{(1)}; \\[2pt]
\#\text{pragma omp atomic} \\[2pt]
x_{(1)k} \mathrel{+\!\!+\!\leftarrow} \phi_{x_k}(x_1,\ldots,x_n)\cdot y_{(1)}; \\[2pt]
x_{(1)k+1} \mathrel{+\!\!+\!\leftarrow} \phi_{x_{k+1}}(x_1,\ldots,x_n)\cdot y_{(1)}; \\[2pt]
\quad\vdots \\[2pt]
x_{(1)n} \mathrel{+\!\!+\!\leftarrow} \phi_{x_n}(x_1,\ldots,x_n)\cdot y_{(1)}; \\[2pt]
y_{(1)} \leftarrow 0;
\end{array}
\right. \tag{4.43}
$$

Proposition 58. *Let $P \in SPLOMP^1$ be a noncritical parallel region. Then the tangent-linear source transformation $\tau(P)$ and the adjoint source transformation $\sigma(P)$ fulfill the closure property from Definition 39.*

Proof. We can built on the properties that we already have shown for the case that P is contained in *SPL*.

Tangent-linear source transformation $\tau(P)$: We showed in Lemma 46 that τ is closed in case that P is contained in *SPL* and it is noncritical. Thus, we assume that $P \in SPLOMP^1$. The transformation rules for a barrier (4.2), the master construct (4.21), the critical construct (4.27), and the atomic construct (4.36) are defined recursively and the synchronization construct is just mapped by the identity mapping. Therefore, we conclude that $\tau(P)$ is noncritical and fulfills the closure property.

Adjoint source transformation $\sigma(P)$: We showed in Proposition 47 that $\sigma(P)$ is noncritical as long as P is in *SPL* and it fulfills the exclusive read property. Therefore, we assume that P does not fulfill the exclusive read property. The statement that does not fulfill the exclusive read property is supposed to be the assignment $s = y \leftarrow \phi(x_1,\ldots,x_n)$. s contains a reference $\&x_k$ where $k \in \{1,\ldots,n\}$ on its right-hand side that is read by multiple threads.

As stated above, we have different possibilities to handle the missing exclusive read property. We can be conservative by transforming each and every assignment from P by the rule (4.41). Another approach is that the source transformation tool is implemented such that it expects the user to augment these assignments by a pragma that do not fulfill the exclusive read property. Then the tool can use the rule (2.40) as before and for the assignments with the customized pragma it uses the rule (4.42). Probably the best way is to let the exclusive read analysis from Chapter 3 try to reveal that s does not fulfill the exclusive read property. In this case, we can use the rule (4.43) to transform s and all the other assignments can be transformed with (2.40). This shows that we use the synchronization construct atomic of OpenMP to resolve the race condition in $\rho(P)$ that occurs when transforming s.

The described method indicates that we obtain an adjoint model $\sigma(P)$ that contains at least one atomic construct. Thus, the adjoint code $\sigma(P)$ is contained in $SPLOMP^1$. We examine what the reapplication of the source transformations τ, σ, and ρ supply.

The reapplication of τ to $\sigma(P)$ results in a second-order forward-over-reverse model that contains again the atomic construct due to rule (4.36). In case that we apply σ to $\sigma(P)$, we obtain the second-order reverse-over-reverse model. The atomic construct that is contained in $\sigma(P)$ becomes a critical construct as shown in (4.39) and (4.40). In addition, we need at least one barrier, as shown in Proposition 56 and Proposition 57. The auxiliary variable a_l in (4.39) and (4.40) must be initialized. If this initialization must be inside the parallel region for a certain reason, a master construct can be used to perform this. Summarizing, the following OpenMP constructs may occur inside a derivative code that is the source transformation result of a code contained in $SPLOMP^1$:

- barrier

- master

- critical

- atomic

The fact that the resulting codes are noncritical can be understood by the fact that the barrier and the master constructs are transformed by the identity mapping and the transformations of the atomic and the critical constructs emit code that checks local data or accesses shared data within a critical region.

The remaining fact to show is that in case we transform P that is contained in $SPLOMP^1$ by $\tau(P)$ or by $\sigma(P)$, we obtain code that is again contained in $SPLOMP^1$.

1. barrier: The identity mapping was used to transform a barrier, as shown in (4.2), (4.3), and (4.4).

2. master: The master construct leads to another master construct as shown in the rules (4.21) and (4.22). The rule (4.23) shows that the master construct does not occur in the reverse section.

3. critical: The rules (4.27), (4.29) and (4.30) are defined with the help of a critical region.

4. atomic: The atomic construct in the input code leads to the same constructs that are listed here.

This shows the claim.   □

The proof exposes that we need all of the possible synchronization constructs that OpenMP provides to achieve the closure property. Assume we have a transformation tool at hand that takes $SPLOMP^1$ code as input and emits $SPLOMP^1$ code as output. With Proposition 58 we know that the tool is able to generate derivative codes or arbitrary high-order through reapplication and in addition these codes can be executed concurrently since they are noncritical.

4.3 *SPLOMP*^{2} - **Worksharing Constructs**

This section extends our $SPLOMP^1$ language further with the worksharing constructs of OpenMP. The language that we achieved with this extension is called $SPLOMP^2$. With the help of worksharing constructs one can distribute a part of a code to a group of threads such that this code is executed concurrently. A typical example is a loop where the corresponding body is referred to as S. In case that there are no data dependencies between the loop iterations we can execute these iterations simultaneously by distributing the iterations to a team of threads. In Section 2.3.3 on page 113, we described how the knowledge about the parallel region can be exploited in the AD source transformation, especially in the adjoint case where we differ between the split and joint reversal schemes. This description can also be applied to the worksharing constructs. The arising question is what scheme one should use to transform a worksharing subsequence, the split reversal or the joint reversal scheme.

The joint reversal scheme for the worksharing region S would be similar to the structure of the adjoint model of Q which is shown on page 116. A worksharing region which is equivalent to the original code S is preceded by a code that is responsible for storing the input values of S. Inside the reverse section at the position where the adjoint code of the worksharing construct is, these input values are restored and the worksharing construct is executed forward ($\sigma(S)$) and then reverse ($\rho(S)$). For example, the following parallel region

```
#pragma omp parallel
{
    S₁
    #pragma omp for
l : for ( k←0 ; k<n ; k←k+1)
    {
        Sᵢ
    }
    Sⱼ
}
```

where S_1, S_i, and S_j are some sequences of statements, is transformed by the source transformation $\sigma(P)$ to the following code:

```
 1 #pragma omp parallel
 2 {
 3     σ(S₁)
 4     #include "store_checkpoint.c"
 5     STACKc . push ( l ) ;
 6     #pragma omp for
 7     for ( k←0 ; k<n ; k←k+1)  {  Sᵢ  }
 8     STACKc . push ( l ) ;
 9     σ(Sⱼ)
10     while (  not STACKc . empty ( )  )  {
11         ρ(Sⱼ)
12         if (STACKc . top ( )=l )  {
13             STACKc . pop ( ) ;
14             #include "restore_checkpoint.c"
15             #pragma omp for
16             for ( k←0 ; k<n ; k←k+1)  {  σ(Sᵢ)  }
17             while (  STACKc . top ( ) ≠ l )  {  ρ(Sᵢ)  }
18             STACKc . pop ( ) ;
19         }
20         ρ(S₁)
21     }
22 }
```

Let us consider the above listing in more detail. The forward section of this adjoint code is shown from line 3 to line 9. The sequences S_1 and S_j are transformed by

σ. The worksharing loop in line 7 occurs in the forward section exactly as in the original code. The execution of the loop is marked on the control flow stack STACK$_c$ by pushing label l twice, before executing the loop and again afterwards. An important fact is that the input data for the worksharing loop has to be stored before the execution in line 7 starts. This is indicated by including some code store_checkpoint.c in line 4. The reverse section is represented by the loop starting in line 10. The adjoint statements of the sequences S_j and S_1 are generated by the transformation ρ (line 11 and line 20).

The interesting part here is the branch statement starting in line 12. The test if there is the label I on top of the control flow stack is true if the execution encounters the label that was pushed on the stack through the statement in line 8. In case that the test is valid, the label I is popped from the stack. The code contained in restore_checkpoint.c is responsible for restoring the input values of the worksharing loop. The execution of $\sigma(S_i)$ in line 16 is distributed among the group of threads of the parallel region. For example, if the worksharing loop has three iterations and these iterations are distributed among two threads, the distribution may be that the first thread gets two iterations and the second thread gets the remaining third iteration. This means that the two threads has a different control flow and therefore also a different content in STACK$_c$. Please note that this different stack content is responsible for the concurrent execution during the reverse section of S_i (line 17). The adjoint execution of the worksharing loop is finished when label I is again on STACK$_c$ that initially was pushed on the stack through the statement in line 5. The concluding statement in line 18 pops this label from the control flow stack.

The joint-reversal scheme that is shown in the above listing results in a minimization of the consumption of memory, but this is paid by a relative high computation overhead. All the computations from S_i are again executed in $\sigma(S_i)$. If we assume that that the runtime is dominated by the execution of the worksharing loop then the runtime overhead of the adjoint code is at least twice as high as in the original code.

In case that there is only one worksharing region inside of P we can assume that most of the computation takes place in this region. Hence, we apply the joint reversal scheme to this worksharing construct. The presence of multiple worksharing constructs means that there are several checkpointing points during the execution of the adjoint code. The corresponding overhead should be prevented and therefore we apply the split reversal scheme to each of the worksharing regions.

We determine for this chapter that we apply the split reversal to all the standalone worksharing constructs. The combined pragmas, containing a parallel and a worksharing directive, will be transformed by the joint reversal scheme.

4.3.1 Loop Construct

The loop associated with a OpenMP loop construct is a counting loop as displayed in OpenMP 3.1 citation 18 (page 37). Such a shape of a loop is not supported by the *SPL* language. However, a counting loop can easily be transformed into a semantically equivalent *SPL* code and vice versa. Therefore, we define the loop construct syntax such that it fits into our *SPL* context with

$$
R_{OMP} = R_{OMP} \cup \left\{
\begin{array}{ll}
c & : & \text{\#pragma omp for} \\
& & \{ \\
& & \quad a_1 \leftarrow e_1; \\
& & \quad \textbf{while } (b)\{ \quad S; a_1 \leftarrow e_2; \quad \} \\
& & \}
\end{array}
\right\} .
$$

Example 31. This example presents how a counting loop written in C looks like in *SPL*. The statements of the loop body are composite as *S*. The C code

```
#pragma omp for
for(k←0;k<n;k←k+1)
{
   S
}
```

is semantically equivalent to the following *SPL* pendant

$$
\begin{array}{l}
\text{\#pragma omp for} \\
\{ \\
\quad k \leftarrow 0; \\
\quad while(k < n) \\
\quad \{ \\
\quad\quad S \\
\quad\quad k \leftarrow k+1; \\
\quad \} \\
\}
\end{array}
$$

whereby one has to take into account the fact that k is a private variable for each thread, see OpenMP 3.1 [63, p. 85].

In addition, we saw in Example 6 that the loop construct distribute the work through implicit data decomposition. This means that the OpenMP enabled compiler is responsible for creating code that portions the value range of the counting variable (variable k in the code above) into intervals such that each thread gets a certain part of the value range. In Example 4, we showed a data decomposition performed in an explicit way.   △

We want to assume that the loop associated with the loop construct fulfills the requirements mentioned in the OpenMP 3.1 [63, p. 40]. Briefly said, the initialization assignment, the update statement, as well as the test expression of the loop counter variable must have a certain well-defined form in C/C++. Furthermore, the number of iterations must be known before the execution of any iteration of this loop starts. We do not distinguish between the cases where a nowait clause is present and where this is not the case. The implicit barrier must be considered by the transformation tool but for the transformation of the loop itself this has no effect. Let us define S as

$$
S := \left\{
\begin{array}{ll}
l_0: & \text{\#pragma omp for} \\
     & \{ \\
l_1: & \quad a_1 \leftarrow e_1; \\
l_2: & \quad \textbf{while } (b)\{ \\
     & \quad\quad \{ \\
l_3: & \quad\quad\quad S'; \\
l_4: & \quad\quad\quad a_1 \leftarrow e_2; \\
     & \quad\quad \} \\
     & \quad \}
\end{array}
\right.
$$

where $S' = (s_1; \ldots; s_q)$. The tangent-linear transformation of S is obtained by

$$
\tau(S) := \left\{
\begin{array}{l}
\text{\#pragma omp for} \\
\{ \\
\quad \tau(a_1 \leftarrow e_1) \\
\quad \tau(\quad \textbf{while } (b) \ \{ \ S'; \quad a_1 \leftarrow e_2; \ \} \quad) \\
\}
\end{array}
\right.
$$

$$\underset{\equiv}{\text{(2.22),(2.21)}} \left\{ \begin{array}{l} \text{\#pragma omp for} \\ \{ \\ \qquad a_1 \leftarrow e_1 \\ \qquad \text{while } (b) \; \{ \; \tau(S'); \quad a_1 \leftarrow e_2; \; \} \\ \} \end{array} \right. \qquad (4.44)$$

The code in (4.44) displays that the tangent-linear source transformation is quite similar to the initial code and only the subsequence S' is transformed. A simple transformation into a valid C/C++ OpenMP loop construct is possible.

The adjoint source transformation takes more effort and needs more attention to the specific shape of the loop construct with its associated loop. One important observation about the loop construct is that the loop iteration variable is private inside the scope of the loop construct, see OpenMP 3.1 [63, p. 40, and p. 85]. The loop iteration variable may only be changed by the update assignment $a_1 \leftarrow e_2$ OpenMP 3.1 [63, p. 40]. In the following, we distinguish between three cases:

1. a_1 is invisible outside of the loop construct.

2. a_1 is visible and private outside of the loop construct.

3. a_1 is visible and shared outside of the loop construct.

Case 1: The loop iteration variable a_1 is invisible outside of the loop construct.

This case is the easiest, because the invisibility outside the loop relieves the transformation from the necessity of storing and restoring data.

$$\sigma(S) := \left\{ \begin{array}{l} \text{\#pragma omp for} \\ \{ \\ \qquad a_1 \leftarrow e_1 \\ \qquad \sigma(\text{while } (b) \; \{ \; S'; \quad a_1 \leftarrow e_2; \; \}) \\ \} \end{array} \right.$$

$$\underset{\equiv}{\text{(2.44)}} \left\{ \begin{array}{l} \text{\#pragma omp for} \\ \{ \\ \qquad a_1 \leftarrow e_1 \\ \qquad \text{while } (b) \; \{ \; \sigma(\quad (S'; a_1 \leftarrow e_2) \quad) \; \} \\ \} \end{array} \right.$$

When we assume that $S' \in SLC$ we obtain:

$$
\equiv \left\{
\begin{array}{l}
\texttt{\#pragma omp for} \\
\{ \\
\quad a_1 \leftarrow e_1 \\
\quad \texttt{while } (b) \ \{ \\
\qquad \texttt{STACK}_{(1)c}.\text{push}(l_3); \\
\qquad \sigma(s_1); \cdots ; \sigma(s_q); \\
\qquad \texttt{STACK}_{(1)i}.\text{push}(a_1); \\
\qquad a_1 \leftarrow e_2; \\
\quad \} \\
\} \\
\end{array}
\right.
\qquad (4.45)
$$

The code for the reverse section is given by:

$$
\rho(S) := \rho(\ (S';a_1 \leftarrow e_2) \ )
$$

$$
\equiv \left\{
\begin{array}{l}
\texttt{if } \left(\texttt{STACK}_{(1)c}.\text{top}() = l_3\right) \ \{ \\
\quad \texttt{STACK}_{(1)c}.\text{pop}(); \\
\quad a_1 \leftarrow \texttt{STACK}_{(1)i}.\text{top}(); \\
\quad \texttt{STACK}_{(1)i}.\text{pop}(); \\
\quad \rho(s_q); \cdots ; \rho(s_1); \\
\} \\
\end{array}
\right.
\qquad (4.46)
$$

When one would examine the stack $\texttt{STACK}_{(1)c}$ after the forward section of P, the only trace that we would have from the loop construct is label l_3. All these labels are consumed during the reverse section.

Please remember that the parallel execution is performed in the reverse section through the work-split during the forward section. The iterations of the work-sharing loop are distributed among the threads and this distribution is also used during the reverse section because the individual control flow has been stored on $\texttt{STACK}_c$. This individual control flow is processed backward during the reverse section. Hence, the worksharing remains valid during the reverse section. We conclude this case by showing that we can transform (4.45) into an OpenMP loop construct written in C/C++ code.

$$
\begin{aligned}
&\text{\#pragma omp for} \\
&\mathbf{for}(a_1 \leftarrow e_1; b; a_1 \leftarrow e_2) \\
&\{ \\
&\qquad \text{STACK}_{(1)c}.\text{push}(l_3); \\
&\qquad \sigma(s_1) \\
&\qquad \vdots \\
&\qquad \sigma(s_q) \\
&\qquad \text{STACK}_{(1)i}.\text{push}(a_1); \\
&\}
\end{aligned}
$$

Case 2: The loop iteration variable a_1 is visible and private outside the loop construct.

Since the value of a_1 is visible outside of the loop, we store the value of a_1 before the program execution enters the loop construct. This value is restored after the adjoint part of the loop construct has been executed.

$$
\sigma(S) := \begin{cases}
\text{STACK}_{(1)c}.\text{push}(l_1) \\
\text{STACK}_{(1)i}.\text{push}(a_1) \\
\text{\#pragma omp for} \\
\{ \\
\qquad a_1 \leftarrow e_1 \\
\qquad \sigma(\text{while } (b) \{ S'; \quad a_1 \leftarrow e_2; \}) \\
\}
\end{cases}
\tag{4.47}
$$

Before we store the value of a_1 onto the integer value stack, we mark the position in the execution where the reverse section should recover the value a_1 by putting the label l_1 onto $\text{STACK}_{(1)c}$. The rewriting of the loop body that we made in (4.45) also holds for (4.47). The code for the reverse section is also similar as shown in

(4.46) and we only display the code that differs:

$$\rho(S) := \begin{cases} \rho(\ (a_1 \leftarrow e_1)\ ) \\ \rho(\ (S';a_1 \leftarrow e_2)\ ) \end{cases}$$

$$\underset{(2.49),(2.55)}{\equiv} \begin{cases} \text{if } \left(\text{STACK}_{(1)c}.\text{top}() = l_1\right) \{ \\ \qquad \text{STACK}_{(1)c}.\text{pop}(); \\ \qquad a_1 \leftarrow \text{STACK}_{(1)i}.\text{top}(); \\ \qquad \text{STACK}_{(1)i}.\text{pop}(); \\ \} \\ \rho(\ (S';a_1 \leftarrow e_2)\ ) \end{cases} \qquad (4.48)$$

After the reverse section has executed the code that is shown in (4.46), the label l_1 is on top of $\text{STACK}_{(1)c}$. Thus, the old value of a_1 is restored from $\text{STACK}_{(1)i}$.

Case 3: The loop iteration variable a_1 is visible and shared outside of the loop construct.

This case takes the most effort in the sense that we have to attend to different things. The value of a_1 shall be stored before the runtime system enters the loop construct, just as in the previous case. There is only one instance of a_1 since it is shared. Thus, only one thread needs to take care of the store and the restore of the value of a_1. The method is illustrated in (4.49).

The position where the store of the value of a_1 takes place is marked by the label l_0. Subsequently, one thread enters the single region and puts the label l_1 onto $\text{STACK}_{(1)c}$ and the value of a_1 onto $\text{STACK}_{(1)i}$. It is worth mentioning that each thread puts label l_0 onto its stack but only one thread puts afterwards the label l_1 onto its stack.

Please note that the implicit barrier of the single construct could be dropped by defining the nowait clause in case that there is an implicit barrier at the end of the loop construct. It must be ensured that the value of a_1 is stored on the stack before any thread continues with the code that follows (4.49) in the forward section. This means that only one of the two worksharing constructs in (4.49) has to have an implicit barrier.

$$\sigma(S) := \begin{cases} \text{STACK}_{(1)c}.\text{push}(l_0) \\ \text{\#pragma omp single} \\ \{ \\ \quad \text{STACK}_{(1)c}.\text{push}(l_1) \\ \quad \text{STACK}_{(1)i}.\text{push}(a_1) \\ \} \quad \text{/* implicit barrier */} \\ \text{\#pragma omp for} \\ \{ \\ \quad a_1 \leftarrow e_1 \\ \quad \sigma(\text{while } (b) \ \{ \ S'; \quad a_1 \leftarrow e_2; \ \}) \\ \} \end{cases} \qquad (4.49)$$

$$\rho(S) := \begin{cases} \text{if } \left(\text{STACK}_{(1)c}.\text{top}() = l_0\right) \ \{ \\ \quad \text{STACK}_{(1)c}.\text{pop}(); \\ \quad \text{\#pragma omp barrier} \\ \} \\ \rho(\ \ (a_1 \leftarrow e_1) \ \ ) \\ \rho(\ \ (S'; a_1 \leftarrow e_2) \ \ ) \end{cases} \qquad (4.50)$$

During the reverse section only one thread has label l_1 on $\text{STACK}_{(1)c}$ and this thread restores the previous value of a_1. All threads have the label l_0 on their stack $\text{STACK}_{(1)c}$ and as soon as this label is on top of the stack the test expression in (4.50) is evaluated true. Therefore, all threads enter at some point the contained barrier construct.

What should be true for all these cases is that the value of the loop counting variable a_1 has to be the same at two points during the adjoint execution. The one point is before the control flow enters the loop during the forward section. The other point is after the control flow has finished the execution of all corresponding adjoint statements. This property is shown by the following lemma.

Lemma 59. *Suppose we have a loop construct with a counting variable a_1. Then the value of a_1 before the loop construct during the forward section is the same as the value after the loop construct during the reverse section.*

Proof. Depending on the visibility and the data-sharing status of a_1, we show the correctness of the previous three cases. If a_1 is not defined outside the loop construct, the claim is obviously true. Thus, we assume that a_1 is visible outside the loop construct.

In case that a_1 is defined as a private reference outside the construct, each thread possesses an instance of a_1. We denote the individual values of a_1 by R^t where t is the unique thread identifier number. In (4.47) each thread stores the value of R^t on its stack $STACK_{(1)i}$ and in addition it keeps track of this storing operation by pushing label l_1 onto the control flow stack $STACK_{(1)c}$. Each time a thread processes a loop iteration, it also puts a label l_3 onto $STACK_{(1)c}$ as shown in (4.45). Hence, after the forward section the trace of the loop construct on $STACK_{(1)c}$ looks like

$$l_1 \quad (l_3)*$$

when we write the trace as a regular expression. This means we have label l_1 and a sequence of labels l_3 potentially empty on the stack $STACK_{(1)c}$. The labels l_3 are consumed during the reverse section by the branch shown in (4.46). Eventually, the label l_1 leads to the execution of the branch shown in (4.48). At this point, the individual value R^t is restored by thread t, which shows the claim.

The remaining case is when a_1 is defined as shared outside the construct. Thus, we assume that the shared variable a_1 has the value R before the loop construct. The code of the forward section looks therefore like in (4.49). Each thread puts a label l_0 onto its $STACK_{(1)c}$ before only one thread, say thread t, is allowed to enter the single construct. Thread t puts the value R onto its $STACK_{(1)i}$. Thereafter, each thread puts a label l_3 onto $STACK_{(1)c}$ each time it processes an iteration of the loop construct. When we write the trace of the loop construct after the forward section as a regular expression, we distinguish between thread t and the other threads. Thread t has

$$l_0 \quad l_1 \quad (l_3)*$$

on its $STACK_{(1)c}$, whereby $STACK_{(1)c}$ of the rest of the group contains

$$l_0 \quad (l_3)*.$$

During the reverse section, all threads execute the adjoint codes associated with label l_3. Subsequently, all threads except thread t have label l_0 on their stack and therefore they enter the branch displayed in (4.50) where they temporary stop execution due to the barrier. Meanwhile, thread t has label l_1 on its stack and enters the branch shown in (4.48) where it restores the value R from its $STACK_{(1)i}$. The next label that thread t has on its $STACK_{(1)c}$ is label l_0 and it also enters the branch with the barrier. Once all threads have reached the barrier shown in (4.50), the execution continues and the shared variable a_1 has its previous value R.   □

4.3.2 sections Construct

The sections construct is a non-iterative construct that contains multiple subsequences, as described in OpenMP 3.1 citation 19 (page 39). All subsequences except the first are opened by a section construct. The set of non-terminal symbols V of our grammar $\mathscr{G}_{OMP}$ needs to be extended by two additional symbols.

$$V = V \cup \{ \text{sequence\_of\_sections}, \text{section} \}$$

We will omit the fact that the first section is optional and therefore we define the syntax of the sections construct in the following way.

$$R_{OMP} = R_{OMP}$$
$$\cup \left\{ \begin{array}{ll} c & : \quad \text{\#pragma omp sections} \\ & \quad \{ \quad \text{sequence\_of\_sections} \quad \} \end{array} \right\}$$
$$\cup \left\{ \begin{array}{ll} \text{sequence\_of\_sections} & : \quad \text{section} \\ & \quad | \quad \text{sequence\_of\_sections} \quad \text{section} \end{array} \right\}$$
$$\cup \left\{ \begin{array}{ll} \text{section} & : \quad \text{\#pragma omp section} \\ & \quad \{ \quad S' \quad \} \end{array} \right\}$$

The OpenMP sections construct is a control flow statement and is therefore contained in *CFSTMT*. Suppose a sections construct S is given as

$$S := \left\{ \begin{array}{l} \text{\#pragma omp sections} \\ \{ \\ \quad \text{\#pragma omp section} \quad \{ \quad S'_1 \quad \} \\ \quad \text{\#pragma omp section} \quad \{ \quad S'_2 \quad \} \\ \quad \vdots \\ \quad \text{\#pragma omp section} \quad \{ \quad S'_n \quad \} \\ \} \end{array} \right. \quad .$$

The source transformation for getting the tangent-linear model and the forward section of the adjoint code are both defined recursively.

$$
\tau(S) := \begin{cases}
\text{\#pragma omp sections} \\
\{ \\
\qquad \text{\#pragma omp section} \quad \{ \quad \tau(S_1') \quad \} \\
\qquad \text{\#pragma omp section} \quad \{ \quad \tau(S_2') \quad \} \\
\qquad \vdots \\
\qquad \text{\#pragma omp section} \quad \{ \quad \tau(S_n') \quad \} \\
\}
\end{cases}
$$

$$
\sigma(S) := \begin{cases}
\text{\#pragma omp sections} \\
\{ \\
\qquad \text{\#pragma omp section} \quad \{ \quad \sigma(S_1') \quad \} \\
\qquad \text{\#pragma omp section} \quad \{ \quad \sigma(S_2') \quad \} \\
\qquad \vdots \\
\qquad \text{\#pragma omp section} \quad \{ \quad \sigma(S_n') \quad \} \\
\}
\end{cases}
$$

The adjoint part for the reverse section is obtained by

$$
\rho(S) := \begin{cases}
\rho(S_1') \\
\rho(S_2') \\
\vdots \\
\rho(S_n')
\end{cases}
$$

4.3.3 single Construct

The subsequence associated with the single construct is only entered by one thread out of the team of threads executing the current parallel region. There is an implicit barrier at the end of this construct when the nowait clause is not given, see OpenMP 3.1 citation 20 (page 40) for details. The syntax of the single construct is

$$
R_{OMP} = R_{OMP} \cup \begin{cases}
c & : & \text{\#pragma omp single} \\
& & \{ \quad S' \quad \}
\end{cases}
$$

As the previous worksharing constructs, the single construct is contained in *CFSTMT*. The single construct itself is transformed recursively by

$$\tau \left(\begin{array}{c} \text{\#pragma omp single} \\ \{ \quad S' \quad \} \end{array} \right) := \left\{ \begin{array}{l} \text{\#pragma omp single} \\ \{ \quad \tau(S') \quad \} \end{array} \right.$$

$$\sigma \left(\begin{array}{c} \text{\#pragma omp single} \\ \{ S' \} \end{array} \right) := \left\{ \begin{array}{l} \text{\#pragma omp single} \\ \{ \quad \sigma(S') \quad \} \end{array} \right. \tag{4.51}$$

$$\rho \left(\begin{array}{c} \text{\#pragma omp single} \\ \{ S' \} \end{array} \right) := \rho(S') \tag{4.52}$$

Suppose thread t encounters a single construct during the forward section. Assuming that thread t enters the single construct, the label of the first statement in S' is put onto $\text{STACK}_{(1)c}$. Only thread t has this label on its stack. During the reverse section, thread t encounters as the only one the label and executes the code given by (4.52). The concluding worksharing constructs are the combined constructs.

4.3.4 Combined Parallel Constructs

The parallel loop construct allows to combine the loop construct with the parallel construct. The syntax of the loop construct is not contained in the language *SPL*. However, as explained in Section 4.3.1 we can easily transform the syntax of the counting loop into *SPL* code. Therefore, we will omit the transformation into *SPL* code and back to C/C++. Instead, we use the syntax written as a counting loop.

$$R_{OMP} = R_{OMP} \cup \left\{ \begin{array}{cl} P : & \text{\#pragma omp parallel for} \\ & \text{for}(a_1 \leftarrow e_1; b; a_1 \leftarrow e_2) \quad \{ \quad D \quad S \quad \} \end{array} \right\}$$

The tangent-linear source transformation is defined recursively and is very similar to the one defined in (2.10).

$$\tau \left(\begin{array}{l} \text{\#pragma omp parallel for} \\ \text{for}(a_1 \leftarrow e_1; b; a_1 \leftarrow e_2) \quad \{ \quad D \quad S \quad \} \end{array} \right)$$
$$:= \left\{ \begin{array}{l} \text{\#pragma omp parallel for} \\ \text{for}(a_1 \leftarrow e_1; b; a_1 \leftarrow e_2) \quad \{ \quad \tau(D) \quad \tau(S) \quad \} \end{array} \right.$$

The adjoint source transformation looks very similar to the one defined for the parallel region in rule (2.32). As explained in Section 2.3.3, the parallel region is transformed in the joint reversal scheme and we apply the joint reversal scheme to the combined loop construct as well.

$$
\sigma(P) := \left\{
\begin{array}{l}
\#\text{pragma omp parallel for} \\
\text{for}\,(a_1 \leftarrow e_1; b; a_1 \leftarrow e_2) \\
\{ \\
\quad \sigma(D) \\
\quad \sigma(S) \\
\quad \text{while}\,\left(\text{not STACK}_{(1)_c}.\text{empty}()\right)\,\{\,\rho(S)\,\} \\
\}
\end{array}
\right.
\tag{4.53}
$$

The second combined parallel construct is the parallel sections construct, which has the following syntax.

$$
R_{OMP} = R_{OMP}
$$

$$
\cup \left\{
\begin{array}{ll}
P \;:\; & \#\text{pragma omp parallel sections} \\
& \{ \\
& \quad \#\text{pragma omp section} \quad \{\quad D_1 \quad S_1 \quad \} \\
& \quad \#\text{pragma omp section} \quad \{\quad D_2 \quad S_2 \quad \} \\
& \quad \vdots \\
& \quad \#\text{pragma omp section} \quad \{\quad D_n \quad S_n \quad \} \\
& \}
\end{array}
\right\}
\tag{4.54}
$$

The tangent-linear source transformation is defined section-wise:

$$
\begin{array}{ll}
\tau(P) \;:=\; & \#\text{pragma omp parallel sections} \\
& \{ \\
& \quad \#\text{pragma omp section} \quad \{\quad \tau(D_1) \quad \tau(S_1) \quad \} \\
& \quad \#\text{pragma omp section} \quad \{\quad \tau(D_2) \quad \tau(S_2) \quad \} \\
& \quad \vdots \\
& \quad \#\text{pragma omp section} \quad \{\quad \tau(D_n) \quad \tau(S_n) \quad \} \\
& \}
\end{array}
$$

The adjoint source transformation takes each section and generates the corresponding forward and reverse section.

$$
\begin{aligned}
\sigma(P) \quad := \quad & \#\text{pragma omp parallel sections} \\
& \{ \\
& \qquad \#\text{pragma omp section} \\
& \qquad \{ \\
& \qquad\qquad \sigma(D_1); \quad \sigma(S_1); \\
& \qquad\qquad \text{while} \left(\text{not STACK}_{(1)_c}.\text{empty}() \right) \{ \rho(S_1) \}, \\
& \qquad \} \\
& \qquad \vdots \\
& \qquad \#\text{pragma omp section} \\
& \qquad \{ \\
& \qquad\qquad \sigma(D_n); \quad \sigma(S_n); \\
& \qquad\qquad \text{while} \left(\text{not STACK}_{(1)_c}.\text{empty}() \right) \{ \rho(S_n) \}, \\
& \qquad \} \\
& \}
\end{aligned}
$$

This finishes the second extension of our *SPL* language where we covered the worksharing constructs of OpenMP. The language that this extension produces is called $SPLOMP^2$. The next extension deals with the data-sharing possibilities of OpenMP.

4.4 *SPLOMP*^{3} - Data-Sharing

There are several clauses and one directive in OpenMP to influence the data-sharing environment during runtime. We present the source transformation of the threadprivate directive, the private clause, the firstprivate clause, the lastprivate clause, and the reduction clause. This is the third extension of *SPL* after we showed the transformation of the synchronization constructs and the worksharing constructs of OpenMP in the previous sections. The language that is produced is called $SPLOMP^3$.

4.4.1 Global Data - threadprivate Directive

In Section 4.1, we saw that there are different possibilities to implement the stacks for the adjoint code. One possibility is to define a stack in the global scope of the program and to declare this stack as threadprivate.

To enable the grammar $\mathscr{G}_{OMP}$ to recognize the threadprivate directive, we extend the set of non-terminal symbols with

$$V = V \cup \{v, l\}$$

where v represents a variable and l is a list of variables separated by a comma. The syntax of the threadprivate directive is:

$$R_{OMP} = R_{OMP} \cup \left\{ \begin{array}{lll} v & : & a \quad | \quad x \\ l & : & v \quad | \quad l \quad , \quad v \\ c & : & \texttt{\#pragma omp threadprivate (} l \texttt{)} \end{array} \right\}$$

Despite the fact that the threadprivate directive is placed in the global scope and not inside of the parallel region P, we have to define how the source transformation influences this directive. Suppose that the adjoint model of the parallel region P makes use of the stacks $\text{STACK}_{(1)c}, \text{STACK}_{(1)i}$, and $\text{STACK}_{(1)f}$. Each source transformation that supplies an higher derivative model leads to another definition of these three stacks. The second-order tangent-linear model uses the stacks $\text{STACK}_c^{(2)}, \text{STACK}_i^{(2)}$, and $\text{STACK}_f^{(2)}$, the second-order adjoint model stores data in $\text{STACK}_{(2)c}, \text{STACK}_{(2)i}$, and $\text{STACK}_{(2)f}$, and so forth.

Is there a floating-point variable x in P that is defined as threadprivate then the derivative code contains a threadprivate directive that lists the variable $x^{(1)}$ as threadprivate in case that we apply the tangent-linear transformation, or it lists the variable $x_{(1)}$ as threadprivate in case of the adjoint transformation, respectively. The tangent-linear transformation of the threadprivate directive is defined as:

$$\tau(\texttt{\#pragma omp threadprivate } (list)) := \texttt{\#pragma omp threadprivate } (list_1)$$

where

$$\begin{aligned} list_1 := \quad & \left\{ \text{STACK}_{(c|i|f)}, \text{STACK}_{(c|i|f)}^{(1)} \,|\, \text{STACK}_{(c|i|f)} \in list \right\} \\ & \cup \; \{ a \,|\, a \in INT_{list} \} \\ & \cup \; \left\{ x, x^{(1)} \,|\, x \in FLOAT_{list} \right\} \end{aligned}$$

The adjoint transformation is:

$$\sigma(\texttt{\#pragma omp threadprivate } (list)) := \texttt{\#pragma omp threadprivate } (list_2)$$

where

$$\begin{aligned} list_2 := \quad & \left\{ \text{STACK}_{(c|i|f)}, \text{STACK}_{(1)(c|i|f)} \,|\, \text{STACK}_{(c|i|f)} \in list \right\} \\ & \cup \; \{ a \,|\, a \in INT_{list} \} \\ & \cup \; \left\{ x, x_{(1)} \,|\, x \in FLOAT_{list} \right\} \end{aligned}$$

The following sections explain the transformation of several clauses which can be used to control the data environment at runtime.

4.4.2 Thread-Local Data - private Clause

The private clause may appear in the following constructs: parallel, for, sections, single, and it contains a list of variables, here referred to as *list*. The status private inside of a construct can be interpreted as a substitution of the variables that occur in *list*. Each occurrence of theses variables inside the construct is substituted by a thread local instance of this variable.

We extend the set of non-terminal symbols V in $\mathcal{G}_{OMP}$ by the symbol cl which represents a clause.

$$V = V \cup \{cl\}$$

The production rules R_{OMP} of $\mathcal{G}_{OMP}$ are adjusted by

$$R_{OMP} = R_{OMP} \cup \left\{ \begin{array}{lll} clauses & : & cl \mid clauses \quad cl \\ cl & : & \texttt{private (} \ l \ \texttt{)} \end{array} \right\}$$

We saw that in the previous section that the threadprivate directive is transformed such that it contains a modified *list* where each floating-point variable is associated with another floating-point variable. This could also be done for the private clause but as we will see in the following sections, as soon as the data-sharing clause defines a dataflow into P or from P out to the surrounding code Q, this approach cannot be applied. We decided to present a more general approach that can be applied to all data-sharing clauses instead of presenting an individual transformation for each clause. From the implementation perspective this is also a convenient approach. Suppose that the parallel region P is defined as

$$P := \left\{ \begin{array}{l} \texttt{\#pragma omp parallel } \textit{[clauses]} \\ \{ \quad D \quad S \quad \} \end{array} \right. \tag{4.55}$$

and inside of *clauses* or somewhere in S there is an appearance of private(*list*), firstprivate(*list*), or lastprivate(*list*). The label of this clause is denoted by l.

Our approach is to replace the data clause with its *list* by a sequence of definitions that extends the transformation of D. This new sequence of definitions is referred to as D_1. The notation that is commonly used for a set is here used and should be read a sequence of definitions.

$$D_1 := \tau(D) \cup \{\texttt{int } a_l \mid a \in INT_{list}\}$$
$$\cup \{\texttt{double } x_l; \texttt{double } x_l^{(1)} \leftarrow 0. \mid x \in FLOAT_{list}\} \tag{4.56}$$

when we apply the forward mode. The application of the reverse mode leads to
the definitions

$$D_2 := \sigma(D) \cup \{\text{int } a_l \,|\, a \in INT_{list}\}$$
$$\cup \{\text{double } x_l; \text{double } x_{(1)l} \leftarrow 0. \,|\, x \in FLOAT_{list}\}. \tag{4.57}$$

For each occurrence of a variable v inside of *list* another instance is defined, namely
v_l, which represents the private instance of v. In addition, the transformation must
perform a substitution of variable v by v_l. We denote the substitution of v by v_l
inside of S with

$$S[v/v_l].$$

The tangent-linear source transformation uses the extended sequence of definitions
D_1 and substitutes the references inside the sequence of statements S:

$$\tau \left(\begin{array}{l} \text{\#pragma omp parallel private(\textit{list}) \textit{[clauses]}} \\ \{\quad D \quad S \quad \} \end{array} \right)$$

$$:= \left\{ \begin{array}{l} \text{\#pragma omp parallel \textit{[clauses}$_t$\textit{]}} \\ \{ \\ \quad D_1 \\ \quad \tau(S) \forall a, x \in VAR_{list} : [a/a_l][x/x_l][x^{(1)}/x_l^{(1)}] \\ \} \end{array} \right.$$

The line

$$\tau(S) \forall a, x \in VAR_{list} : [a/a_l][x/x_l][x^{(1)}/x_l^{(1)}]$$

means that the sequence S is transformed by $\tau(S)$ and afterwards a substitution
of all variables a and x contained in *list* takes place where a is substituted by a_l,
x is substituted by x_l and the tangent-linear component $x^{(1)}$ is replaced by $x_l^{(1)}$.
The *clauses*$_t$ in the derivative code is changed such that there is no private clause

present. The adjoint version of a parallel region with a private clause is given by:

$$
\sigma \left(
\begin{array}{l}
\text{\#pragma omp parallel private}(\textit{list})\ [\textit{clauses}] \\
\{ \quad D \quad S \quad \}
\end{array}
\right)
$$

$$
:= \left\{
\begin{array}{l}
\text{\#pragma omp parallel } [\textit{clauses}_a] \\
\{ \\
\quad D_2 \\
\quad \sigma(S) \forall a, x \in \textit{VAR}_{\textit{list}} : [a/a_l][x/x_l] \\
\quad \text{while} \left(\text{not STACK}_{(1)c}.\text{empty}() \right) \{ \\
\quad\quad \rho(S) \forall a, x \in \textit{VAR}_{\textit{list}} : [a/a_l][x/x_l][x_{(1)}/x_{(1)l}] \\
\quad \} \\
\}
\end{array}
\right.
$$

We substitute all variables a and x that are part of $\textit{VAR}_{\textit{list}}$. The adjoint values for x_l are stored in $x_{(1)l}$. The sequence $\textit{clauses}_a$ does not contain a private clause.

Example 32. In this example we assume that the following parallel region is given.

$$
\begin{array}{l}
\text{\#pragma omp parallel private}(a,x) \\
\{ \quad D \quad S \quad \}
\end{array}
$$

The variable a is an integer variable and x is a floating-point variable. The label of the private clause is l. In the tangent-linear code we define a private instance for 'a' called 'a_l' and analogous for x. Inside the derivative code, we substitute each occurrence of a by a_l and all x become x_l.

$$
\begin{array}{l}
\text{\#pragma omp parallel} \\
\{ \\
\quad \tau(D) \\
\quad \text{int } a_l; \quad \text{double } x_l; \quad \text{double } x_l^{(1)} \leftarrow 0.; \\
\quad \tau(S) \forall a, x \in \textit{VAR}_{\textit{list}} : [a/a_l][x/x_l][x^{(1)}/x_l^{(1)}] \\
\}
\end{array}
$$

The adjoint code is

```
#pragma omp parallel
{
```
$$\sigma(D)$$
$$\text{int } a_l; \quad \text{double } x_l; \quad \text{double } x_{(1)l} \leftarrow 0.;$$
$$\sigma(S)\forall a,x \in VAR_{list} : [a/a_l][x/x_l]$$
$$\text{while } \left(\text{not STACK}_{(1)c}.\text{empty}()\right) \{$$
$$\rho(S)\forall a,x \in VAR_{list} : [a/a_l][x/x_l][x_{(1)}/x_{(1)l}]$$
```
    }
}
```

where we only replace the variables a and x inside the forward section. Inside the reverse section we have to replace a, x, and $x_{(1)}$. $\triangle$

In the following, we consider the case where the private clause is part of a worksharing construct W. This means W is either for, sections, or single. We define the sequence S in the parallel region P as:

$$S := \begin{cases} \text{\#pragma omp } W \text{ private}(list) \\ \{ \\ \quad S' \\ \} \end{cases}$$

The private clause is labeled by l what allows us to take (4.56) as sequence of definitions for the tangent-linear code. The tangent-linear source transformation of S is:

$$\tau(S) := \begin{cases} \forall a \in INT_{list} : a_l \leftarrow 0; \\ \forall x \in FLOAT_{list} : x_l \leftarrow 0.;x_l^{(1)} \leftarrow 0.; \\ \text{\#pragma omp } W \\ \{ \\ \quad \tau(S')\forall a,x \in VAR_{list} : [a/a_l][x/x_l][x^{(1)}/x_l^{(1)}] \\ \} \end{cases} \quad (4.58)$$

The description of the private clause in the OpenMP standard does not determine that the private variables has to be initialized by a value. However, we perform an initialization with zero since we reuse this code pattern in the next section where we cover the firstprivate clause.

We determined in Section 4.3 that we apply the split reversal scheme to a stand-alone worksharing construct. The split reversal scheme of W means that we have two separated parts that belong to the transformation of W. The first part is contained in the forward section of P and the second part is in the reverse section of P. The private variables have therefore two different scopes. Let us first consider the case where we did not use the above approach of substituting every variable that occurs in the *list* of the private clause. In this case and the fact that the scope of the private variables is left unchanged, means that their values must be stored on the stack at the end of the first scope and at the begin of the second scope the values have to be restored. Since we substitute each variable v in *list* by a variable v_l and the variable v_l is defined in the scope of the parallel region P, the value keeps alive in between the two scopes and therefore we do not need to store and restore the values.

$$
\sigma(S) := \left\{
\begin{array}{l}
\forall a \in INT_{list} : a_l \leftarrow 0; \\[4pt]
\forall x \in FLOAT_{list} : x_l \leftarrow 0.; \\[4pt]
\#\text{pragma omp } W \\[4pt]
\{ \\[4pt]
\qquad \sigma(S')\forall a,x \in VAR_{list} : [a/a_l][x/x_l] \\[4pt]
\}
\end{array}
\right.
\tag{4.59}
$$

$$
\rho(S) := \left\{
\begin{array}{l}
\text{if} \left(\quad STACK_{(1)c}.top() = LABEL(S') \quad \right) \{ \\[4pt]
\qquad \rho(S')\forall a,x \in VAR_{list} : [a/a_l][x/x_l][x_{(1)}/x_{(1)l}] \\[4pt]
\}
\end{array}
\right.
\tag{4.60}
$$

This concludes the source transformation for the private clause. The next two sections cover the firstprivate and the lastprivate clause. These two clauses can be seen as a superset of the functionality provided by the private clause. Hence, the definitions of the current section are necessary for these clauses as well.

4.4.3 firstprivate Construct

We consider again a parallel region P structured as shown in (4.55). The clause firstprivate(*list*) has the same functionality as the private(*list*) clause. In addition, it defines that each variable v in *list* is to be initialized with the value of the global instance of v when the runtime execution enters the associated construct. The production rule for this clause is given by

$$R_{OMP} = R_{OMP} \cup \big\{ \ cl \ : \ \texttt{firstprivate} \ (\ l \ ) \ \big\}$$

In case that the firstprivate(*list*) clause associated with a parallel directive, we define a sequence of definitions D_3 such that it creates private copies of all *list* items. The label of the firstprivate clause is assumed to be l. Each variable v_l is initialized by the value of the global instance of v.

$$D_3 := \tau(D) \cup \{\text{int } a_l \leftarrow a; \, | \, a \in INT_{list}\}$$
$$\cup \{\text{double } x_l \leftarrow x; \text{double } x_l^{(1)} \leftarrow x^{(1)}; \, | \, x \in FLOAT_{list}\}$$

With this sequence of definitions, we define the tangent-linear source transformation by

$$\tau \left(\begin{array}{c} \texttt{\#pragma omp parallel firstprivate(\textit{list}) [clauses]} \\ \{ \quad D \quad S \quad \} \end{array} \right)$$

$$:= \left\{ \begin{array}{l} \texttt{\#pragma omp parallel [\textit{clauses}_t]} \\ \{ \\ \quad D_3 \\ \quad \tau(S) \forall a, x \in VAR_{list} : [a/a_l][x/x_l][x^{(1)}/x_l^{(1)}] \\ \} \end{array} \right.$$

where we substitute each occurrence of a, x, or $x^{(1)}$ in $\tau(S)$ with its private instance a_l, x_l, or $x_l^{(1)}$. In *clauses*$_t$ there must not be a firstprivate clause. This is similar in case of the adjoint source transformation. We define the sequence of definitions D_4 such that

$$D_4 := \sigma(D) \cup \{\text{int } a_l \leftarrow a; | a \in INT_{list}\}$$
$$\cup \{\text{double } x_l \leftarrow x; \text{double } x_{(1)l} \leftarrow 0.; | x \in FLOAT_{list}\}$$

With the definition of D_4, we obtain a dataflow from outside of P into the parallel region, namely from variables a and x to variables a_l and x_l. This dataflow has to be reversed after the code from $\rho(P)$ has been executed. Hence, we define a sequence of statements S_1 which is responsible to reverse the dataflow of all floating-point elements in VAR_{list}.

$$S_1 := \left\{ \left[\begin{array}{l} \texttt{\#pragma omp atomic} \\ x_{(1)} \, +\!\!\leftarrow x_{(1)l}; \end{array} \right] \Big| x \in FLOAT_{list} \right\} \tag{4.61}$$

The scope of instance $x_{(1)}$ is valid inside the master thread and this instance is a scalar reference. For each thread in the group of executing threads there is a private copy $x_{(1)l}$ of this instance. The initialization from one scalar variable into p private copies in the forward section is represented in the reverse section by adding up the p values of the adjoint private copies and to put this value into the adjoint associate of x. Therefore, we have to use the atomic construct to prevent a race condition. The adjoint source transformation for P with a firstprivate clause may look like

$$\sigma \left(\begin{array}{c} \text{\#pragma omp parallel firstprivate}(\textit{list}) \ [\textit{clauses}] \\ \{ \quad D \quad S \quad \} \end{array} \right)$$

$$:= \left\{ \begin{array}{l} \text{\#pragma omp parallel } [\textit{clauses}_a] \\ \{ \\ \quad D_4 \\ \quad \sigma\,(S) \forall a, x \in \textit{VAR}_{\textit{list}} : [a/a_l][x/x_l] \\ \quad \text{while} \left(\text{not } \textbf{STACK}_{(1)c}.\text{empty()} \right) \ \{ \\ \qquad \rho\,(S) \forall a, x \in \textit{VAR}_{\textit{list}} : [a/a_l][x/x_l][x_{(1)}/x_{(1)l}] \\ \quad \} \\ \quad S_1 \\ \} \end{array} \right. \qquad (4.62)$$

where $\textit{clauses}_a$ must not contain a firstprivate clause. The OpenMP experienced reader may make an objection to S_1 in (4.62) and could argue that this also can be done by another OpenMP construct, namely the reduction clause. Therefore, we display another definition which is semantically equivalent to (4.62) and contains a reduction clause with a $\textit{list}_r$ composed of all adjoint floating-point elements from $\textit{list}$. Since the reduction clause implicitly defines private variables for all variables contained in $list_r$, we exclude these definitions from D_4.

$$\sigma \left(\begin{array}{l} \text{\#pragma omp parallel firstprivate}(\textit{list}) \ [\textit{clauses}] \\ \{ \quad D \quad S \quad \} \end{array} \right)$$

$$:= \left\{ \begin{array}{l} \text{\#pragma omp parallel reduction}(\textit{list}_r) \ [\textit{clauses}_a] \\ \{ \\ \quad D_4 \setminus \{\text{double } x_{(1)l} \leftarrow 0.; | x \in FLOAT_{list}\} \\ \quad \sigma\,(S) \forall a, x \in VAR_{list} : [a/a_l][x/x_l] \\ \quad \text{while } \left(\text{not STACK}_{(1)c}.\text{empty}() \right) \ \{ \\ \qquad \rho\,(S) \forall a, x \in VAR_{list} : [a/a_l][x/x_l][x_{(1)}/x_{(1)l}] \\ \quad \} \\ \} \end{array} \right.$$

with $list_r := \{x_{(1)} | x \in FLOAT_{list}\}$.

For the case that the firstprivate clause is defined inside of P, we assume S to have the following shape where the label of the firstprivate clause is l.

$$S := \left\{ \begin{array}{l} \text{\#pragma omp } W \text{ firstprivate}(\textit{list}) \\ \{ \\ \qquad S' \\ \} \end{array} \right.$$

The tangent-linear source transformation of the firstprivate clause is similar to the one of the private clause in (4.58). The only difference is that the private copies are not initialized by zero but rather with the value of the global instance.

$$\tau\,(S) := \left\{ \begin{array}{l} \forall a \in INT_{list} : a_l \leftarrow a; \\ \forall x \in FLOAT_{list} : x_l \leftarrow x; x_l^{(1)} \leftarrow x^{(1)}; \\ \text{\#pragma omp } W \\ \{ \\ \qquad \tau\,(S') \forall a, x \in VAR_{list} : [a/a_l][x/x_l][x^{(1)}/x_l^{(1)}] \\ \} \end{array} \right.$$

The source transformation that emits the code for the forward section initializes the local copies with the values of the global instance of a and x not with zero. This dataflow must be reversed after the execution has finished the reverse section part of the worksharing construct W. For this reason, we push the label of the firstprivate clause onto the control flow stack $STACK_{(1)c}$ to mark that at this point the dataflow has to be reversed as soon as the reverse section has reached this label l.

$$\sigma(S) := \begin{cases} STACK_{(1)c}.push(l) \\ \forall a \in INT_{list} : a_l \leftarrow a; \\ \forall x \in FLOAT_{list} : x_l \leftarrow x; \\ \#pragma\ omp\ W \\ \{ \\ \quad \sigma(S')\forall a,x \in VAR_{list} : [a/a_l][x/x_l] \\ \} \end{cases}$$

The transformation $\rho(S)$ for the reverse section contains a branch that checks if the label l is on top of the control flow stack. If this is the case, the local values of each thread are summarized and put into the global instance that is associated with the private variable. This code is defined in S_1 (4.61). The other branch statement in $\rho(S)$ contains the adjoint statements for S' together with the substitutions of the private variables.

$$\rho(S) := \begin{cases} if\ \left(\quad STACK_{(1)c}.top() = l \quad \right) \{ \quad S_1 \quad \} \\ if\ \left(\quad STACK_{(1)c}.top() = LABEL(S') \quad \right) \{ \\ \quad \rho(S')\forall a,x \in VAR_{list} : [a/a_l][x/x_l][x_{(1)}/x_{(1)l}] \\ \} \end{cases}$$

The third kind of a private clause is the lastprivate clause.

4.4.4 lastprivate Construct

The semantic of the lastprivate(*list*) allows to define a dataflow from, for example, a loop construct to the code that follows the loop. The production rule is

$$R_{OMP} = R_{OMP} \cup \{\ cl\ :\ \texttt{lastprivate}\ (\ l\ )\ \}.$$

Let us consider v as an element in *list*. This means that there are two instances of v. One global instance $v^{\{t\}}$ that can be read by all threads, and v^t as a private variable of thread t. At the end of the construct associated with the lastprivate clause, the global instance of v is assigned to with the current value of a certain v^t.

The decision what private v^t is assigned to the global instance depends on the kind of worksharing construct associated with the lastprivate clause. In case that the clause is associated with a loop then the thread that executes the sequentially last iteration of the loop writes its thread local value to the global instance of $v^{\{t\}}$. The other possible worksharing construct is the sections construct. There, the thread that executes the code from the lexically last section assigns its private value v^t to the global instance $v^{\{t\}}$.

Combined parallel for Construct with lastprivate

Let us first consider the case where a combined parallel loop construct is associated with a lastprivate clause. Therefore, the following code structure is given where the counting variable runs from 0 to $N - 1$ with $N \in \mathbb{N}$.

$$P \quad : \quad \text{\#pragma omp parallel for lastprivate}(list) \; [clauses]$$
$$\text{for}\,(a_2 \leftarrow 0; a_2 < N; a_2 \leftarrow a_2 + 1) \quad \{ \quad D \quad S \quad \}$$

The test for entering the loop is $a_2 < N$ with a_2 being the counting variable of the loop. The easiest approach to transform the lastprivate(*list*) clause in the forward mode is to adjust the *list* such that each occurrence of a floating-point variable x in *list* leads to two elements x and $x^{(1)}$ in the corresponding *list* occurring in the tangent-linear model. However, in case of the adjoint code we cannot use this approach and therefore we present a more common approach for the forward mode as well to ensure a similar implementation of these source transformations.

Our approach is to check in each iteration if the last iteration is currently executed. In case that there are several hundreds or even millions of loop iterations this test becomes a bottleneck and the lastprivate variant would probably the better choice. Since we only consider the source transformation on the inner parallel region and we do not want to define code that is located outside of the parallel region, we keep the above transformation as it is and the software engineer who implements these techniques can decide which solution fits her or his demands.

The tangent-linear transformation of P is shown in (4.63). Inside of the sequence *clauses*$_t$ is no lastprivate clause present and D_1 is defined as given in (4.56).

The adjoint transformation of P which is shown in (4.64), is performed in the joint reversal scheme. The body of the loop starts with the sequence of definitions D_2 from (4.57). The forward section with the substitution of the variables a and x from *VAR*$_{list}$ follows D_2. Subsequently, a branch statement checks if the current iteration is the last iteration. If this is the case, we perform the dataflow that is defined through the lastprivate clause. This means we assign the values of the thread local variables a_l and x_l to the global instances a and x. We do not need an atomic statement since only one thread performs these assignments. Please note the assignment $x_{(1)l} \leftarrow x_{(1)}$ inside the branch statement with the test expression $(a_2 = N - 1)$ shown in (4.64).

$$\tau(P) := \begin{cases} \text{\#pragma omp parallel for } [clauses_t] \\ \text{for } (a_2 \leftarrow 0; a_2 < N; a_2 \leftarrow a_2 + 1) \\ \{ \\ \quad D_1 \\ \quad \tau(S) \forall a, x \in VAR_{list} : [a/a_l][x/x_l][x^{(1)}/x_l^{(1)}] \\ \quad \textbf{if } (a_2 = N - 1)\{ \\ \qquad \forall a \in INT_{list} : a \leftarrow a_l; \\ \qquad \forall x \in FLOAT_{list} : x \leftarrow x_l; x^{(1)} \leftarrow x_l^{(1)}; \\ \quad \} \\ \} \end{cases} \qquad (4.63)$$

$$\sigma(P) := \begin{cases} \text{\#pragma omp parallel for } [clauses_a] \\ \text{for } (a_2 \leftarrow 0; a_2 < N; a_2 \leftarrow a_2 + 1) \\ \{ \\ \quad D_2 \\ \quad \sigma(S) \forall a, x \in VAR_{list} : [a/a_l][x/x_l] \\ \quad \textbf{if } (a_2 = N - 1)\{ \\ \qquad \forall a \in INT_{list} : a \leftarrow a_l; \\ \qquad \forall x \in FLOAT_{list} : x \leftarrow x_l; x_{(1)l} \leftarrow x_{(1)}; \\ \quad \} \\ \quad \text{while } \left(\text{not STACK}_{(1)c}.\text{empty}() \right) \{ \\ \qquad \rho(S) \forall a, x \in VAR_{list} : [a/a_l][x/x_l][x_{(1)}/x_{(1)l}] \\ \quad \} \\ \} \end{cases} \qquad (4.64)$$

This assignment is an adjoint assignment that occurs here before the code of the reverse section. We perform with this assignment the reversal of the lastprivate clause. At the beginning of the reverse section all adjoint variables have to be zero. For the variable $x_{(1)l}$ this means it would be initialized with zero and the thread that performed the lastprivate dataflow in the forward section would increment the adjoint $x_{(1)l}$ by the value of $x_{(1)}$. These two steps are semantically equivalent with the assignment $x_{(1)l} \leftarrow x_{(1)}$. We could perform this reversal dataflow similar to the solution that we used in the previous section where we used a label l to mark the position where the reverse section has to execute the reversal of the dataflow. But, we select the simpler solution despite the commonly used code pattern that we only have adjoint assignments inside the reverse section.

Combined parallel sections Construct with lastprivate

In this section, the parallel region P is supposed to have the following structure:

$$P \; := \quad \text{\#pragma omp parallel sections lastprivate}(list) \; [clauses]$$
$$\{$$
$$\qquad \text{\#pragma omp section} \quad \{ \quad D_{l,1} \quad S_{l,1} \quad \}$$
$$\qquad \text{\#pragma omp section} \quad \{ \quad D_{l,2} \quad S_{l,2} \quad \}$$
$$\qquad \vdots$$
$$\qquad \text{\#pragma omp section} \quad \{ \quad D_{l,N} \quad S_{l,N} \quad \}$$
$$\}$$

The definition for the syntax in (4.54) determines that each section has its own sequence with definitions and its own sequence of statements. The label of the lastprivate clause is assumed to be l. The sequence of definitions are referred to as $D_{l,\{1,...,N\}}$ and the sequence of statements as $S_{l,\{1,...,N\}}$ where $N \in \mathbb{N}$. The index l should only avoid misunderstandings with the already defined sequences D_1, D_2, and so forth. At this point, there is no connection between $D_{l,1}$ and the lastprivate clause with label l.

We define two sequences of definitions D_5 and D_6 that consist of the variables listed in the lastprivate clause together with their tangent-linear or adjoint associates. The sequence of definitions for the tangent-linear code is

$$D_5 := \{\text{int } a_l | a \in INT_{list}\} \cup \{\text{double } x_l; \text{double } x_l^{(1)} \leftarrow 0. | x \in FLOAT_{list}\}$$

whereby the sequence of definitions for the adjoint code is

$$D_6 := \{\text{int } a_l | a \in INT_{list}\} \cup \{\text{double } x_l; \text{double } x_{(1)l} \leftarrow 0. | x \in FLOAT_{list}\}.$$

The tangent-linear source transformation is

$$
\tau(P) := \begin{cases}
\text{\#pragma omp parallel sections } [clauses_l] \\
\{ \\
\quad \text{\#pragma omp section } \{ \\
\quad\quad \tau(D_{l,1}) \\
\quad\quad D_5 \\
\quad\quad \tau(S_{l,1}) \forall a, x \in VAR_{list} : \\
\quad\quad\quad [a/a_l][x/x_l][x^{(1)}/x_l^{(1)}] \\
\quad \} \\
\quad \vdots \\
\quad \text{\#pragma omp section } \{ \\
\quad\quad \tau(D_{l,N}) \\
\quad\quad D_5 \\
\quad\quad \tau(S_{l,N}) \forall a, x \in VAR_{list} : \\
\quad\quad\quad [a/a_l][x/x_l][x^{(1)}/x_l^{(1)}] \\
\quad\quad \forall a \in INT_{list} : a \leftarrow a_l; \\
\quad\quad \forall x \in FLOAT_{list} : x \leftarrow x_l; x^{(1)} \leftarrow x_l^{(1)}; \\
\quad \} \\
\}
\end{cases} \tag{4.65}
$$

where $clauses_l$ may not contain a lastprivate clause. Each section defines the private variables that are listed in the lastprivate clause. The code inside the last section contains in addition a sequence of assignments that is semantically equivalent to the dataflow performed by the lastprivate clause.

The adjoint transformation of the combined parallel sections construct supplies a joint reversal scheme inside each section. This is shown in (4.66) and means that each section contains its own sequence of definition D_6 and its own forward and reverse section. The dataflow from the lastprivate clause is inherited by the additional code in the last section just before the reverse section. In addition, there is an adjoint assignment that performs the reversal of the dataflow of the lastprivate clause. This is the same solution as in the combined parallel loop construct.

$\sigma(P) :=$

$$
\left\{
\begin{aligned}
&\text{\#pragma omp parallel sections } \textit{[clauses}_a\textit{]} \\
&\{ \\
&\qquad \text{\#pragma omp section } \quad \{ \\
&\qquad\quad \sigma\left(D_{l,1}\right) \\
&\qquad\quad D_6 \\
&\qquad\quad \sigma\left(S_{l,1}\right)\forall a,x \in \textit{VAR}_{list} : [a/a_l][x/x_l] \\
&\qquad\quad \text{while } \left(\text{not STACK}_{(1)_c}.\text{empty}()\right) \{ \\
&\qquad\qquad \rho\left(S_{l,1}\right)\forall a,x \in \textit{VAR}_{list} : [a/a_l][x/x_l][x_{(1)}/x_{(1)l}] \\
&\qquad\quad \} \\
&\qquad \} \\
&\qquad \vdots \\
&\qquad \text{\#pragma omp section } \quad \{ \\
&\qquad\quad \sigma\left(D_{l,N}\right) \\
&\qquad\quad D_6 \\
&\qquad\quad \sigma\left(S_{l,N}\right)\forall a,x \in \textit{VAR}_{list} : [a/a_l][x/x_l] \\
&\qquad\quad \forall a \in \textit{INT}_{list} : a \leftarrow a_l; \\
&\qquad\quad \forall x \in \textit{FLOAT}_{list} : x \leftarrow x_l; \quad x_{(1)l} \leftarrow x_{(1)}; \\
&\qquad\quad \text{while } \left(\text{not STACK}_{(1)_c}.\text{empty}()\right) \{ \\
&\qquad\qquad \rho\left(S_{l,1}\right)\forall a,x \in \textit{VAR}_{list} : [a/a_l][x/x_l][x_{(1)}/x_{(1)l}] \\
&\qquad\quad \} \\
&\qquad \} \\
&\} \\
\end{aligned}
\right. \tag{4.66}
$$

Worksharing Loop Construct with lastprivate

What remains is the transformation of a worksharing construct that contains a lastprivate clause. The parallel region P is structured as shown in (4.55). Let us first explain the transformation of the loop construct where we assume that the loop is structured as shown in (4.67). We have rewritten the loop again as semantically equivalent while-loop since our *SPL* language does not contain a for-loop statement.

$$S := \begin{cases} \#\text{pragma omp for lastprivate}(\textit{list}) \ [\textit{clauses}] \\ \{ \\ \qquad a_2 \leftarrow 0; \\ \qquad \textbf{while } (a_2 < N)\{ \quad S'; a_2 \leftarrow a_2 + 1; \quad \} \\ \} \end{cases} \tag{4.67}$$

To register the thread that executes the sequentially last iteration, we use an auxiliary variable $a_{l,l}$ that is indexed twice with label l to avoid misunderstandings with the variable a_l. a_l is used as identifier for the local copy of variable a contained in the *list* of lastprivate variables, $a_{l,l}$ is used to mark the thread that executes the sequentially last iteration.

According to (4.67), variable $a_{l,l}$ stores the current value of the counting variable a_2. Once the loop has been executed, only one thread has a private variable $a_{l,l}$ that contains the value $N - 1$. This method prevents that the test that checks if the current iteration is the last iteration must be performed by each iteration. This should be a performance gain when assuming that the loop has millions of iterations. The sequence of definitions has to be adjusted such that it contains the auxiliary variable $a_{l,l}$. Thus, the tangent-linear transformation $\tau(P)$ is supposed to be D_7 where

$$D_7 := \tau(D) \cup \{\text{int } a_l | a \in \textit{INT}_{list}\}$$
$$\cup \{\text{double } x_l; \text{double } x_l^{(1)} \leftarrow 0. | x \in \textit{FLOAT}_{list}\} \cdot \tag{4.68}$$
$$\cup \{\text{int } a_{l,l} \leftarrow 0\}$$

The adjoint source transformation $\sigma(P)$ is assumed to have the following sequence of definitions:

$$D_8 := \sigma(D) \cup \{\text{int } a_l | a \in \textit{INT}_{list}\}$$
$$\cup \{\text{double } x_l; \text{double } x_{(1)l} \leftarrow 0. | x \in \textit{FLOAT}_{list}\} \tag{4.69}$$
$$\cup \{\text{int } a_{l,l} \leftarrow 0\}$$

The tangent-linear source transformation is defined as shown in (4.70). Inside of *clauses*$_t$ is no lastprivate allowed. The auxiliary variable $a_{l,l}$ is initialized with zero before the loop construct and is assigned to during the execution of the loop with the current value of the counting variable a_2. The branch statement that follows

$$
\tau(S) := \begin{cases}
a_{l,l} \leftarrow 0; \\
\text{\#pragma omp for } [clauses_t] \\
\{ \\
\quad a_2 \leftarrow 0; \\
\quad \textbf{while } (a_2 < N)\{ \\
\quad\quad \tau(S')\forall a,x \in VAR_{list} : [a/a_l][x/x_l][x^{(1)}/x_l^{(1)}] \\
\quad\quad a_{l,l} \leftarrow a_2; \\
\quad\quad a_2 \leftarrow a_2 + 1; \\
\quad \} \\
\} \\
\textbf{if } (a_{l,l} = N - 1)\{ \\
\quad \forall a \in INT_{list} : a \leftarrow a_l; \\
\quad \forall x \in FLOAT_{list} : x \leftarrow x_l; \; x^{(1)} \leftarrow x_l^{(1)}; \\
\}
\end{cases}
\tag{4.70}
$$

$$
\sigma(S) := \begin{cases}
a_{l,l} \leftarrow 0; \\
\text{\#pragma omp for } [clauses_a] \\
\{ \\
\quad a_2 \leftarrow 0; \\
\quad \textbf{while } (a_2 < N)\{ \\
\quad\quad \sigma(S')\forall a,x \in VAR_{list} : [a/a_l][x/x_l] \\
\quad\quad a_{l,l} \leftarrow a_2; \\
\quad\quad \sigma(a_2 \leftarrow a_2 + 1) \\
\quad \} \\
\} \\
\textbf{if } (a_{l,l} = N - 1)\{ \\
\quad \forall a \in INT_{list} : a \leftarrow a_l; \\
\quad \forall x \in FLOAT_{list} : x \leftarrow x_l; \\
\quad \text{STACK}_{(1)_c}.\text{push}(l) \\
\}
\end{cases}
\tag{4.71}
$$

the loop construct checks whether or not the local variable has the value $N-1$. The thread where this check is valid must perform the dataflow connected with lastprivate clause.

The source transformation for the forward section of the adjoint code is defined in (4.71). The label of the lastprivate clause is assumed to be l. The structure of the transformation is similar to the tangent-linear but, in addition, we push label l onto the control flow stack to indicate that the thread with this label on its stack has to reverse the dataflow associated with the lastprivate clause.

The code for the reverse section

$$\rho\left(S\right) := \begin{cases} \text{if } \left(\text{STACK}_{(1)c}.\text{top}() = l\right) \ \{ \\ \quad \forall x \in FLOAT_{list} : x_{(1)l} \leftarrow x_{(1)}; \\ \} \\ \rho\left(S';a_2 \leftarrow a_2 + 1\right)\forall a,x \in VAR_{list} : [a/a_l][x/x_l][x_{(1)}/x_{(1)l}] \end{cases}$$

contains two parts. The first part is a branch where the test expression is valid if the executing thread has the label l on its control flow stack. In case that the test is successful, the executing thread has to reverse the dataflow connected with the lastprivate clause. We use an assignment and not an incremental assignment to set the value of the adjoint variable $x_{(1)l}$. This is correct since this adjoint variable is a local variable inside the worksharing construct and its value is therefore zero at the beginning of the adjoint part. The second part is the part that contains the adjoint statements of the loop body.

Worksharing sections Construct with lastprivate

We denote the sequence of statements contained in the first section with $S_{l,1}$, the sequence of the second section with $S_{l,2}$. The last section is $S_{l,N}$ with $N \in \mathbb{N}$.

$$S := \begin{cases} \text{\#pragma omp sections lastprivate}(list) \ [clauses] \\ \{ \\ \quad \text{\#pragma omp section} \quad \{ \quad S_{l,1} \quad \} \\ \quad \text{\#pragma omp section} \quad \{ \quad S_{l,2} \quad \} \\ \quad \vdots \\ \quad \text{\#pragma omp section} \quad \{ \quad S_{l,N} \quad \} \\ \} \end{cases}$$

We assume that the parallel region has the sequence of definitions as defined for the worksharing loop in (4.68) and (4.69). The tangent-linear transformation is similar to the one of the combined parallel sections construct shown in (4.65). The only difference is the missing sequences of definitions D_5 and $\tau(D_{l,1,...,N})$. For this reason we do not display the definition of $\tau(S)$ here and refer to (4.65).

The adjoint source transformation differs from (4.66) because we used the joint reversal scheme there and we use here the split reversal scheme for the stand-alone worksharing construct. The transformation for the forward section part is

$$
\sigma(S') := \left\{
\begin{array}{l}
\texttt{\#pragma omp sections } [clauses_t] \\
\{ \\
\quad \texttt{\#pragma omp section} \\
\quad \{ \quad \sigma(S_{l,1}) \forall a,x \in VAR_{list} : [a/a_l][x/x_l] \quad \} \\
\quad \vdots \\
\quad \texttt{\#pragma omp section} \\
\quad \{ \quad \sigma(S_{l,N-1}) \forall a,x \in VAR_{list} : [a/a_l][x/x_l] \quad \} \\
\quad \texttt{\#pragma omp section} \\
\quad \{ \\
\qquad \sigma(S_{l,N}) \forall a,x \in VAR_{list} : [a/a_l][x/x_l] \\
\qquad \forall a \in INT_{list} : a \leftarrow a_l; \\
\qquad \forall x \in FLOAT_{list} : x \leftarrow x_l; \\
\qquad STACK_{(1)c}.\text{push}(l) \\
\quad \} \\
\} \\
\end{array}
\right.
$$

where we perform the dataflow connected with the lastprivate clause in the last section. We assume that the label of the lastprivate clause is l. Therefore, we push label l onto the control flow stack to indicate that the dataflow connected to the lastprivate clause has to be reversed during the reverse section.

The transformation $\rho\left(S'\right)$ for the reverse section, where

$$\rho\left(S'\right) := \begin{cases} \rho\left(S_{l,1}\right) \forall a, x \in VAR_{list} : [a/a_l][x/x_l][x_{(1)}/x_{(1)l}] \\ \rho\left(S_{l,2}\right) \forall a, x \in VAR_{list} : [a/a_l][x/x_l][x_{(1)}/x_{(1)l}] \\ \quad \vdots \\ \rho\left(S_{l,N}\right) \forall a, x \in VAR_{list} : [a/a_l][x/x_l][x_{(1)}/x_{(1)l}] \\ \text{if } \left(\text{STACK}_{(1)c}.\text{top}() = l\right) \{ \\ \qquad \forall x \in FLOAT_{list} : x_{(1)l} \leftarrow x_{(1)}; \\ \} \end{cases}$$

contains the adjoint statements for all the sections $S_{l,1}$ to $S_{l,N}$. In addition, it contains a branch statement that checks if there is the label l on the top of the control flow stack. In this case, the executing thread has to reverse the dataflow that is connected with lastprivate clause.

4.4.5 reduction Clause

This section covers the reduction$(\otimes:\ list)$ clause where $\otimes$ is the addition operator '+' or the multiplication operator '*' and $list$ contains a list of variables. We extend the set of non-terminal symbols in our grammar $\mathcal{G}_{OMP}$ by

$$V = V \cup \{op\}$$

and the production rules by

$$R_{OMP} = R_{OMP} \cup \left\{ \begin{array}{lll} op & : & \texttt{*} \mid \texttt{+} \\ cl & : & \texttt{reduction (} op \texttt{ : } l \texttt{)} \end{array} \right\}.$$

Each OpenMP construct with an associated reduction clause can be expressed in a semantically equivalent form. For example, let us consider the following example of a parallel region P.

$$P := \left\{ \begin{array}{l} \textit{\#pragma omp parallel reduction}(\otimes\textit{: list) [clauses]} \\ \{ \quad D \quad S \quad \} \end{array} \right.$$

For simplicity reasons, we assume that all variables in $list$ are floating-point variables. According to the description in OpenMP 3.1 citation 33 (page 52), P can be expressed semantically equivalent by

$$
\begin{aligned}
&\#\text{pragma omp parallel } [clauses_r] \\
&\{ \\
&\qquad D \\
&\qquad \forall x \in FLOAT_{list} : \text{double } x_l \leftarrow ID(\otimes); \\
&\qquad S \quad \forall x \in FLOAT_{list} : [x/x_l] \\
&\qquad \#\text{pragma omp critical} \\
&\qquad \{ \\
&\qquad\qquad \forall x \in FLOAT_{list} : x \leftarrow x \otimes x_l; \\
&\qquad \} \\
&\}
\end{aligned}
\qquad (4.72)
$$

where $clauses_r$ does not contain a reduction clause and the label of the reduction clause in P is assumed to be l. For each element $x \in FLOAT_{list}$ each thread creates a private instance x_l that is initialized with $ID(\otimes)$. ID maps the given operator to its associated identity element. This means in case of an addition, $ID(+)$ is zero, and a multiplication operator as argument of $ID(*)$ yields the value one.

The reader may ask why the reduction clause has been introduced in OpenMP when it can be expressed by the above method. The answer is that a reduction is a widely used operation in scientific computing and the clause hides the underlying synchronization need from the user. Another advantage is that the back-end compiler can decide how this synchronization is being implemented on a lower level. The fact that the reduction clause often occurs in scientific computing motivates this section although we already have the tools at hand to transform (4.72).

Reduction of a Sum

Let us assume that P has the shape

$$
P := \left\{
\begin{aligned}
&\#\text{pragma omp parallel reduction}(+:list) \\
&\{ \quad D \quad S \quad \}
\end{aligned}
\right.
$$

where $list$ only contains floating-point variables. We define for each variable in $list$ a private variable that is initialized with zero. In addition we need a derivative component for this private variable. The sequence of definitions for the tangent-linear code is

$$
D_9 := \tau(D) \cup \{\text{double } x_l \leftarrow 0.; \text{ double } x_l^{(1)} \leftarrow 0.; | x \in FLOAT_{list}\}
$$

whereby the adjoint code has

$$D_{10} := \sigma(D) \cup \{\text{double } x_l \leftarrow 0.; \text{double } x_{(1)l} \leftarrow 0. | x \in FLOAT_{list}\}$$

as the sequence of definitions. The tangent-linear transformation is

$$\tau(P) := \left\{ \begin{array}{l} \text{\#pragma omp parallel} \\ \{ \\ \quad D_9 \\ \quad \tau(S) \forall x \in FLOAT_{list} : [x/x_l][x^{(1)}/x_l^{(1)}] \\ \quad S_4 \\ \} \end{array} \right. \tag{4.73}$$

where S_4 is a sequence of statements with

$$S_4 := \left\{ \left[\begin{array}{l} \text{\#pragma omp atomic} \\ x^{(1)} \mathrel{+\!\!\!+\!\!\leftarrow} x_l^{(1)}; \\ \text{\#pragma omp atomic} \\ x \mathrel{+\!\!\!+\!\!\leftarrow} x_l; \end{array} \right] \middle| x \in FLOAT_{list} \right\} \tag{4.74}$$

that perform the reduction operation.
The adjoint code of P is obtained by applying the joint reversal scheme:

$$\sigma(P) := \left\{ \begin{array}{l} \text{\#pragma omp parallel} \\ \{ \\ \quad D_{10} \\ \quad \sigma(S) \forall x \in FLOAT_{list} : [x/x_l] \\ \quad S_5 \\ \quad \text{while} \left(\text{not STACK}_{(1)c}.\text{empty}() \right) \{ \\ \quad\quad \rho(S) \forall x \in FLOAT_{list} : [x/x_l][x_{(1)}/x_{(1)l}] \\ \quad \} \\ \} \end{array} \right. \tag{4.75}$$

where

$$S_5 := \left\{ \left[\begin{array}{l} \text{\#pragma omp atomic} \\ x \mathrel{+\!\!\!+\!\!\leftarrow} x_l; \\ x_{(1)l} \leftarrow x_{(1)}; \end{array} \right] \middle| x \in FLOAT_{list} \right\}$$

is the sequence of statements that is responsible for the reduction. More precisely, the first statement in S_5 performs the reduction with an atomic assignment, whereby the second assignment is the adjoint counterpart of the first assignment. We do not use an atomic construct for the second assignment since the left-hand side reference $x_{(1)l}$ is a private variable. As in the section where we covered the lastprivate clause we set here the adjoint variable despite the fact that the execution is still in the forward section.

Suppose that a worksharing construct W inside of the parallel region P is augmented with a reduction clause as shown in (4.76). We expect in *list* again only floating-point variables.

$$S := \left\{ \; \#\text{pragma omp } W \text{ reduction}(+: list) \;\; \{ \;\; S' \;\; \} \right. \tag{4.76}$$

Since we substitute the variables contained in *list* we assume that $\tau(P)$ contains D_9, and $\sigma(P)$ includes D_{10} as their sequence of definitions, see (4.73) and (4.75). The tangent-linear source transformation is defined as

$$\tau(S) := \begin{cases} \#\text{pragma omp } W \\ \{ \;\; \tau(S')\forall x \in FLOAT_{list} : [x/x_l][x^{(1)}/x_l^{(1)}] \;\; \} \\ S_4 \end{cases} \tag{4.77}$$

where we again use S_4 from (4.74).

The split reversal scheme prevents us from taking S_5 to compute the reduction and its adjoint computation. During the forward section we add up consecutively all private variables x_l when x is a variable occurring in *list*. In addition, we have to store the value that x has before the program execution enters W for the first time in the forward section. This value would otherwise be lost through the reduction operation but it is necessary when the reverse section amount of W has been completed. We use a shared auxiliary variable $z_{l,x}$ for each variable x contained in *list* where l is assumed to be the label of the reduction clause. The code for the forward section is obtained by

$$\sigma(S) := \begin{cases} \#\text{pragma omp single} \\ \{ \;\; \forall x \in FLOAT_{list} : z_{l,x} \leftarrow x; \;\; \} \\ \#\text{pragma omp } W \\ \{ \;\; \sigma(S')\forall x \in FLOAT_{list} : [x/x_l] \;\; \} \\ \text{STACK}_{(1)c}.\text{push}(l) \\ S_6 \\ \#\text{pragma omp barrier} \end{cases} \tag{4.78}$$

with

$$S_6 := \left\{ \left[\begin{array}{l} \text{\#pragma omp atomic} \\ x \mathrel{+\!\!\leftarrow} x_l; \end{array} \right] \middle| x \in FLOAT_{list} \right\}$$

The single construct stores the values of the variables from *list* in the auxiliary variables $z_{l,x}$. The implicit barrier ensures that no thread enters the worksharing construct before all the values has been stored. Before the reduction is started by executing the sequence S_6 we mark this position with the label l on the control flow. This label indicates the reverse section that the adjoint pendant of the reduction must be executed. In S_6, we use an atomic assignment for each x in *list* to perform the reduction operation. The barrier at the end of $\sigma(S)$ ensures that all threads have finished the reduction before its adjoint pendant is executed. In case that there are multiple reduction clauses a better solution would be to emit one barrier between the forward and the reverse section.

The code that belongs to the reverse section is

$$\rho(S) := \left\{ \begin{array}{l} \text{if } \left(\text{STACK}_{(1)_c}.\text{top}() = LABEL(S') \right) \{ \\ \qquad \rho(S') \forall x \in FLOAT_{list} : [x/x_l][x_{(1)}/x_{(1)l}] \\ \} \\ \text{if } \left(\text{STACK}_{(1)_c}.\text{top}() = l \right) \{ \\ \qquad \forall x \in FLOAT_{list} : x_{(1)l} \leftarrow x_{(1)}; \\ \qquad \text{\#pragma omp single} \\ \qquad \{ \quad \forall x \in FLOAT_{list} : x \leftarrow z_{l,x}; \quad \} \\ \} \end{array} \right. \qquad (4.79)$$

where the first branch statement contains the adjoint statements for the body of the worksharing construct. The second branch statement is responsible for two things. On the one hand, the reversal of the dataflow inherited from the reduction is performed. On the other hand, the single construct restores the values of the global instances of all elements of *list*. One could ask why this happens at this point where the reverse section part of the worksharing construct is lying ahead. But this does not influence the adjoint code of the worksharing construct since all the instances of x from *list* are substituted by x_l.

Reduction of a Multiplication

The parallel region P is defined as

$$P := \begin{cases} \text{\#pragma omp parallel reduction(*:\textit{list})} \\ \{ \quad D \quad S \quad \} \end{cases} .$$

Again, we assume that *list* only contains floating-point variables. We have to initialize the private copies of variable x contained in *list* with the identity element of the multiplication. The sequence of definitions for the tangent-linear source transformation is

$$D_{11} := \tau(D) \cup \{\text{double } x_l \leftarrow 1.; \text{double } x_l^{(1)} \leftarrow 0.; | x \in FLOAT_{list}\}$$

and the sequence for the adjoint code looks like

$$D_{12} := \sigma(D) \cup \{\text{double} x_l \leftarrow 1.; \text{double} x_{(1)l} \leftarrow 0.; | x \in FLOAT_{list}\}$$
$$\cup \{\text{double} z_{1,x,0,\dots,p-1}, z_{2,x} | x \in FLOAT_{list}\}$$

where for each variable in *list* the following auxiliary variables are necessary. $z_{1,x}$ is a vector with p elements whereby $z_{2,x}$ is a scalar variable.

The tangent-linear transformation of P is

$$\tau(P) := \begin{cases} \text{\#pragma omp parallel} \\ \{ \\ \quad D_{11} \\ \quad \tau(S) \forall x \in FLOAT_{list} : [x/x_l][x^{(1)}/x_l^{(1)}] \qquad (4.80) \\ \quad \text{\#pragma omp critical} \\ \quad \{ \quad S_7 \quad \} \\ \} \end{cases}$$

where

$$S_7 := \left\{ \left[\begin{array}{l} x^{(1)} \leftarrow x^{(1)} * x_l + x_l^{(1)} * x; \\ x \leftarrow x * x_l; \end{array} \right] \middle| x \in FLOAT_{list} \right\} \qquad (4.81)$$

The reduction is performed explicitly by the sequence S_7. Since x is shared among the threads, we have to put S_7 into a critical region. We cannot use atomic assignments in S_7 since the *SPL* language does not contain an assignment with the shape $x *\!\!\leftarrow e$. OpenMP, on the other side, knows this shape and allows to declare this shape of assignment to be declared as atomic. The atomic assignment is for sure the better alternative as to put it in a critical region but here we do not have a choice.

In Section 4.2.3, we saw that situations can occur where it is necessary to store the order of the threads that execute a certain part of P. The reason for this is the fact that a reference occurs on both sides of an assignment. We can avoid the overhead connected to the storing of the thread order in case of a multiplication as reduction operation. This is done by calculating the derivative of the reduction result with respect to all its input references. The result of a reduction that is connected with a multiplication of a floating-point variable x is

$$x_a \leftarrow x_b \cdot \prod_{t=0}^{p-1} x^t \qquad (4.82)$$

where x_a denotes the state of reference x after the reduction, and x_b is the state of reference x before the reduction. This means that the references $\&x_a$ and $\&x_b$ are the same. We need this notation to differ between these two instances of x. The private copies of each thread are referred to as $x^{0,\dots,p-1}$. For illustration reasons, we assume for a moment that all $x^{0,\dots,p-1}$ are not zero. Then the adjoint model of (4.82) can be expressed by:

$$x^0_{(1)} +\!\!\!\leftarrow x_{(1)a} \cdot \frac{x_a}{x^0}; \qquad (4.83)$$

$$x^1_{(1)} +\!\!\!\leftarrow x_{(1)a} \cdot \frac{x_a}{x^1};$$

$$\vdots$$

$$x^{p-1}_{(1)} +\!\!\!\leftarrow x_{(1)a} \cdot \frac{x_a}{x^{p-1}}; \qquad (4.84)$$

$$x_{(1)b} \leftarrow x_{(1)a} \cdot \frac{x_a}{x_b}; \qquad (4.85)$$

Since the adjoint references of all thread local references are zero at the beginning of the reverse section, we can rewrite the assignments from (4.83) to (4.84) as common assignments instead of incremental assignments. (4.85) indicates that the result before the reduction as well as the result after the reduction are necessary. In order to cover the special case that an intermediate result can be zero, we use several auxiliary variables. The number of threads is denoted by p.

The joint reversal scheme of P is shown in (4.86). There, we assume that the reduction(*list*) clause is labeled with l and *list* contains the floating-point variable *x*. D_{12} defines the local copy x_l of x. Before the augmented forward section of S starts, a single thread stores the value of x in $z_{2,x}$. The other threads wait at the implicit barrier of the single construct. After the forward section of S, the intermediate result of each thread is stored by

$$z_{1,x,t} \leftarrow x_l;$$

The reduction is performed with help of a critical region. This critical region ends the forward section of P and a barrier ensures that all threads have finished their forward section.

The reverse section starts by setting the local adjoint $x_{(1)l}$. This assignment corresponds to the assignments shown from (4.83) to (4.84) where the fraction $\frac{x_a}{x^l}$ is represented here by the product. The product multiplies all intermediate values except the one from the current thread (omp_get_thread_num $\neq t$). Eventually, the variable x is restored, and the adjoint $x_{(1)}$ is computed inside of a single construct.

Now, we explain how a reduction connected with a worksharing construct W is transformed. Let S be defined as

$$S := \text{\#pragma omp } W \text{ reduction}(*:list) \quad \{ \quad S' \quad \}$$

where worksharing construct W is associated with a reduction(*:*list*) clause.

The tangent-linear transformation, defined in (4.87), uses the sequence S_7 from (4.81) to perform the reduction. The source transformation of S for the forward section is presented in (4.88) where the value of x is stored in $z_{2,x}$ before the reduction (x_b in (4.82)). This is performed by a single thread while the rest of the threads wait at the implicit barrier. The intermediate result of each thread is stored after the worksharing construct. Label l is then pushed onto the control flow stack to indicate that the reduction is about to start. After all threads have left the critical region, they wait at an explicit barrier to ensure that the reduction has been finished by all threads before any thread can start the corresponding reverse section part. This barrier can be placed between the forward and the reverse section of P. In case of several worksharing constructs this is the better choice to minimize the number of barrier constructs.

The transformation of S that arises in the reverse section is shown in (4.89) and consists of two branch statements. The first branch contains the adjoint statements for the body of the worksharing construct. The second branch is responsible for performing the reversal of the dataflow connected with the reduction process.

All threads encounter label l on their control flow stack, and when this happens each thread enters the corresponding branch statement and pops the topmost label from the stack. Subsequently, the adjoints of the local copies $x_{(1)l}$ are computed. This assignment corresponds to the assignments shown from (4.83) to (4.84). The product consists of the multiplication of all intermediate results except the one from the current thread (omp_get_thread_num $\neq t$). Afterwards, the single construct is used to restore the value of x and to compute the adjoint $x_{(1)}$. The value of the adjoint is given by the product of all intermediate values $z_{1,x,0,\ldots,p-1}$.

$$\sigma(P) :=$$

```
#pragma omp parallel
{
    D_12
    #pragma omp single
    {   ∀x ∈ FLOAT_list : z_{2,x} ← x;   }
    σ(S)∀x ∈ FLOAT_list : [x/x_l]
    ∀x ∈ FLOAT_list : z_{1,x,t} ← x_l;
    #pragma omp critical
    {   ∀x ∈ FLOAT_list : x ← x * x_l;   }
    #pragma omp barrier
    ∀x ∈ FLOAT_list :
```

$$x_{(1)l} \leftarrow x_{(1)} \cdot z_{2,x} \cdot$$
$$\prod_{t=0,\text{omp\_get\_thread\_num}\neq t}^{p-1} z_{1,x,t};$$

```
    while (not STACK_{(1)c}.empty()) {
        ρ(S)∀x ∈ FLOAT_list : [x/x_l][x_{(1)}/x_{(1)l}]
    }
    #pragma omp single
    {
```

$$\forall x \in FLOAT_{list} : x \leftarrow z_{2,x};$$
$$\forall x \in FLOAT_{list} : x_{(1)} \leftarrow x_{(1)} \cdot \prod_{t=0}^{p-1} z_{1,x,t};$$

```
    }
}
```

$$(4.86)$$

$$\tau(S) := \begin{cases} \text{\#pragma omp } W \\ \quad \{ \quad \tau(S')\forall x \in FLOAT_{list} : [x/x_l][x^{(1)}/x_l^{(1)}] \quad \} \\ \text{\#pragma omp critical} \\ \quad \{ \quad S_7 \quad \} \end{cases} \qquad (4.87)$$

$$\sigma(S) := \begin{cases} \text{\#pragma omp single} \\ \quad \{ \quad \forall x \in FLOAT_{list} : \quad z_{2,x} \leftarrow x; \quad \} \\ \text{\#pragma omp } W \\ \quad \{ \quad \sigma(S')\forall x \in FLOAT_{list} : [x/x_l] \quad \} \\ \forall x \in FLOAT_{list} : z_{1,x,t} \leftarrow x_l; \\ \text{STACK}_{(1)c}.\text{push}(l) \\ \text{\#pragma omp critical} \\ \quad \{ \quad \forall x \in FLOAT_{list} : \quad x \leftarrow x * x_l; \quad \} \\ \text{\#pragma omp barrier} \end{cases} \qquad (4.88)$$

$$\rho(S) := \begin{cases} \text{if } \left(\quad \text{STACK}_{(1)c}.\text{top}() = LABEL(S') \quad \right) \{ \\ \quad \rho(S')\forall x \in FLOAT_{list} : [x/x_l][x_{(1)}/x_{(1)l}] \\ \} \\ \text{if } \left(\quad \text{STACK}_{(1)c}.\text{top}() = l \quad \right) \{ \\ \quad \text{STACK}_{(1)c}.\text{pop}(); \\ \quad x_{(1)l} \leftarrow x_{(1)} \cdot z_{2,x} \cdot \\ \qquad \Pi_{t=0,\text{omp\_get\_thread\_num} \neq t}^{p-1} z_{1,x,t}; \\ \text{\#pragma omp single} \\ \{ \\ \qquad \forall x \in FLOAT_{list} : \\ \qquad x \leftarrow z_{2,x}; \\ \qquad x_{(1)} \leftarrow x_{(1)} \cdot \Pi_{t=0}^{p-1} z_{1,x,t}; \\ \} \\ \} \end{cases} \qquad (4.89)$$

4.5 Summary

In this chapter we covered the case that the input code for the AD source transformation consists of an OpenMP parallel region. This region can consist of the different possible directives and constructs described in the OpenMP 3.1 standard. In order to achieve rules for an AD source transformation for the possible OpenMP pragmas, we took the context-free grammar for the language *SPL* together with the transformation rules for *SPL* from Chapter 2 as a starting point. The productive rules of the grammar were extended step-by-step during the current chapter. In addition, the corresponding tangent-linear and adjoint source transformation rules were defined.

The first thing that a software engineer who wants to implement an adjoint source transformation has to decide is how the values are stored that potentially can be overwritten during the execution of the forward section. We presented two possibilities in Section 4.1 where both solutions have their assets and drawbacks.

In Section 4.2 we introduced the transformations of the synchronization constructs barrier, master, critical, and atomic of OpenMP. We call the language that is produced by the context-free grammar $\mathscr{G}_{OMP}$ together with rules for recognizing the syntax of the synchronization constructs $SPLOMP^1$. This was a very important section because with help of the synchronization constructs we were able to show the closure property of our source transformations in Section 4.2.5.

The two most important worksharing constructs of OpenMP are the loop construct and the sections construct. We covered these constructs in Section 4.3. The language that contains the worksharing constructs of OpenMP is denoted by $SPLOMP^2$. The combined constructs parallel for and the parallel sections were also part of this section. The adjoint transformation of the combined constructs did differ from the adjoint transformation of the stand-alone worksharing constructs. For the first, we used the joint reversal scheme, for the latter we used the split reversal scheme.

OpenMP allows to define several clauses to control the data-sharing among threads. These data-sharing options were covered in Section 4.4. The language that arises from applying all the production rules of this chapter, is referred to as $SPLOMP^3$. We introduced the transformations of the threadprivate directive and in addition the transformations of the private, firstprivate, lastprivate, and reduction clauses. Again, we had to differ between the joint reversal and the split reversal scheme depending on whether the data-sharing clause was defined for a combined construct or for a worksharing construct.

5 Experimental Results

The previous chapter presented source code transformation rules that take an parallel region as input and provide the tangent-linear or the adjoint model as output code. These source transformation rules have been implemented in a tool called SPLc which is described in Appendix A. The current chapter presents runtime results of several derivative codes and of applications where derivative codes are used. The corresponding derivative codes were all generated by SPLc.

Our target computer architecture was a shared memory system that is a node contained in the high performance computing cluster of the RWTH Aachen University. This cluster was placed as 37th in the TOP 500 supercomputer list[1] when the system was established in the year 2011. The node of the system that we use for testing our approach is a compound of four physical nodes where the connection between these physical nodes is provided by the proprietary Bull Coherent Switch (BCS). Each physical node is equipped with a board carrying four Intel X7550 processors. The X7550 processors are clocked at 2 GHz and consist of eight cores. In total, the system node has 16 sockets with 8 cores per socket which gives the user the possibility to run programs which exploit 128 cores and a maximum of memory that amounts to one terabyte. Detailed information about the architecture can be found in Table 5.1.

To measure the performance of the tangent-linear and the adjoint models that are the result of the source transformation rules from Chapter 4, we will present a test suite that contains example codes with different OpenMP pragmas. The derivative codes from these code examples are obtained by applying our tool SPLc. Two compilers are used to build the binaries containing the derivative codes. The first compiler is from the *GNU Compiler Collection* in version 4.8.0 and the second compiler is the Intel Compiler in version 14.0. The detailed runtime results from the test suite are presented in Section 5.1. To investigate the scaling properties of a second derivative code, we will choose one test case from the test suite. By reapplying SPLc to the first derivative code of this certain test case, we obtain the second derivative code. The scaling properties of the second derivative codes are pointed out in Section 5.2.

In Section 5.3, we investigate the performance that we gain from applying the

---

[1]www.top500.org

Architecture:	x86_64
CPU op-mode(s):	32-bit, 64-bit
Byte Order:	Little Endian
CPU(s):	128
On-line CPU(s) list:	0-127
Thread(s) per core:	1
Core(s) per socket:	8
Socket(s):	16
NUMA node(s):	16
Vendor ID:	GenuineIntel
CPU family:	6
Model:	46
Stepping:	6
CPU MHz:	2000.175
BogoMIPS:	3999.51
Virtualization:	VT-x
L1d cache:	32K
L1i cache:	32K
L2 cache:	256K
L3 cache:	18432K

Table 5.1: The output of the shell command `lscpu` provides a detailed description about the test environment's architecture.

exclusive read analysis (ERA). This static program analysis was introduced in Chapter 3. We apply the source transformation twice, on the one hand we generate the derivative code without using the ERA. On the other hand, we exploit the additional information of the static analysis to generate the derivative code. The difference between the two source transformation results is the number of atomic statements.

The least-squares problem from the introduction (Section 1.1.1) was implemented and its scaling properties are presented in Section 5.4. The second motivating example in the introduction was the nonlinear constrained optimization problem that we introduced in Section 1.2.3. The runtime results of the implementation are shown in Section 5.5.

5.1 Test Suite

In this section, we present several example codes with varying OpenMP pragmas inside of a parallel region P. We examine the scaling properties of the corresponding derivative codes of P. In order to compare this with the scaling properties of the original code P, we present the runtime results of the derivative codes together with the one from P. The test suite consists of the following example codes which comprise of

1. a pure parallel region with no further pragmas,

2. a parallel region with a barrier construct,

3. a parallel region with a master construct,

4. a parallel region with a critical region, and

5. a parallel region with an atomic statement.

The listings of the example codes can be found in Appendix B. The computation kernel in each example code makes usage of two dynamically allocated arrays, the one-dimensional array x, and the two-dimensional array A. The size of x is around one megabyte, the size of A is about 125 gigabytes.

We used two compilers for the runtime tests, the Intel compiler version 14.0, and the GNU Compiler Collection (GCC) version 4.8.0. Subsequently, we abbreviate these two compilers by icpc and g++. Both of these compilers support a certain set of static program analysis techniques. Since each of the compilers support a command line option for using no program optimization (-O0) and for a sophisticated program optimization (-O3), we will use these two options to obtain two different binaries. In case we note that optimization level 0 has been used for a test, we compile the source files with the command line option -O0. Alternatively, we write optimization level 3 when the corresponding binary was compiled with -O3 as command line option. Please note that these optimization levels have nothing to do with the exclusive read analysis.

In order to achieve precise results in addition to only measuring the runtime, we use hardware counters to measure the CPU cycles and the floating-point operations. For accessing these hardware counters, we use the *Performance API* (PAPI) in version $5.2.0$[2]. For a given runtime test we display the runtime in seconds, CPU cycles, million floating-point operations per second (MFLOPS), and the speedup. MFLOPS is a commonly used measure of computer performance, typically used

---

[2]The PAPI project website is http://icl.cs.utk.edu/papi/software/.

in the domain of scientific computing. The speedup is the ratio between two executions, one execution is performed with only one thread, the other execution uses a group of, for example, p threads. The number of CPU cycles of the run with one thread divided by the number of CPU cycles with p threads yields the speedup result.

The scale abilities of the three different codes, namely the original code P, the tangent-linear code $\tau(P)$, and the adjoint code $\sigma(P)$, are measured by using an increasing number of threads ranging from one to 128 threads. All the measured values are displayed in a table and for better illustration reasons we show for each table a bar chart that summarizes the speedup and MFLOPS results. An example bar chart is shown Figure 5.1. The x-axis represents different executions with an increasing number of threads, starting with only one thread and ending with the maximum number of threads that the test architecture provides (for example 16 in Figure 5.1). The bar chart has two y-axes, the left y-axis illustrates speedup values (light gray bars), the right y-axis clarifies the MFLOPS values (dark gray bars) in the chart.

The results from each individual execution is shown in a group of six bars. The six bars can be seen as three pairs of light and dark gray bars. The first pair displays the results of the original code, the second pair corresponds to the results of the tangent-linear code, and the last pair displays the adjoint code results. This fact is explained for the execution with four threads below the Figure 5.1.

5.1.1 Pure Parallel Region

This test case is called 'plain-parallel' and the example code consists of a pure parallel region. This means that the only OpenMP construct inside the code is the parallel directive to open a parallel code region. The original code is shown in Listing B.1 on page 343, the first-order tangent-linear code is presented in Listing B.2 on page 343, and Listing B.6 on page 363 contains the first-order adjoint code. The next section describes the results that we achieved by using the Intel C/C++ compiler to assemble a binary. Afterwards, we demonstrate the runtime results of the binary that results from using the GNU Compiler Collection.

Intel compiler 14.0:

The detailed results are presented in Table 5.2 and Table 5.3. The reader may find a quick overview in the corresponding bar charts in Figure 5.2 and Figure 5.3. In Table 5.2, we see the results for the test run without any optimization (-O0). Comparing the CPU cycles of the original code and its corresponding derivatives

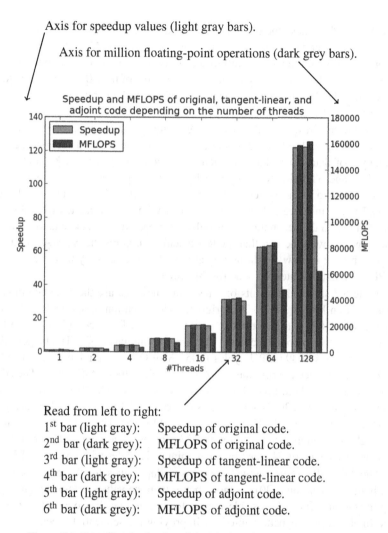

Figure 5.1: Example plot for describing the bar charts in this chapter.

codes, one recognizes that the tangent-linear code evaluation takes about twice as much cycles as the evaluation of the original code. The execution of the adjoint code needs about five times more CPU cycles as the evaluation of the original parallel region. These factors fit into our experiences with AD source transformation and similar factors are presented in [57]. Comparing the columns that show the MFLOPS values, one recognizes that the values for the original code and the tangent-linear code are very similar. If we compare the MFLOPS values of the original code with the one from the adjoint code then the adjoint code execution has a rate that lies between one third and two thirds of the MFLOPS values for the original code. This is probably the impact of the stack operations which store and restore data of a size of 250 gb.

The lower part of the table shows the memory consumption of the evaluation of the adjoint code. In overall, the evaluation of the forward section needs 250 gigabyte of data where half the memory is consumed in $STACK_f$ and each of $STACK_i$ and $STACK_c$ need one fourth of the 250 gigabyte. These numbers are divided by two when we redouble the number of threads since the data for each thread is cut in halves. For example, when we use 128 threads to evaluate the parallel region then each thread has only two megabyte of stack data to store. This fact together with the fact that the test machine supports the NUMA architecture (see Table 5.1), allows each individual thread to use memory that is close to the CPU core that executes the code for this thread.

Table 5.3 shows the results of a test run where we use the Intel compiler and optimization level 3 (-O3). The original code execution needs 9590 CPU cycles (1032 MFLOPS) with one thread and it takes 170 CPU cycles (58149 MFLOPS) when the computation is made by 128 threads simultaneously. The tangent-linear code consumes 10494 CPU cycles (1056 MFLOPS) when 1 thread is used and 128 threads reduce the runtime to 308 CPU cycles (36040 MFLOPS). The adjoint code take about three times longer than the original code. If only one thread is used, the runtime is 30964 CPU cycles (1265 MFLOPS) and with 128 threads the computation lasts 580 CPU cycles (67510 MFLOPS). The original code achieves its highest speedup value of 56 with 128 threads. The tangent-linear code only reaches a speedup of 34 whereby the adjoint code scales with a value of 53.

The MFLOPS values as well as the speedup values for the binary compiled without optimization (-O0) are much better than the values obtained with enabled optimization (-O3). A possible reason for this is that the code optimization of the Intel compiler is better suited for improving the sequential execution and not the parallel execution. One fact is that the number of CPU cycles of the tangent-linear code in Table 5.3 is almost the same as the number for the original code but in Table 5.2 we see that the number of operations is almost doubled through the computation of the derivative values. The adjoint code needs without optimization almost four times more CPU cycles than the original code. With optimization the difference between the CPU cycles of the original code and the adjoint code is about three. This means the code optimization of the Intel compiler improves the tangent-linear code such that the derivative computation comes almost for free.

		Original code		
#Threads	Runtime (sec)	CPU cycles ($\cdot 10^{-9}$)	Speedup	MFLOPS
1	1299	25635	1.0	1296
2	661	13043	1.97	2547
4	329	6521	3.93	5094
8	162	3213	7.98	10339
16	82	1615	15.86	20562
32	41	817	31.38	40667
64	21	409	62.59	81128
128	11	209	122.33	158559

		Tangent-linear code		
#Threads	Runtime (sec)	CPU cycles ($\cdot 10^{-9}$)	Speedup	MFLOPS
1	2629	52031	1.0	1319
2	1336	26430	1.97	2597
4	674	13326	3.9	5152
8	336	6618	7.86	10374
16	164	3258	15.97	21069
32	83	1642	31.68	41810
64	42	818	63.57	83899
128	21	425	122.38	161508

		Adjoint code		
#Threads	Runtime (sec)	CPU cycles ($\cdot 10^{-9}$)	Speedup	MFLOPS
1	6032	113644	1.0	898
2	2997	57681	1.97	1771
4	1586	29332	3.87	3483
8	807	14693	7.73	6955
16	388	7245	15.68	14102
32	201	3701	30.71	27625
64	121	2130	53.35	47993
128	96	1639	69.3	62330

	Sizes of stacks in gigabyte per thread used in the adjoint code			
#Threads	$STACK_c$	$STACK_i$	$STACK_f$	Overall
1	62.5	62.5	125.0	250.0
2	31.3	31.3	62.5	125.0
4	15.6	15.6	31.3	62.5
8	7.8	7.8	15.6	31.3
16	3.9	3.9	7.8	15.6
32	2.0	2.0	3.9	7.8
64	1.0	1.0	2.0	3.9
128	0.5	0.5	1.0	2.0

Table 5.2: Runtime results of the original, the tangent-linear, and the adjoint code of test **'plain-parallel'**, compiled by the **Intel C++ compiler** with optimization level **0**. In addition, the size of the adjoint code stacks are shown in gigabytes. The speedup is computed with respect to CPU cycles not runtime. The test was run with vector $x \in \mathbb{R}^{129536}$ with a memory size of 1012 kilobytes. The matrix $A \in \mathbb{R}^{129536 \times 129536}$ consumes 125 gigabytes of memory.

Original code

#Threads	Runtime (sec)	CPU cycles $(\cdot 10^{-9})$	Speedup	MFLOPS
1	486	9590	1.0	1032
2	261	5149	1.86	1923
4	151	2916	3.29	3400
8	75	1497	6.4	6618
16	37	748	12.81	13243
32	19	380	25.22	26068
64	11	230	41.63	43048
128	8	170	56.24	58149

Tangent-linear code

#Threads	Runtime (sec)	CPU cycles $(\cdot 10^{-9})$	Speedup	MFLOPS
1	533	10494	1.0	1056
2	270	5327	1.97	2081
4	138	2739	3.83	4050
8	69	1372	7.65	8078
16	34	682	15.37	16246
32	20	411	25.52	26983
64	18	366	28.62	30318
128	15	308	34.01	36040

Adjoint code

#Threads	Runtime (sec)	CPU cycles $(\cdot 10^{-9})$	Speedup	MFLOPS
1	1733	30964	1.0	1265
2	864	15497	2.0	2528
4	441	7995	3.87	4901
8	218	3951	7.84	9917
16	109	1994	15.52	19642
32	56	1039	29.79	37705
64	45	805	38.43	48635
128	31	580	53.33	67510

Sizes of stacks in gigabyte per thread used in the adjoint code

#Threads	$STACK_c$	$STACK_i$	$STACK_f$	Overall
1	64.0	64.0	128.0	256.0
2	32.0	32.0	64.0	128.0
4	16.0	16.0	32.0	64.0
8	8.0	8.0	16.0	32.0
16	4.0	4.0	8.0	16.0
32	2.0	2.0	4.0	8.0
64	1.0	1.0	2.0	4.0
128	0.5	0.5	1.0	2.0

Table 5.3: Runtime results of the original, the tangent-linear, and the adjoint code of test **'plain-parallel'**, compiled by the **Intel C++ compiler** with optimization level **3**. In addition, the size of the adjoint code stacks are shown in gigabytes. The speedup is computed with respect to CPU cycles not runtime. The test was run with vector $x \in \mathbb{R}^{131072}$ with a memory size of 1 megabytes. The matrix $A \in \mathbb{R}^{131072 \times 131072}$ consumes 128 gigabytes of memory.

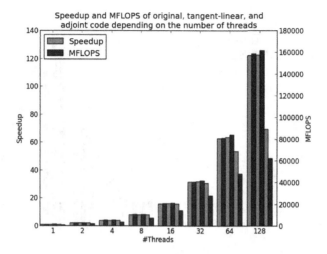

Figure 5.2: Plot for the test 'plain parallel region O0 icpc'. The actual values are shown in Table 5.2. The original as well as the derivative codes are presented in Appendix B.1.

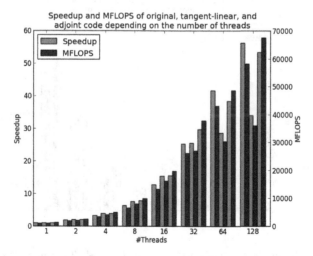

Figure 5.3: Plot for the test 'plain parallel region O3 icpc'. Table 5.3 presents the values of this chart. The original as well as the derivative codes are presented in Appendix B.1.

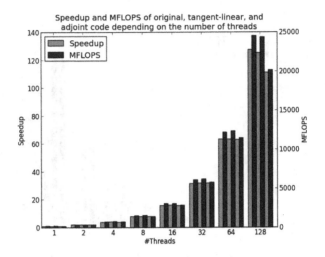

Figure 5.4: Plot for the test 'plain parallel region O0 g++'. Table 5.4 displays the corresponding values. The original as well as the derivative codes are presented in Appendix B.1.

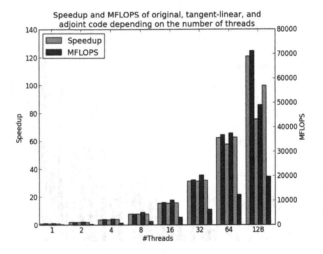

Figure 5.5: Plot for the test 'plain parallel region O3 g++'. The values illustrated in this chart can be found in Table 5.5. The original as well as the derivative codes are presented in Appendix B.1.

Original code

#Threads	Runtime (sec)	CPU cycles ($\cdot 10^{-9}$)	Speedup	MFLOPS
1	6074	120129	1.0	192
2	3142	61862	1.94	373
4	1568	30932	3.88	747
8	770	15251	7.88	1515
16	382	7570	15.87	3053
32	192	3807	31.55	6068
64	95	1894	63.43	12201
128	48	942	127.44	24515

Tangent-linear code

#Threads	Runtime (sec)	CPU cycles ($\cdot 10^{-9}$)	Speedup	MFLOPS
1	12185	240801	1.0	194
2	6208	122263	1.97	383
4	3104	60946	3.95	769
8	1533	30301	7.95	1548
16	765	15144	15.9	3098
32	383	7560	31.85	6205
64	193	3800	63.36	12345
128	97	1926	125.0	24357

Adjoint code

#Threads	Runtime (sec)	CPU cycles ($\cdot 10^{-9}$)	Speedup	MFLOPS
1	19658	379776	1.0	181
2	10007	195077	1.95	353
4	5009	96957	3.92	710
8	2468	48416	7.84	1423
16	1223	23933	15.87	2880
32	612	11964	31.74	5761
64	311	6027	63.01	11436
128	192	3417	111.12	20170

Sizes of stacks in gigabyte per thread used in the adjoint code

#Threads	$STACK_c$	$STACK_i$	$STACK_f$	Overall
1	62.5	62.5	125.0	250.0
2	31.3	31.3	62.5	125.0
4	15.6	15.6	31.3	62.5
8	7.8	7.8	15.6	31.3
16	3.9	3.9	7.8	15.6
32	2.0	2.0	3.9	7.8
64	1.0	1.0	2.0	3.9
128	0.5	0.5	1.0	2.0

Table 5.4: Runtime results of the original, the tangent-linear, and the adjoint code of test **'plain-parallel'**, compiled by the **GCC (g++)** with optimization level **0**. In addition, the size of the adjoint code stacks are shown in gigabytes. The speedup is computed with respect to CPU cycles not runtime. The test was run with vector $x \in \mathbb{R}^{129536}$ with a memory size of 1012 kilobytes. The matrix $A \in \mathbb{R}^{129536 \times 129536}$ consumes 125 gigabytes of memory.

Original code

#Threads	Runtime (sec)	CPU cycles ($\cdot 10^{-9}$)	Speedup	MFLOPS
1	1155	22833	1.0	589
2	604	11859	1.93	1134
4	301	5935	3.85	2266
8	148	2938	7.77	4577
16	74	1470	15.53	9149
32	37	732	31.18	18369
64	19	365	62.51	36824
128	10	189	120.74	71125

Tangent-linear code

#Threads	Runtime (sec)	CPU cycles ($\cdot 10^{-9}$)	Speedup	MFLOPS
1	1233	24295	1.0	650
2	617	12185	1.99	1296
4	320	6284	3.87	2514
8	154	3072	7.91	5143
16	78	1559	15.58	10130
32	39	775	31.34	20384
64	21	420	57.75	37559
128	16	322	75.42	49055

Adjoint code

#Threads	Runtime (sec)	CPU cycles ($\cdot 10^{-9}$)	Speedup	MFLOPS
1	12814	245440	1.0	197
2	6426	125100	1.96	387
4	3204	62222	3.94	779
8	1594	31052	7.9	1561
16	800	15455	15.88	3137
32	401	7744	31.69	6260
64	208	3917	62.66	12377
128	143	2457	99.85	19729

Sizes of stacks in gigabyte per thread used in the adjoint code

#Threads	$STACK_c$	$STACK_i$	$STACK_f$	Overall
1	62.6	62.6	125.3	250.5
2	31.3	31.3	62.6	125.3
4	15.7	15.7	31.3	62.6
8	7.8	7.8	15.7	31.3
16	3.9	3.9	7.8	15.7
32	2.0	2.0	3.9	7.8
64	1.0	1.0	2.0	3.9
128	0.5	0.5	1.0	2.0

Table 5.5: Runtime results of the original, the tangent-linear, and the adjoint code of test **'plain-parallel'**, compiled by the **GCC (g++)** with optimization level **3**. In addition, the size of the adjoint code stacks are shown in gigabytes. The speedup is computed with respect to CPU cycles not runtime. The test was run with vector $x \in \mathbb{R}^{129664}$ with a memory size of 1013 kilobytes. The matrix $A \in \mathbb{R}^{129664 \times 129664}$ consumes 125 gigabytes of memory.

The adjoint code can be improved by code optimization but by far not as good as the tangent-linear code can be improved. This fact indicates that more performance of the adjoint code is only achieved when we use parallel programming. Indeed, the adjoint code executed with 128 threads achieves with 67510 a better result than the tangent-linear code (36040).

GCC 4.8.0:

The results for the test run without optimization (-O0) are presented as a bar chart in Figure 5.4 and as a table in Table 5.4. The tangent-linear code needs about twice as much CPU cycles (240801) as the original code (120129). The 192 MFLOPS for the original code and the 194 MFLOPS for the tangent-linear code shows that both codes provide quite similar performance. The CPU cycles consumed by the adjoint code are about three times higher than the values from the original code and the MFLOPS values are slightly below the ones from the original code. If we compare Table 5.4 with the results from the Intel compiler (Table 5.2), it is obvious that the Intel compiler supplies a binary that is more than twice as fast as the binary created by g++.

Table 5.5 contains the details of the test run with a binary that is provided by g++ with optimization level 3 (-O3). The corresponding bar chart is Figure 5.5. The original code evaluation takes 22833 CPU cycles with one thread and 189 cycles with 128 threads. The corresponding MFLOPS values are 589 and 71125. The tangent-linear code spends 24295 CPU cycles with one thread (650 MFLOPS) and 322 cycles with 128 threads (49055 MFLOPS).

An interesting fact is that the difference between the original code and the adjoint code is about a factor 10. This factor is three when we compare the original code and the adjoint code obtained without using code optimization. As in the Intel case, the adjoint code seems not very suited for applying code optimization. The execution with 128 threads consumes 2457 CPU cycles (19729 MFLOPS) which is 13 times slower than the original code. The speedup value with 128 threads of the adjoint is around 100 and therefore bigger than 75 which is the speedup value for the tangent-linear execution.

5.1.2 Parallel Region with a Barrier

The listings for this test case are contained in Appendix B.2, Listing B.7 on page 374 shows the original code, Listing B.8 on page 375 contains the first-order tangent-linear code, and the first-order adjoint code is shown in Listing B.9 on page 376.

The original code is basically a matrix-vector product with a succeeding reduction performed by the master thread. Before this reduction can be performed, each thread must have finished its computation. Therefore, a barrier is placed after the matrix-vector product. After each thread has reached the barrier, the master thread starts the reduction.

Intel compiler 14.0:

The reader finds details about the test run with optimization level 0 (-O0) in Table 5.6 and the corresponding bar chart is Figure 5.6. The sequential execution of the original code consumes 4163 CPU cycles (165 MFLOPS). In case we use all possible 128 cores the runtime decreases to 239 CPU cycles (2880 MFLOPS). The tangent-linear code achieves a sequential runtime of 7944 CPU cycles (259 MFLOPS) in contrast to the execution with 128 threads which takes 387 CPU cycles (5318 MFLOPS). This means again, the tangent-linear execution needs about twice as long as the original code. A maximal speedup value of 20 with 128 is all but efficient and is probably the result of the overhead connected with the synchronization introduced by the barrier.

The sequential execution of the adjoint code needs 37875 cycles (64 MFLOPS) whereby the execution of the same code with 128 threads consumes 812 cycles (2997 MFLOPS). One barrier in the original code leads to two barriers in the adjoint code, one in the forward section and one in the reverse section. This means the overhead for synchronization increases with each application of the adjoint source transformation. However, we achieve a maximal speedup of 46 with 128 threads which is better than the maximal speedup for the original and the tangent-linear code. While considering the columns where the speedup values are listed, one observes that the adjoint code scales are acceptable until 64 threads. The original and the tangent-linear code on the other hand have a good speedup until they reach eight threads.

The bar chart in Figure 5.7 summarizes the detailed information from Table 5.7 where the underlying test run was compiled with optimization level 3 (-O3). The original code execution needs 2863 CPU cycles (144 MFLOPS) with one thread and when the execution uses 128 threads it takes 170 CPU cycles (2365 MFLOPS). The tangent-linear evaluation consumes with one thread 4656 CPU cycles (235 MFLOPS) and with 128 threads 284 CPU cycles (3777 MFLOPS) are necessary. The runtime of the adjoint code with one thread is 15633 CPU cycles (143 MFLOPS) and with 128 threads the CPU takes 624 cycles (3605 MFLOPS).

The speedup values 47 (-O0) and 25 (-O3) are better than the results from the tangent-linear codes and even better than the speedup values from the original codes. This is a fact one would not expect because the adjoint transformation of a barrier ((4.3), (4.4)) introduces a barrier construct in the forward section as well as in the reverse section. But the impact of the synchronization seems to be small against the improvement of smaller stack sizes when using an higher number of threads. The smaller stack sizes can be stored locally what on the one hand means on a memory chip near the CPU socket where the thread is executed, and on the other side this means on a physical node where the thread is executed. We always have to keep in mind that the machine with 128 threads is a compound of four physical nodes where each node has 32 cores.

GCC 4.8.0:

The results from the test run without any optimization (-O0) is shown in Table 5.8 and in Figure 5.8. The original code execution takes 4195 CPU cycles (163 MFLOPS) with one thread and 195 CPU cycles (3523 MFLOPS) with 128 threads. The related tangent-linear code runs with one thread 8411 CPU cycles (245 MFLOPS) long and it runs 356 CPU cycles (5789 MFLOPS) when 128 threads are used. The speedup of original and tangent-linear code is almost linear until 16 threads are used. Further increase in number of threads does not reduce the run time much, the synchronization overhead seams to be the major factor.

The adjoint computation takes with one thread 27284 CPU cycles (88 MFLOPS) and with 128 threads it needs 772 CPU cycles (3126 MFLOPS). The adjoint source transformation doubles the number of barriers in the output code. Therefore, the synchronization overhead of the adjoint execution is twice as high as in the original execution. This is reflected by low MFLOPS values and the speedup does not grow much when we use more than 8 threads.

If the compilation is done with optimization level 3 (-O3), we achieve the results presented in Table 5.9 and the corresponding bar chart is shown in Figure 5.9. The original code execution is performed in 3222 CPU cycles (213 MFLOPS) when only one thread is used and it takes 207 CPU cycles (3316 MFLOPS) with 128 threads. The tangent-linear execution finishes after 5346 CPU cycles (387 MFLOPS) with one thread and when all 128 threads are used, the runtime is down at 376 CPU cycles (5503 MFLOPS). The adjoint code consumes 12429 CPU cycles (168 MFLOPS) with one thread whereby the execution with 128 threads reduces the runtime to 750 CPU cycles (2806 MFLOPS). All three codes have a speedup value of about 13 with 16 threads. Afterwards, further increase in the number of threads shows a stagnation of the speedup.

Original code

#Threads	Runtime (sec)	CPU cycles ($\cdot 10^{-9}$)	Speedup	MFLOPS
1	214	4163	1.0	165
2	108	2137	1.95	321
4	53	1053	3.95	653
8	27	536	7.75	1281
16	15	307	13.55	2240
32	11	225	18.44	3048
64	11	232	17.9	2963
128	12	239	17.38	2880

Tangent-linear code

#Threads	Runtime (sec)	CPU cycles ($\cdot 10^{-9}$)	Speedup	MFLOPS
1	405	7944	1.0	259
2	203	4010	1.98	514
4	102	1996	3.98	1032
8	51	1018	7.8	2024
16	29	580	13.67	3548
32	21	430	18.44	4785
64	20	397	19.98	5186
128	19	387	20.48	5318

Adjoint code

#Threads	Runtime (sec)	CPU cycles ($\cdot 10^{-9}$)	Speedup	MFLOPS
1	2030	37875	1.0	64
2	1062	20027	1.89	121
4	545	10214	3.71	238
8	282	5353	7.08	454
16	159	3038	12.47	800
32	86	1649	22.96	1474
64	46	881	42.95	2759
128	43	812	46.61	2997

Sizes of stacks in gigabyte per thread used in the adjoint code

#Threads	$STACK_c$	$STACK_i$	$STACK_f$	Overall
1	64.0	64.0	128.0	256.0
2	32.0	32.0	64.0	128.0
4	16.0	16.0	32.0	64.0
8	8.0	8.0	16.0	32.0
16	4.0	4.0	8.0	16.0
32	2.0	2.0	4.0	8.0
64	1.0	1.0	2.0	4.0
128	0.5	0.5	1.0	2.0

Table 5.6: Runtime results of the original, the tangent-linear, and the adjoint code of test **'barrier'**, compiled by the **Intel C++ compiler** with optimization level **0**. In addition, the size of the adjoint code stacks are shown in gigabytes. The speedup is computed with respect to CPU cycles not runtime. The test was run with vector $x \in \mathbb{R}^{131072}$ with a memory size of 1 megabytes. The matrix $A \in \mathbb{R}^{131072 \times 131072}$ consumes 128 gigabytes of memory.

#Threads	Runtime (sec)	Original code CPU cycles ($\cdot 10^{-9}$)	Speedup	MFLOPS
1	146	2863	1.0	144
2	71	1380	2.07	297
4	40	803	3.56	515
8	22	442	6.48	937
16	11	236	12.1	1754
32	8	165	17.26	2496
64	8	175	16.28	2328
128	8	170	16.8	2365

#Threads	Runtime (sec)	Tangent-linear code CPU cycles ($\cdot 10^{-9}$)	Speedup	MFLOPS
1	237	4656	1.0	235
2	117	2298	2.03	469
4	65	1291	3.6	849
8	34	686	6.79	1585
16	22	436	10.67	2488
32	19	377	12.34	2868
64	14	289	16.1	3732
128	14	284	16.38	3777

#Threads	Runtime (sec)	Adjoint code CPU cycles ($\cdot 10^{-9}$)	Speedup	MFLOPS
1	941	15633	1.0	143
2	493	7880	1.98	283
4	230	4044	3.87	554
8	143	2128	7.35	1052
16	71	1140	13.71	1965
32	46	834	18.74	2691
64	40	647	24.16	3477
128	35	624	25.01	3605

#Threads	Sizes of stacks in gigabyte per thread used in the adjoint code $STACK_c$	$STACK_i$	$STACK_f$	Overall
1	64.0	64.0	128.0	256.0
2	32.0	32.0	64.0	128.0
4	16.0	16.0	32.0	64.0
8	8.0	8.0	16.0	32.0
16	4.0	4.0	8.0	16.0
32	2.0	2.0	4.0	8.0
64	1.0	1.0	2.0	4.0
128	0.5	0.5	1.0	2.0

Table 5.7: Runtime results of the original, the tangent-linear, and the adjoint code of test **'barrier'**, compiled by the **Intel C++ compiler** with optimization level **3**. In addition, the size of the adjoint code stacks are shown in gigabytes. The speedup is computed with respect to CPU cycles not runtime. The test was run with vector $x \in \mathbb{R}^{131072}$ with a memory size of 1 megabytes. The matrix $A \in \mathbb{R}^{131072 \times 131072}$ consumes 128 gigabytes of memory.

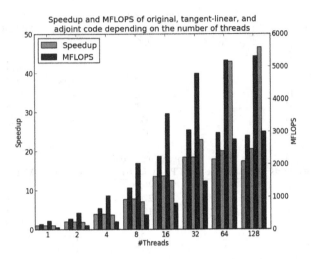

Figure 5.6: Plot for the test 'Barrier O0 icpc'. We refer to Table 5.6 for details.

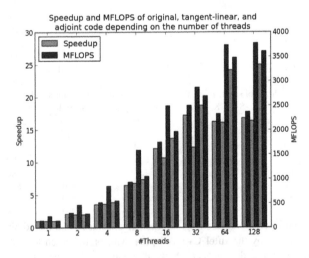

Figure 5.7: Plot for the test 'Barrier O3 icpc'. In Table 5.7, the reader can find the connected values.

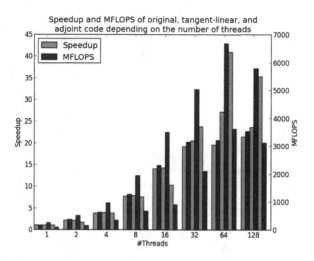

Figure 5.8: Plot for the test 'Barrier O0 g++'. The actual values are shown in Table 5.8.

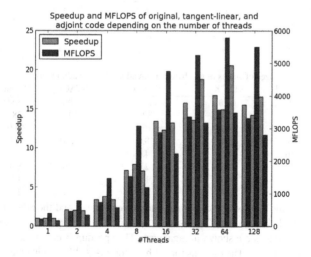

Figure 5.9: Plot for the test 'Barrier O3 g++'. Table 5.9 presents the values of this chart.

Original code

#Threads	Runtime (sec)	CPU cycles ($\cdot 10^{-9}$)	Speedup	MFLOPS
1	215	4195	1.0	163
2	98	1903	2.2	361
4	55	1092	3.84	629
8	28	541	7.75	1270
16	15	298	14.06	2304
32	11	218	19.22	3149
64	10	213	19.61	3214
128	9	195	21.49	3523

Tangent-linear code

#Threads	Runtime (sec)	CPU cycles ($\cdot 10^{-9}$)	Speedup	MFLOPS
1	426	8411	1.0	245
2	206	4081	2.06	505
4	109	2156	3.9	956
8	54	1065	7.89	1934
16	29	589	14.26	3495
32	20	410	20.51	5027
64	15	309	27.22	6672
128	17	356	23.61	5789

Adjoint code

#Threads	Runtime (sec)	CPU cycles ($\cdot 10^{-9}$)	Speedup	MFLOPS
1	1498	27284	1.0	88
2	857	15830	1.72	152
4	388	7116	3.83	338
8	195	3582	7.62	672
16	140	2654	10.28	908
32	61	1144	23.84	2106
64	36	665	40.97	3621
128	40	772	35.34	3126

Sizes of stacks in gigabyte per thread used in the adjoint code

#Threads	$STACK_c$	$STACK_i$	$STACK_f$	Overall
1	64.0	64.0	128.0	256.0
2	32.0	32.0	64.0	128.0
4	16.0	16.0	32.0	64.0
8	8.0	8.0	16.0	32.0
16	4.0	4.0	8.0	16.0
32	2.0	2.0	4.0	8.0
64	1.0	1.0	2.0	4.0
128	0.5	0.5	1.0	2.0

Table 5.8: Runtime results of the original, the tangent-linear, and the adjoint code of test **'barrier'**, compiled by the **GCC (g++)** with optimization level **0**. In addition, the size of the adjoint code stacks are shown in gigabytes. The speedup is computed with respect to CPU cycles not runtime. The test was run with vector $x \in \mathbb{R}^{131072}$ with a memory size of 1 megabytes. The matrix $A \in \mathbb{R}^{131072 \times 131072}$ consumes 128 gigabytes of memory.

Original code

#Threads	Runtime (sec)	CPU cycles ($\cdot 10^{-9}$)	Speedup	MFLOPS
1	166	3222	1.0	213
2	80	1533	2.1	448
4	47	945	3.41	726
8	22	451	7.13	1521
16	12	239	13.45	2870
32	10	204	15.77	3365
64	9	192	16.76	3578
128	10	207	15.52	3316

Tangent-linear code

#Threads	Runtime (sec)	CPU cycles ($\cdot 10^{-9}$)	Speedup	MFLOPS
1	274	5346	1.0	387
2	136	2664	2.01	777
4	71	1411	3.79	1467
8	34	673	7.94	3071
16	21	435	12.29	4754
32	19	393	13.58	5250
64	18	357	14.96	5788
128	19	376	14.21	5503

Adjoint code

#Threads	Runtime (sec)	CPU cycles ($\cdot 10^{-9}$)	Speedup	MFLOPS
1	737	12429	1.0	168
2	367	6206	2.0	336
4	209	3682	3.38	568
8	100	1750	7.1	1193
16	54	939	13.23	2226
32	36	660	18.81	3170
64	32	603	20.59	3486
128	39	750	16.57	2806

Sizes of stacks in gigabyte per thread used in the adjoint code

#Threads	$STACK_c$	$STACK_i$	$STACK_f$	Overall
1	64.0	64.0	128.0	256.0
2	32.0	32.0	64.0	128.0
4	16.0	16.0	32.0	64.0
8	8.0	8.0	16.0	32.0
16	4.0	4.0	8.0	16.0
32	2.0	2.0	4.0	8.0
64	1.0	1.0	2.0	4.0
128	0.5	0.5	1.0	2.0

Table 5.9: Runtime results of the original, the tangent-linear, and the adjoint code of test **'barrier'**, compiled by the **GCC (g++)** with optimization level **3**. In addition, the size of the adjoint code stacks are shown in gigabytes. The speedup is computed with respect to CPU cycles not runtime. The test was run with vector $x \in \mathbb{R}^{131072}$ with a memory size of 1 megabytes. The matrix $A \in \mathbb{R}^{131072 \times 131072}$ consumes 128 gigabytes of memory.

Similar to the binary obtained by Intel compiler, the adjoint code provides the best speedup values. But the best speedup value is not provided with 128 threads but with 64 threads. It seams that the code provided by the Intel compiler can better handle the overhead that is involved by connecting the four physical nodes.

5.1.3 Parallel Region with a master Construct

The listings of this test case can be found in Appendix B.3. The original code is shown Listing B.10 on page 380, the first-order tangent-linear code is contained in Listing B.11 on page 381, the first-order adjoint code is presented in Listing B.12 on page 383.

The original code is almost the same code as the one used in the 'barrier' test case. The computation is a matrix-vector product with subsequent reduction of the thread's results. After the matrix-vector a barrier ensures that all threads have finished their work before the master thread starts the reduction. The only change in this code is that we use the master directive to choose the master thread in contrast to testing that the thread ID is zero.

Intel compiler 14.0:

The optimization level 0 (-O0) provides the results shown in Table 5.10. The corresponding bar chart is shown in Figure 5.10. Comparing Figure 5.10 with the bar chart from the test case 'barrier' it is conspicuous that the current results are much better in terms of scalability. The original code spends 25878 CPU cycles (1315 MFLOPS) processing the data with one thread. In contrast to this, 128 threads reduce the runtime to 226 CPU cycles (150308 MFLOPS). The tangent-linear execution needs 52606 CPU cycles (1381 MFLOPS) with one thread and 452 cycles (160541 MFLOPS) with 128 threads. The runtime of the adjoint code is 118448 CPU cycles (924 MFLOPS) with one thread. In case that 128 threads are processing the data, the execution finishes after 1485 CPU cycles (73755 MFLOPS). Speaking about the scalability, the original and tangent-linear code scale almost the same in terms of speedup values. The adjoint code scales almost linear up to 32 threads. Afterwards, it ends up with a speedup value of 80 with 128 threads.

In Table 5.11, the reader finds the results for the test case with the optimization level 3 (-O3). Figure 5.11 summarizes these results as a bar chart. The original code consumes 8461 CPU cycles (1179 MFLOPS) during the runtime when one thread is used. In case that 128 threads are used the runtime is reduced to 151 CPU cycles (66049 MFLOPS). The tangent-linear code needs 21702 CPU cycles (1761 MFLOPS) for the computation with one thread and with 128 threads it takes 302

CPU cycles (126363 MFLOPS). The adjoint computation with one thread finishes after 48470 CPU cycles (1210 MFLOPS) and 128 threads provide the adjoints after 683 CPU cycles (85070 MFLOPS). According to the scalability one recognizes that the original code and the tangent-linear code scale well until 32 threads. The execution of the adjoint code achieves a speedup factor of 27 with 32 threads and the highest speedup of 71 is achieved with 128 threads.

GCC 4.8.0:

The details of a test run where the binary was compiled without using optimization (-O0) are shown in Table 5.12 and in Figure 5.12. The runtime of the original code is 122932 CPU cycles (182 MFLOPS) with one thread and it takes 961 CPU cycles (23359 MFLOPS) with 128 threads. The tangent-linear execution lasts 250849 CPU cycles (187 MFLOPS) with one thread and 1962 CPU cycles (23923 MFLOPS) with 128 threads. 407296 CPU cycles (172 MFLOPS) are necessary to provide the adjoints with one thread and 128 threads are able to provide the same values in 4570 CPU cycles (15358 MFLOPS).

When the optimization level 3 (-O3) is used to compile the binary we achieve the results displayed in Table 5.13 whereas the corresponding bar chart is shown in Figure 5.13. The original code execution consumes 23719 CPU cycles (581 MFLOPS) with one thread where 128 threads are performing the same computation in 222 CPU cycles (62090 MFLOPS). The tangent-linear runtime with one thread takes 24469 CPU cycles (726 MFLOPS) and 370 cycles (48019 MFLOPS) with 128 threads. The adjoint computation lasts 61907 CPU cycles (521 MFLOPS) with one thread and 767 CPU cycles (42020 MFLOPS) with 128 threads. While comparing the columns where the speedup values are listed, one recognizes that both the derivative codes cannot reach the speedup that the original code achieves. The next test case is a parallel region containing a critical region. This section starts on page 291.

5.1.4 Parallel Region with a Critical Region

The corresponding listings of this test case are contained in Appendix B.4, more precisely the original code is presented in Listing B.13 on page 387, the first-order tangent-linear code can be found in Listing B.14 on page 388, and the first-order adjoint code is presented in Listing B.15 on page 389. The original code implements a matrix-vector product with a succeeding reduction operation that is contained in a critical region.

Original code

#Threads	Runtime (sec)	CPU cycles $(\cdot 10^{-9})$	Speedup	MFLOPS
1	1309	25878	1.0	1315
2	672	13272	1.95	2571
4	339	6569	3.94	5184
8	167	3310	7.82	10304
16	84	1668	15.51	20438
32	42	833	31.04	40901
64	21	423	61.13	80520
128	11	226	114.06	150308

Tangent-linear code

#Threads	Runtime (sec)	CPU cycles $(\cdot 10^{-9})$	Speedup	MFLOPS
1	2674	52606	1.0	1381
2	1396	27408	1.92	2651
4	677	13350	3.94	5442
8	345	6654	7.9	10916
16	168	3331	15.79	21807
32	85	1673	31.44	43420
64	43	855	61.52	84968
128	24	452	116.23	160541

Adjoint code

#Threads	Runtime (sec)	CPU cycles $(\cdot 10^{-9})$	Speedup	MFLOPS
1	6312	118448	1.0	924
2	3227	60542	1.96	1808
4	1603	30545	3.88	3584
8	815	15423	7.68	7099
16	416	7925	14.95	13817
32	218	4196	28.22	26097
64	119	2276	52.03	48112
128	81	1485	79.75	73755

Sizes of stacks in gigabyte per thread used in the adjoint code

#Threads	$STACK_c$	$STACK_i$	$STACK_f$	Overall
1	64.0	64.0	128.0	256.0
2	32.0	32.0	64.0	128.0
4	16.0	16.0	32.0	64.0
8	8.0	8.0	16.0	32.0
16	4.0	4.0	8.0	16.0
32	2.0	2.0	4.0	8.0
64	1.0	1.0	2.0	4.0
128	0.5	0.5	1.0	2.0

Table 5.10: Runtime results of the original, the tangent-linear, and the adjoint code of test **'master'**, compiled by the **Intel C++ compiler** with optimization level **0**. In addition, the size of the adjoint code stacks are shown in gigabytes. The speedup is computed with respect to CPU cycles not runtime. The test was run with vector $x \in \mathbb{R}^{131072}$ with a memory size of 1 megabytes. The matrix $A \in \mathbb{R}^{131072 \times 131072}$ consumes 128 gigabytes of memory.

#Threads	Runtime (sec)	Original code CPU cycles ($\cdot 10^{-9}$)	Speedup	MFLOPS
1	429	8461	1.0	1179
2	210	4171	2.03	2392
4	110	2179	3.88	4579
8	63	1106	7.65	9016
16	28	556	15.21	17941
32	14	282	29.92	35285
64	9	193	43.66	51536
128	7	151	55.89	66049

#Threads	Runtime (sec)	Tangent-linear code CPU cycles ($\cdot 10^{-9}$)	Speedup	MFLOPS
1	1104	21702	1.0	1761
2	553	10937	1.98	3496
4	277	5472	3.97	6988
8	142	2810	7.72	13606
16	69	1375	15.78	27801
32	35	700	30.99	54613
64	21	434	49.99	88100
128	15	302	71.69	126363

#Threads	Runtime (sec)	Adjoint code CPU cycles ($\cdot 10^{-9}$)	Speedup	MFLOPS
1	2604	48470	1.0	1210
2	1366	25054	1.93	2330
4	659	12448	3.89	4678
8	335	6405	7.57	9081
16	173	3305	14.66	17582
32	93	1789	27.09	32480
64	53	989	48.97	58709
128	42	683	70.95	85070

Sizes of stacks in gigabyte per thread used in the adjoint code

#Threads	$STACK_c$	$STACK_i$	$STACK_f$	Overall
1	64.0	64.0	128.0	256.0
2	32.0	32.0	64.0	128.0
4	16.0	16.0	32.0	64.0
8	8.0	8.0	16.0	32.0
16	4.0	4.0	8.0	16.0
32	2.0	2.0	4.0	8.0
64	1.0	1.0	2.0	4.0
128	0.5	0.5	1.0	2.0

Table 5.11: Runtime results of the original, the tangent-linear, and the adjoint code of test **'master'**, compiled by the **Intel C++ compiler** with optimization level **3**. In addition, the size of the adjoint code stacks are shown in gigabytes. The speedup is computed with respect to CPU cycles not runtime. The test was run with vector $x \in \mathbb{R}^{131072}$ with a memory size of 1 megabytes. The matrix $A \in \mathbb{R}^{131072 \times 131072}$ consumes 128 gigabytes of memory.

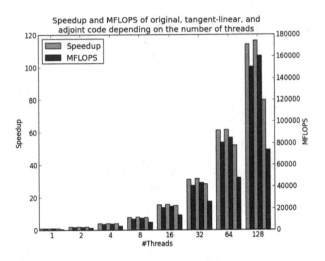

Figure 5.10: Plot for the test 'Master O0 icpc'. The detailed results can be found in Table 5.10. The original code for this test can be found in Appendix B.3 on page 380.

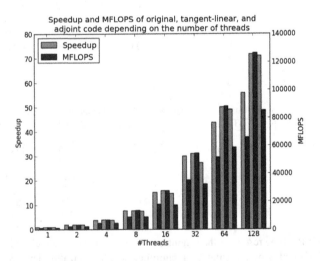

Figure 5.11: Plot for the test 'Master O3 icpc'. The values of this chart are presented in Table 5.11. Appendix B.3 on page 380 displays the corresponding original code for this test case.

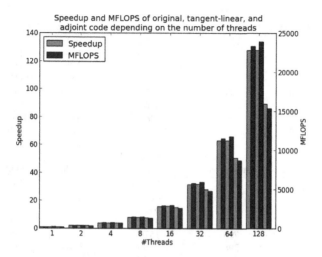

Figure 5.12: Plot for the test 'Master O0 g++'. Table 5.12 displays the corresponding values. The original code for this test can be found in Appendix B.3 on page 380.

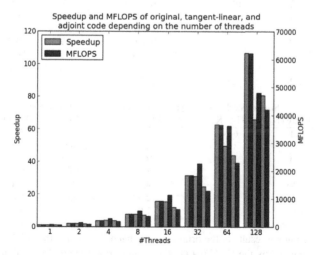

Figure 5.13: Plot for the test 'Master O3 g++'. The values illustrated in this chart can be found in Table 5.13. Appendix B.3 on page 380 displays the corresponding original code for this test case.

| | | Original code | | |
#Threads	Runtime (sec)	CPU cycles $(\cdot 10^{-9})$	Speedup	MFLOPS
1	6216	122932	1.0	182
2	3217	63520	1.94	353
4	1596	31609	3.89	710
8	797	15669	7.85	1433
16	391	7737	15.89	2903
32	196	3892	31.58	5772
64	98	1948	63.1	11535
128	49	961	127.81	23359

| | | Tangent-linear code | | |
#Threads	Runtime (sec)	CPU cycles $(\cdot 10^{-9})$	Speedup	MFLOPS
1	12670	250849	1.0	187
2	6430	127496	1.97	367
4	3250	64232	3.91	730
8	1606	31824	7.88	1474
16	796	15762	15.91	2976
32	398	7918	31.68	5928
64	202	3987	62.91	11774
128	101	1962	127.85	23923

| | | Adjoint code | | |
#Threads	Runtime (sec)	CPU cycles $(\cdot 10^{-9})$	Speedup	MFLOPS
1	20676	407296	1.0	172
2	10673	209124	1.95	335
4	5369	105019	3.88	667
8	2727	53087	7.67	1320
16	1371	26986	15.09	2597
32	738	14574	27.95	4812
64	409	8072	50.46	8691
128	232	4570	89.12	15358

| Sizes of stacks in gigabyte per thread used in the adjoint code | | | | |
#Threads	$STACK_c$	$STACK_i$	$STACK_f$	Overall
1	64.0	64.0	128.0	256.0
2	32.0	32.0	64.0	128.0
4	16.0	16.0	32.0	64.0
8	8.0	8.0	16.0	32.0
16	4.0	4.0	8.0	16.0
32	2.0	2.0	4.0	8.0
64	1.0	1.0	2.0	4.0
128	0.5	0.5	1.0	2.0

Table 5.12: Runtime results of the original, the tangent-linear, and the adjoint code of test **'master'**, compiled by the **GCC (g++)** with optimization level **0**. In addition, the size of the adjoint code stacks are shown in gigabytes. The speedup is computed with respect to CPU cycles not runtime. The test was run with vector $x \in \mathbb{R}^{131072}$ with a memory size of 1 megabytes. The matrix $A \in \mathbb{R}^{131072 \times 131072}$ consumes 128 gigabytes of memory.

		Original code		
#Threads	Runtime (sec)	CPU cycles ($\cdot 10^{-9}$)	Speedup	MFLOPS
1	1198	23719	1.0	581
2	623	12254	1.94	1125
4	315	6158	3.85	2240
8	152	3021	7.85	4565
16	77	1505	15.75	9161
32	38	751	31.58	18365
64	19	377	62.81	36530
128	11	222	106.75	62090

		Tangent-linear code		
#Threads	Runtime (sec)	CPU cycles ($\cdot 10^{-9}$)	Speedup	MFLOPS
1	1239	24469	1.0	726
2	637	12583	1.94	1413
4	315	6215	3.94	2861
8	159	3132	7.81	5678
16	79	1567	15.61	11350
32	39	788	31.04	22564
64	24	491	49.75	36169
128	18	370	66.04	48019

		Adjoint code		
#Threads	Runtime (sec)	CPU cycles ($\cdot 10^{-9}$)	Speedup	MFLOPS
1	3256	61907	1.0	521
2	1880	36147	1.71	892
4	860	16406	3.77	1966
8	448	8568	7.22	3763
16	266	5149	12.02	6260
32	130	2502	24.74	12881
64	73	1409	43.93	22863
128	40	767	80.69	42020

	Sizes of stacks in gigabyte per thread used in the adjoint code			
#Threads	$STACK_c$	$STACK_i$	$STACK_f$	Overall
1	64.0	64.0	128.0	256.0
2	32.0	32.0	64.0	128.0
4	16.0	16.0	32.0	64.0
8	8.0	8.0	16.0	32.0
16	4.0	4.0	8.0	16.0
32	2.0	2.0	4.0	8.0
64	1.0	1.0	2.0	4.0
128	0.5	0.5	1.0	2.0

Table 5.13: Runtime results of the original, the tangent-linear, and the adjoint code of test **'master'**, compiled by the **GCC (g++)** with optimization level **3**. In addition, the size of the adjoint code stacks are shown in gigabytes. The speedup is computed with respect to CPU cycles not runtime. The test was run with vector $x \in \mathbb{R}^{131072}$ with a memory size of 1 megabytes. The matrix $A \in \mathbb{R}^{131072 \times 131072}$ consumes 128 gigabytes of memory.

Intel compiler 14.0:

In Table 5.14 are shown the details of a test run without using compiler optimization (-O0). The corresponding bar chart is Figure 5.14. The processing time for the original code is 4301 CPU cycles (159 MFLOPS) with one thread and 156 CPU cycles (4406 MFLOPS) with 128 threads. The tangent-linear execution needs 7614 CPU cycles (270 MFLOPS) with one thread and 263 CPU cycles (7830 MFLOPS) with 128 threads. The adjoint code consumes during runtime 33951 CPU cycles (71 MFLOPS) with one thread and 1141 CPU cycles (2116 MFLOPS) with 128 threads. These numbers emphasize that the critical section is an expensive construct according to runtime. The synchronization overhead grows with an increasing number of threads. The scalability of the original and the tangent-linear code is well up to 16 threads. Subsequently, a further increase of the number of threads does not improve the runtime much.

In case of the adjoint code we should remind ourselves that the source transformation, described in Section 4.2.3, is defined in a way such that the order in which the threads enter the critical section during the forward section must be tracked and reversed during the reverse section. This makes things even worse in terms of synchronization overhead what one can recognize clearly in the bad speedup results of the adjoint code. A noticeable fact is that the adjoint code reaches 30 as peak speedup value what is better than the speedup value of the original and tangent-linear code with 28 and 29, respectively. Another thing that should be noted is that the gain in performance from doubling the number of threads is similar in case of the adjoint code until we reach 128 threads. For the two other codes it holds that the gain is marginal when more than 32 threads are used.

The bar chart shown in Figure 5.15 summarizes the details of Table 5.15. In the table we can recognize the results that we achieve when one uses the compiler provided optimization (-O3). The original code needs 1679 CPU cycles (243 MFLOPS) for its computation with one thread and it takes 145 CPU cycles (2816 MFLOPS) when 128 threads execute the same code.

The tangent-linear code execution consumes 4112 CPU cycles (264 MFLOPS) with one thread and 277 CPU cycles (3896 MFLOPS) with 128 threads. The adjoint needs the most CPU cycles, namely 13904 CPU cycles (160 MFLOPS) when the code is processed by one thread and 623 CPU cycles (3605 MFLOPS) when 128 threads are used. As in the non-optimized case (-O0) the scalability is poor with a maximal speedup value of 15 for the original and the tangent-linear code. The maximal speedup value of the adjoint code is 22 and therefore better than the values from the original and the tangent-linear code.

The overhead for synchronizing the access to the critical region seems to have the most influence on the parallel execution. The performance of the adjoint code surprises a bit when one thinks of the adjoint code of a critical region. The source transformation rules (4.29) and (4.30) introduce a lot overhead but this overhead does not make things worse what is recognizable from the speedup value of 29.73 and 22.29.

GCC 4.8.0:

Table 5.16 illustrates the detailed results from the non-optimized compilation (-O0). For getting a first impression, the bar chart in Figure 5.16 illustrates the detailed results. The original code needs 3471 CPU cycles (198 MFLOPS) for processing its computation with one thread whereby the same computation takes 155 CPU cycles (4419 MFLOPS) with 128 threads. The runtime of the tangent-linear code consumes 8018 CPU cycles (257 MFLOPS) with one thread and 278 CPU cycles (7402 MFLOPS) with 128 threads. In the adjoint case, we achieve 22351 CPU cycles (107 MFLOPS) when we execute the binary with one thread and if 128 threads are used, the CPU performs 41832 cycles (57 MFLOPS).

The original code and the tangent-linear code provide an acceptable speedup value until the number of threads is 32 but afterwards the speedup stagnates. The data for the adjoint code reveals that an execution with more than eight threads is slower than the sequential execution. This is probably due to the fact that we have to synchronize the order of the threads that enter the critical region.

Table 5.17 displays the results of an execution where we used the optimization level 3 (-O3) to build the binary. Figure 5.17 shows the corresponding bar chart. The original code execution lasts 2900 CPU cycles (237 MFLOPS) with one thread and it consumes 144 CPU cycles (4782 MFLOPS) when 128 threads are used. The computation of the tangent-linear code takes 4483 CPU cycles (461 MFLOPS) with one thread and 270 CPU cycles (7658 MFLOPS) with 128 threads. The adjoint code results are 12420 CPU cycles (170 MFLOPS) with one thread and 20448 CPU cycles (104 MFLOPS) with 128 threads.

Similar to the non-optimized case, the speedup values grow for the original and the tangent-linear code up to 20 and 17, respectively. More than 32 threads does not impact the performance much. The speedup values for the adjoint code are even worse. Until the usage of 16 threads the speedup grows slowly and afterwards the speedup decreases until it ends up with a slowdown of the runtime with 128 threads compared to the sequential execution.

Comparing the results for the adjoint code for the four different runs it is conspicuous that the Intel compiler provides a binary that scales far better than the code supplied by g++. The binary from the Intel compiler scales until 128 threads and even has better speedup values with 128 threads than the original and the tangent-linear code. The MFLOPS values are acceptable when we think of the effort that the adjoint code has to carry out. The binary that the g++ produces seems to have not the best implementation for the critical region. More precisely the scalability drops when the number of threads is higher than eight (Table 5.16) or 16 (Table 5.17). Afterwards, the performance falls down and we achieve even a slowdown comparing to the sequential execution. This shows that the performance of an OpenMP implementation can be very different depending on which compiler one uses.

5.1.5 Parallel Region with atomic Construct

The example code for this test case can be found in Appendix B.5 on page 392. The parallel computation is the same as in the test case called 'critical' except that the reduction operation here is an addition whereby the critical region contained a multiplication. This fact allows us to use atomic construct that is often implemented with hardware support and is therefore more efficient than the software implementation of the critical construct.

Intel compiler 14.0:

Table 5.18 shows the detailed results when we do not use any optimization provided by the compiler (-O0). A bar chart that summarizes the details can be found in Figure 5.18. The original code execution spends 3438 CPU cycles (200 MFLOPS) for the sequential computation while 128 threads finish the computation after 143 CPU cycles (4792 MFLOPS). For its computation, the tangent-linear code needs twice as long, compared to the original code. In detail, this means the CPU consumes 7991 CPU cycles (258 MFLOPS) with one thread and 268 CPU cycles (7674 MFLOPS) with 128 threads. The adjoint code, on the other side, needs 30951 CPU cycles (78 MFLOPS) to compute the adjoint values with one thread. This means that the adjoint code execution is 9 times slower than the execution of the original code. The computation with 128 threads lasts 621 CPU cycles (3932 MFLOPS) which means that the adjoint computation needs just four times as long as the original function evaluation needs with 128 threads.

Due to our source transformation rule, each atomic directive in the original code results in two atomic directives in the tangent-linear code. However, the performance values for the tangent-linear code are better than the values obtained by the original code execution. The peak speedup value for the original code is 24 whereby the tangent-linear code achieves 30.

#Threads	Runtime (sec)	Original code CPU cycles ($\cdot 10^{-9}$)	Speedup	MFLOPS
1	219	4301	1.0	159
2	106	2076	2.07	331
4	55	1090	3.94	630
8	29	581	7.4	1182
16	15	299	14.35	2293
32	10	210	20.47	3272
64	8	175	24.49	3915
128	7	156	27.54	4406

#Threads	Runtime (sec)	Tangent-linear code CPU cycles ($\cdot 10^{-9}$)	Speedup	MFLOPS
1	386	7614	1.0	270
2	203	3994	1.91	516
4	105	2085	3.65	988
8	55	1076	7.07	1915
16	29	586	12.99	3518
32	17	352	21.61	5852
64	16	326	23.32	6318
128	13	263	28.89	7830

#Threads	Runtime (sec)	Adjoint code CPU cycles ($\cdot 10^{-9}$)	Speedup	MFLOPS
1	2011	33951	1.0	71
2	1179	20701	1.64	116
4	759	14123	2.4	170
8	357	6840	4.96	352
16	254	4746	7.15	508
32	159	2984	11.38	808
64	102	1861	18.24	1296
128	63	1141	29.75	2116

Sizes of stacks in gigabyte per thread used in the adjoint code

#Threads	$STACK_c$	$STACK_i$	$STACK_f$	Overall
1	64.0	64.0	128.0	256.0
2	32.0	32.0	64.0	128.0
4	16.0	16.0	32.0	64.0
8	8.0	8.0	16.0	32.0
16	4.0	4.0	8.0	16.0
32	2.0	2.0	4.0	8.0
64	1.0	1.0	2.0	4.0
128	0.5	0.5	1.0	2.0

Table 5.14: Runtime results of the original, the tangent-linear, and the adjoint code of test **'critical'**, compiled by the **Intel C++ compiler** with optimization level **0**. In addition, the size of the adjoint code stacks are shown in gigabytes. The speedup is computed with respect to CPU cycles not runtime. The test was run with vector $x \in \mathbb{R}^{131072}$ with a memory size of 1 megabytes. The matrix $A \in \mathbb{R}^{131072 \times 131072}$ consumes 128 gigabytes of memory.

Original code

#Threads	Runtime (sec)	CPU cycles ($\cdot 10^{-9}$)	Speedup	MFLOPS
1	86	1679	1.0	243
2	73	1408	1.19	291
4	39	772	2.17	536
8	22	445	3.77	933
16	10	202	8.29	2039
32	6	122	13.7	3389
64	7	141	11.86	2923
128	7	145	11.57	2816

Tangent-linear code

#Threads	Runtime (sec)	CPU cycles ($\cdot 10^{-9}$)	Speedup	MFLOPS
1	210	4112	1.0	264
2	117	2306	1.78	468
4	65	1274	3.23	853
8	35	701	5.86	1556
16	21	419	9.81	2592
32	12	243	16.9	4455
64	14	280	14.65	3853
128	14	277	14.83	3896

Adjoint code

#Threads	Runtime (sec)	CPU cycles ($\cdot 10^{-9}$)	Speedup	MFLOPS
1	954	13904	1.0	160
2	447	7557	1.84	295
4	262	4697	2.96	476
8	148	2673	5.2	839
16	90	1673	8.31	1341
32	60	1098	12.66	2044
64	40	722	19.25	3114
128	36	623	22.29	3605

Sizes of stacks in gigabyte per thread used in the adjoint code

#Threads	$STACK_c$	$STACK_i$	$STACK_f$	Overall
1	64.0	64.0	128.0	256.0
2	32.0	32.0	64.0	128.0
4	16.0	16.0	32.0	64.0
8	8.0	8.0	16.0	32.0
16	4.0	4.0	8.0	16.0
32	2.0	2.0	4.0	8.0
64	1.0	1.0	2.0	4.0
128	0.5	0.5	1.0	2.0

Table 5.15: Runtime results of the original, the tangent-linear, and the adjoint code of test 'critical', compiled by the **Intel C++ compiler** with optimization level **3**. In addition, the size of the adjoint code stacks are shown in gigabytes. The speedup is computed with respect to CPU cycles not runtime. The test was run with vector $x \in \mathbb{R}^{131072}$ with a memory size of 1 megabytes. The matrix $A \in \mathbb{R}^{131072 \times 131072}$ consumes 128 gigabytes of memory.

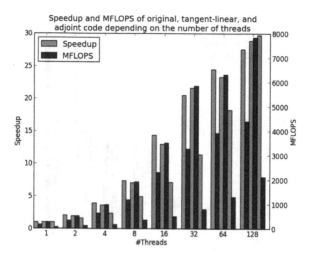

Figure 5.14: Plot for the test 'Critical O0 intel'. The corresponding values can be found in Table 5.14. The original code for this test can be found in Appendix B.4 on page 387.

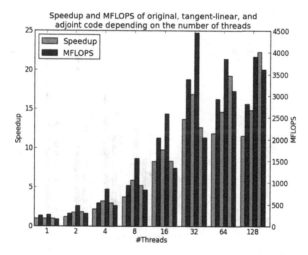

Figure 5.15: Plot for the test 'Critical O3 intel'. The reader can find the details in Table 5.15 useful. Appendix B.4 on page 387 displays the corresponding original code for this test case.

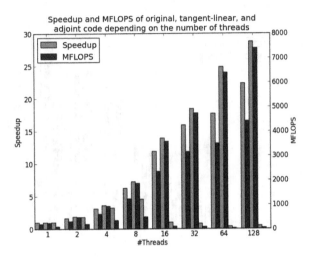

Figure 5.16: Plot for the test 'Critical O0 g++'. We refer to Table 5.16 for details. The original code for this test can be found in Appendix B.4 on page 387.

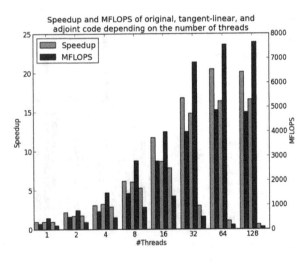

Figure 5.17: Plot for the test 'Critical O3 g++'. Table 5.17 represents the actual values of this chart. Appendix B.4 on page 387 displays the corresponding original code for this test case.

Original code

#Threads	Runtime (sec)	CPU cycles ($\cdot 10^{-9}$)	Speedup	MFLOPS
1	176	3471	1.0	198
2	109	2143	1.62	320
4	56	1116	3.11	615
8	28	558	6.22	1231
16	15	292	11.88	2354
32	11	218	15.92	3154
64	9	196	17.69	3505
128	7	155	22.29	4419

Tangent-linear code

#Threads	Runtime (sec)	CPU cycles ($\cdot 10^{-9}$)	Speedup	MFLOPS
1	407	8018	1.0	257
2	218	4292	1.87	480
4	111	2210	3.63	932
8	56	1101	7.28	1872
16	29	576	13.92	3579
32	22	435	18.42	4736
64	16	322	24.85	6393
128	14	278	28.76	7402

Adjoint code

#Threads	Runtime (sec)	CPU cycles ($\cdot 10^{-9}$)	Speedup	MFLOPS
1	1406	22351	1.0	107
2	676	12201	1.83	197
4	372	6821	3.28	353
8	275	4867	4.59	495
16	1153	22495	0.99	107
32	1439	28304	0.79	85
64	3226	63135	0.35	38
128	2288	41832	0.53	57

Sizes of stacks in gigabyte per thread used in the adjoint code

#Threads	STACK$_c$	STACK$_i$	STACK$_f$	Overall
1	64.0	64.0	128.0	256.0
2	32.0	32.0	64.0	128.0
4	16.0	16.0	32.0	64.0
8	8.0	8.0	16.0	32.0
16	4.0	4.0	8.0	16.0
32	2.0	2.0	4.0	8.0
64	1.0	1.0	2.0	4.0
128	0.5	0.5	1.0	2.0

Table 5.16: Runtime results of the original, the tangent-linear, and the adjoint code of test **'critical'**, compiled by the **GCC (g++)** with optimization level **0**. In addition, the size of the adjoint code stacks are shown in gigabytes. The speedup is computed with respect to CPU cycles not runtime. The test was run with vector $x \in \mathbb{R}^{131072}$ with a memory size of 1 megabytes. The matrix $A \in \mathbb{R}^{131072 \times 131072}$ consumes 128 gigabytes of memory.

Original code

#Threads	Runtime (sec)	CPU cycles ($\cdot 10^{-9}$)	Speedup	MFLOPS
1	150	2900	1.0	237
2	67	1325	2.19	518
4	47	943	3.08	728
8	23	470	6.16	1461
16	12	247	11.73	2782
32	8	172	16.78	3981
64	7	141	20.46	4859
128	7	144	20.1	4782

Tangent-linear code

#Threads	Runtime (sec)	CPU cycles ($\cdot 10^{-9}$)	Speedup	MFLOPS
1	228	4483	1.0	461
2	134	2629	1.71	786
4	70	1381	3.25	1498
8	37	740	6.05	2792
16	26	519	8.62	3978
32	15	302	14.8	6834
64	13	274	16.35	7551
128	13	270	16.56	7658

Adjoint code

#Threads	Runtime (sec)	CPU cycles ($\cdot 10^{-9}$)	Speedup	MFLOPS
1	832	12420	1.0	170
2	412	6965	1.78	304
4	244	4341	2.86	489
8	131	2353	5.28	904
16	86	1585	7.83	1343
32	210	4089	3.04	520
64	590	11538	1.08	184
128	1142	20448	0.61	104

Sizes of stacks in gigabyte per thread used in the adjoint code

#Threads	$STACK_c$	$STACK_i$	$STACK_f$	Overall
1	64.0	64.0	128.0	256.0
2	32.0	32.0	64.0	128.0
4	16.0	16.0	32.0	64.0
8	8.0	8.0	16.0	32.0
16	4.0	4.0	8.0	16.0
32	2.0	2.0	4.0	8.0
64	1.0	1.0	2.0	4.0
128	0.5	0.5	1.0	2.0

Table 5.17: Runtime results of the original, the tangent-linear, and the adjoint code of test **'critical'**, compiled by the **GCC (g++)** with optimization level **3**. In addition, the size of the adjoint code stacks are shown in gigabytes. The speedup is computed with respect to CPU cycles not runtime. The test was run with vector $x \in \mathbb{R}^{131072}$ with a memory size of 1 megabytes. The matrix $A \in \mathbb{R}^{131072 \times 131072}$ consumes 128 gigabytes of memory.

This is different if we consider the adjoint code. As we have shown in Section 4.2.4, the source transformation of an atomic construct introduces two critical regions, one in the forward section and the other one in the reverse section. Therefore, one may assume that the synchronization overhead is even higher than in the tangent-linear code. But the adjoint code has a maximal speedup value of 50 if the number of threads is 128. This is quite good before the background of the synchronization overhead of the two critical regions.

When we use the optimization level 3 (-O3), we achieve the results illustrated in Table 5.19. The corresponding bar chart is shown in Figure 5.19. The execution of the original code leads to 2769 CPU cycles (149 MFLOPS) if we use one thread and 148 CPU cycles (2735 MFLOPS) are necessary when we use 128 threads. The tangent-linear execution with one thread lasts 2624 CPU cycles (402 MFLOPS). In case we use 128 threads, the CPU needs 272 cycles (3891 MFLOPS) to process the tangent-linear code. The computation of the adjoint values with one thread takes 12585 CPU cycles (180 MFLOPS) and the parallel execution with 128 threads needs 602 CPU cycles (3837 MFLOPS). The columns where the speedup results are shown reveal a maximal speedup value of 19 for the original code. The peak speedup for tangent-linear code is only 10 and the adjoint code achieves 21. It should be mentioned that the speedup values for the adjoint code are better than the speedup values from the original and the tangent-linear code.

GCC 4.8.0:

If we do not use any optimization (-O0) for compiling the binary, we obtain the results displayed in Table 5.20. These results are additionally illustrated as a bar chart in Figure 5.20. The original code needs 3859 CPU cycles (178 MFLOPS) for finishing its computation when one thread is used. The parallel computation with 128 threads lasts 150 CPU cycles (4576 MFLOPS). The tangent-linear code takes 8309 CPU cycles (248 MFLOPS) if we use one thread and 256 CPU cycles (8043 MFLOPS) in case that we use 128 threads. The execution of the adjoint code with one thread consumes 21889 CPU cycles (110 MFLOPS) whereby the parallel computation with 128 threads reduces the runtime to 687 CPU cycles (3518 MFLOPS). All three codes have their peak speedup value with 64 threads.

The optimization level 3 of GCC (-O3) provides the runtime results shown in Table 5.21. The corresponding figure in form of a bar chart is shown in Figure 5.21. The original code execution lasts 1779 CPU cycles (386 MFLOPS) with one thread and the same computation performed by 128 threads reduces the runtime to 153 CPU cycles (4495 MFLOPS). The tangent-linear computation with one thread finishes after 3094 CPU cycles (667 MFLOPS) whereby 128 threads finish the execution after 239 CPU cycles (8660 MFLOPS). The adjoint code needs 11929 CPU cycles (175 MFLOPS) if it is executed with one thread. With 128 threads the computation consumes 737 CPU cycles (2918 MFLOPS).

It is striking that the speedup values for the adjoint code is clearly above the speedup values from the original and the tangent-linear code. A possible explanation is that the critical region introduced by the source transformation[3], does not have a big impact but the fact that we do not store all the intermediate results obviously saves a lot of stack operations and improves therefore the performances.

This concludes the test suite and we continue with presenting the runtime results of three second derivative codes in Section 5.2 on page 315. As an example, we use the 'plain-parallel' test case from Section 5.1.1. In detail, we will see results for the forward-over-forward mode, the forward-over-reverse mode, and the reverse-over-forward mode.

---

[3] See source transformation rules (4.39) and (4.40).

#Threads	Runtime (sec)	Original code CPU cycles ($\cdot 10^{-9}$)	Speedup	MFLOPS
1	174	3438	1.0	200
2	107	2108	1.63	326
4	55	1092	3.15	629
8	27	540	6.36	1272
16	14	275	12.48	2496
32	9	186	18.46	3693
64	8	159	21.58	4319
128	7	143	23.93	4792

#Threads	Runtime (sec)	Tangent-linear code CPU cycles ($\cdot 10^{-9}$)	Speedup	MFLOPS
1	406	7991	1.0	258
2	207	4075	1.96	505
4	103	2022	3.95	1019
8	53	1054	7.58	1956
16	26	522	15.28	3943
32	18	367	21.74	5611
64	13	268	29.76	7683
128	13	268	29.72	7674

#Threads	Runtime (sec)	Adjoint code CPU cycles ($\cdot 10^{-9}$)	Speedup	MFLOPS
1	1815	30951	1.0	78
2	965	16157	1.92	151
4	432	7934	3.9	307
8	231	4067	7.61	600
16	125	2328	13.29	1048
32	65	1180	26.21	2067
64	39	694	44.54	3514
128	34	621	49.81	3932

Sizes of stacks in gigabyte per thread used in the adjoint code

#Threads	STACK_c	STACK_i	STACK_f	Overall
1	64.0	64.0	128.0	256.0
2	32.0	32.0	64.0	128.0
4	16.0	16.0	32.0	64.0
8	8.0	8.0	16.0	32.0
16	4.0	4.0	8.0	16.0
32	2.0	2.0	4.0	8.0
64	1.0	1.0	2.0	4.0
128	0.5	0.5	1.0	2.0

Table 5.18: Runtime results of the original, the tangent-linear, and the adjoint code of test **'atomic'**, compiled by the **Intel C++ compiler** with optimization level **0**. In addition, the size of the adjoint code stacks are shown in gigabytes. The speedup is computed with respect to CPU cycles not runtime. The test was run with vector $x \in \mathbb{R}^{131072}$ with a memory size of 1 megabytes. The matrix $A \in \mathbb{R}^{131072 \times 131072}$ consumes 128 gigabytes of memory.

Original code

#Threads	Runtime (sec)	CPU cycles ($\cdot 10^{-9}$)	Speedup	MFLOPS
1	141	2769	1.0	149
2	75	1491	1.86	278
4	41	823	3.36	504
8	19	390	7.1	1064
16	11	217	12.71	1909
32	7	153	18.05	2702
64	7	147	18.78	2797
128	7	148	18.68	2735

Tangent-linear code

#Threads	Runtime (sec)	CPU cycles ($\cdot 10^{-9}$)	Speedup	MFLOPS
1	134	2624	1.0	402
2	121	2370	1.11	451
4	66	1318	1.99	812
8	33	653	4.02	1635
16	19	393	6.67	2710
32	16	329	7.96	3221
64	13	264	9.91	4008
128	13	272	9.64	3891

Adjoint code

#Threads	Runtime (sec)	CPU cycles ($\cdot 10^{-9}$)	Speedup	MFLOPS
1	789	12585	1.0	180
2	395	6560	1.92	346
4	201	3459	3.64	657
8	99	1718	7.32	1324
16	56	981	12.82	2320
32	36	640	19.65	3565
64	33	586	21.45	3906
128	32	602	20.9	3837

Sizes of stacks in gigabyte per thread used in the adjoint code

#Threads	$STACK_c$	$STACK_i$	$STACK_f$	Overall
1	64.0	64.0	128.0	256.0
2	32.0	32.0	64.0	128.0
4	16.0	16.0	32.0	64.0
8	8.0	8.0	16.0	32.0
16	4.0	4.0	8.0	16.0
32	2.0	2.0	4.0	8.0
64	1.0	1.0	2.0	4.0
128	0.5	0.5	1.0	2.0

Table 5.19: Runtime results of the original, the tangent-linear, and the adjoint code of test **'atomic'**, compiled by the **Intel C++ compiler** with optimization level **3**. In addition, the size of the adjoint code stacks are shown in gigabytes. The speedup is computed with respect to CPU cycles not runtime. The test was run with vector $x \in \mathbb{R}^{131072}$ with a memory size of 1 megabytes. The matrix $A \in \mathbb{R}^{131072 \times 131072}$ consumes 128 gigabytes of memory.

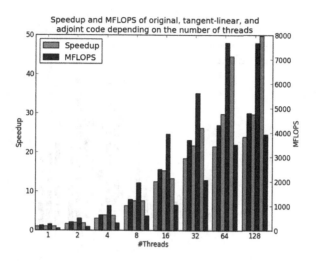

Figure 5.18: Plot for the test 'Atomic O0 intel'. The details are illustrated in Table 5.18. The original code for this test can be found in Appendix B.5 on page 392.

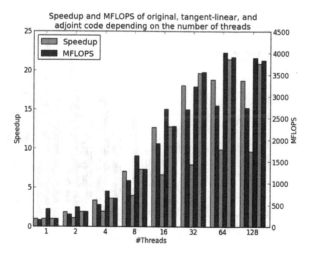

Figure 5.19: Plot for the test 'Atomic O3 intel'. Table 5.19 shows the details of this test. Appendix B.5 on page 392 displays the corresponding original code for this test case.

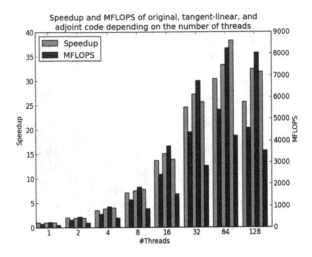

Figure 5.20: Plot for the test 'Atomic O0 g++'. The details are illustrated in Table 5.20. The original code for this test can be found in Appendix B.5 on page 392.

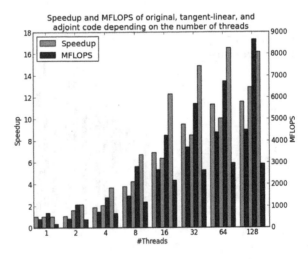

Figure 5.21: Plot for the test 'Atomic O3 g++'. Table 5.21 shows the details of this test. Appendix B.5 on page 392 displays the corresponding original code for this test case.

Original code

#Threads	Runtime (sec)	CPU cycles ($\cdot 10^{-9}$)	Speedup	MFLOPS
1	196	3859	1.0	178
2	99	1967	1.96	349
4	55	1108	3.48	620
8	27	542	7.12	1268
16	14	281	13.7	2440
32	7	156	24.59	4381
64	6	126	30.43	5421
128	7	150	25.67	4576

Tangent-linear code

#Threads	Runtime (sec)	CPU cycles ($\cdot 10^{-9}$)	Speedup	MFLOPS
1	421	8309	1.0	248
2	219	4303	1.93	479
4	109	2160	3.85	954
8	56	1113	7.46	1852
16	27	550	15.08	3741
32	15	305	27.22	6753
64	12	250	33.23	8247
128	13	256	32.4	8043

Adjoint code

#Threads	Runtime (sec)	CPU cycles ($\cdot 10^{-9}$)	Speedup	MFLOPS
1	1321	21889	1.0	110
2	634	11221	1.95	214
4	309	5522	3.96	436
8	157	2783	7.86	865
16	87	1575	13.9	1530
32	49	854	25.63	2823
64	33	572	38.22	4214
128	37	687	31,85	3518

Sizes of stacks in gigabyte per thread used in the adjoint code

#Threads	$STACK_c$	$STACK_i$	$STACK_f$	Overall
1	64.0	64.0	128.0	256.0
2	32.0	32.0	64.0	128.0
4	16.0	16.0	32.0	64.0
8	8.0	8.0	16.0	32.0
16	4.0	4.0	8.0	16.0
32	2.0	2.0	4.0	8.0
64	1.0	1.0	2.0	4.0
128	0.5	0.5	1.0	2.0

Table 5.20: Runtime results of the original, the tangent-linear, and the adjoint code of test **'atomic'**, compiled by the **GCC (g++)** with optimization level **0**. In addition, the size of the adjoint code stacks are shown in gigabytes. The speedup is computed with respect to CPU cycles not runtime. The test was run with vector $x \in \mathbb{R}^{131072}$ with a memory size of 1 megabytes. The matrix $A \in \mathbb{R}^{131072 \times 131072}$ consumes 128 gigabytes of memory.

Original code

#Threads	Runtime (sec)	CPU cycles ($\cdot 10^{-9}$)	Speedup	MFLOPS
1	92	1779	1.0	386
2	86	1692	1.05	406
4	48	951	1.87	722
8	23	469	3.79	1464
16	13	257	6.91	2670
32	9	186	9.54	3687
64	7	157	11.32	4384
128	7	153	11.58	4495

Tangent-linear code

#Threads	Runtime (sec)	CPU cycles ($\cdot 10^{-9}$)	Speedup	MFLOPS
1	158	3094	1.0	667
2	98	1921	1.61	1075
4	75	1484	2.08	1393
8	37	733	4.22	2821
16	24	486	6.36	4253
32	18	362	8.53	5706
64	15	307	10.06	6733
128	12	239	12.91	8660

Adjoint code

#Threads	Runtime (sec)	CPU cycles ($\cdot 10^{-9}$)	Speedup	MFLOPS
1	719	11929	1.0	175
2	442	5635	2.12	371
4	191	3234	3.69	650
8	111	1772	6.73	1185
16	63	969	12.3	2171
32	49	800	14.9	2644
64	43	720	16.55	2964
128	45	737	16.18	2918

Sizes of stacks in gigabyte per thread used in the adjoint code

#Threads	$STACK_c$	$STACK_i$	$STACK_f$	Overall
1	64.0	64.0	128.0	256.0
2	32.0	32.0	64.0	128.0
4	16.0	16.0	32.0	64.0
8	8.0	8.0	16.0	32.0
16	4.0	4.0	8.0	16.0
32	2.0	2.0	4.0	8.0
64	1.0	1.0	2.0	4.0
128	0.5	0.5	1.0	2.0

Table 5.21: Runtime results of the original, the tangent-linear, and the adjoint code of test **'atomic'**, compiled by the **GCC (g++)** with optimization level **3**. In addition, the size of the adjoint code stacks are shown in gigabytes. The speedup is computed with respect to CPU cycles not runtime. The test was run with vector $x \in \mathbb{R}^{131072}$ with a memory size of 1 megabytes. The matrix $A \in \mathbb{R}^{131072 \times 131072}$ consumes 128 gigabytes of memory.

5.2 Second Derivative Codes

We showed the experimental results from the test suite in a very detailed form. This section presents the second derivative codes from the test case 'plain-parallel'. The example code is presented in Appendix B.1 on page 342 and the results of the first derivative code can be found in Section 5.1.1.

We used the tool SPLc (see Appendix A) that implements the approach of this work to achieve the second derivative code. SPLc applied to a parallel region P generates the first derivative code of P. Afterwards, we reapply the SPLc to its own output to obtain the second derivative code. To recap, our tangent-linear source transformation from Section 2.3 is referred to as τ and the adjoint source transformation is denoted by σ. The forward-over-forward mode of a parallel region P is obtained by $\tau \circ \tau(P)$ where $\circ$ represents the composition of two source transformations. $\tau \circ \sigma(P)$ results in the second-order adjoint code in forward-over-reverse mode. The second-order adjoint code in reverse-over-forward mode is obtained by $\sigma \circ \tau(P)$.

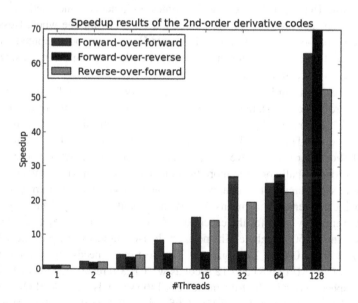

Figure 5.22: This bar chart shows the speedup results for the second derivative codes of the test suite example 'plain-parallel', see Section 5.1.1. The original code of this test case is shown in Appendix B.1. The Intel compiler was used with optimization level 3 (-O3).

Figure 5.22 summarizes the runtime results for the second derivative codes of the 'plain-parallel' example code. The binary for this test was created by the Intel compiler with optimization level 3. The corresponding values are presented in Table 5.22. Each bar represents one execution of the corresponding second derivative code. The x-axis shows the number of threads, the y-axis shows the speedup values of the execution. The red bar stands for one evaluation of the second-order tangent-linear code. The blue bar represents the second-order adjoint execution in forward-over-reverse mode while the yellow bar illustrates the reverse-over-forward mode. We did not consider the reverse-over-reverse mode due to its huge memory usage.

In order to set the results from Figure 5.22 and Table 5.22 into the right context, we remind the reader that the second-order models have different output dimensions. Let us assume that the original function is $F : \mathbb{R}^n \to \mathbb{R}^m$. According to Definition 10, the second-order tangent-linear model has the output dimension m. The second-order adjoint model, on the other side, has the output dimension n as determined in Definition 11. Depending on the actual values for n and m, the choice of the second-order model can have a big impact. For example, in case that m is one, the Hessian is a matrix with n times n elements. One needs n^2 evaluations of the second-order tangent-linear model to compute the whole Hessian. With the second-order adjoint model the whole Hessian can be computed with n evaluations. These remarks emphasize that the first impression when considering Figure 5.22 is no reason to jump to conclusions.

The second and third column in Table 5.22 show the results for the forward-over-forward mode (f-o-f), the third and fourth column contains the results for the forward-over-reverse mode (f-o-r), and the last two columns illustrates the reverse-over-forward (r-o-f) results. The input size n for the f-o-f and the f-o-r mode is 100.000. For the r-o-f code we had to reduce the size to 32.000 because otherwise the process shall consume more than one terabyte of memory what exceeds the system's resource. The code example uses a two-dimensional array A with 10^{10} double elements. This means A consumes about 75 gigabyte memory. The maximum memory peak for the execution of the f-o-f mode is 300 gigabyte, the f-o-r mode needs 500 gigabyte. The peak for the r-o-f mode where we reduced the input size n to 32.000, is 200 gigabyte. The f-o-f mode needs 758 seconds with one thread (819 MFLOPS) and 12 seconds with 128 threads (72727 MLFOPS). This is a speedup of 63. The f-o-r mode lasts 1466 seconds (1268 MFLOPS) with one thread and 21 seconds (68425 MFLOPS) with 128 threads which represents a speedup factor of 70. Due to the reduced input size of the r-o-f mode, we cannot really compare this mode with the other results. Nevertheless, we present the results for the sake of completeness. With one thread the execution of the r-o-f mode

#Threads	Forward-over-forward (n=100.000)		Forward-over-reverse (n=100.000)		Reverse-over-forward (n=32.000)	
	MFLOPS	Time	MFLOPS	Time	MFLOPS	Time
1	819	758	1268	1466	1436	158
2	1602	352	2467	752	2807	80
4	3192	177	4008	421	5668	39
8	6352	90	5275	324	11399	21
16	12842	50	6220	295	22530	11
32	25941	28	7168	279	35065	8
64	22465	30	27627	53	32716	7
128	72727	12	68425	21	89549	3

Table 5.22: This tables shows the runtime results for the second-order derivative code of the 'plain-parallel' test case. The backend compiler was Intel and the optimization flag we used was O3.

needs 158 seconds (1436 MFLOPS), with 128 threads the runtime is reduced to 3 seconds (89549 MFLOPS). This is a speedup value of 52.

This clearly favors the f-o-r mode instead of the r-o-f mode. The memory usage is in the r-o-f case much bigger. The best performance in terms of MFLOPS achieves the f-o-f mode. But one has to keep in mind that the complexity of the second-order tangent-linear model is different to the second-order adjoint model. Therefore, the actual performance can be quite different depending on the number of input and output variables of the original function.

Another fact that arise when one considers Figure 5.22 is that it makes quite a difference when the execution uses more than 32 threads. In this case there are more than only one physical node involved in the execution. The speedup values of the f-o-r code does not grow much when using more than eight and less or equal to 32 threads. As soon as one uses 64 or 128 threads the speedup grows fast what probably is connected with the distribution of the data among more than one physical node.

5.3 Exclusive Read Analysis

In Section 2.4.2 we introduced the exclusive read property to reach closure of the adjoint source transformation. The AD tool has to be conservative in the sense that without any additional analysis information it has to assume that its input code P does not fulfill the exclusive read property. This fact leads to an adjoint code that probably contains unnecessary atomic constructs. For this reason, we developed the exclusive read analysis (ERA) in Chapter 3. The information provided by the static analysis supports the AD tool in the decision whether an adjoint statement has to be synchronized or not. In the current section we compare the runtime results of the adjoint code created with the support of the ERA information and the adjoint code that is generated by the conservative assumption that every adjoint assignment has to be synchronized.

As example code for this section serves Listing 3.1 on page 139. In this code we consider the two-dimensional array A as a data structure for storing coefficients of polynomials of degree four. The code consist of two loops, the outer loop is responsible for computing different polynomial values, the inner loop takes the current value for x[i] and calculates the value of the polynomial. The method to compute the polynomial value is a very simple one and some multiplications can be spared by using Horners method instead.

The read access to x[i] is an exclusive read for each thread as the value of index variable i is achieved by data decomposition. This means that no component of x is read by more than one thread. If the source transformation tool does not use the ERA, it cannot obtain the information whether x[i] is only used by one or by more threads. Therefore, the tool assumes conservatively that multiple threads read the memory reference x[i]. The lack of the exclusive read property for the original code leads to a storage operation of multiple threads to the same memory location in the reverse section of the adjoint code. The corresponding race condition has to be synchronized by an atomic construct.

As for the test suite, we used the compiler g++ and icpc to obtain the binaries for this test. In addition, the two common options O0 and O3 were used to create two binaries, one without using any code optimization and one with using it. This results in four different binaries. The setups for these binaries are once applied to the code that SPLc provides while using the ERA and once where SPLc does not use the static analysis. Eventually, this results in eight different binaries. Figure 5.23 shows the results as a bar chart and the detailed runtime results are explained in the following two paragraphs.

Intel compiler 14.0:   Let us first consider the runtime results for the binary obtained without backend code optimization (O0). The results where we did not use ERA are displayed in Table 5.23 whereby Table 5.24 shows the corresponding results of using ERA before the source transformation. The last column shows improvement between these two tables. The improvement values lies between 1.51

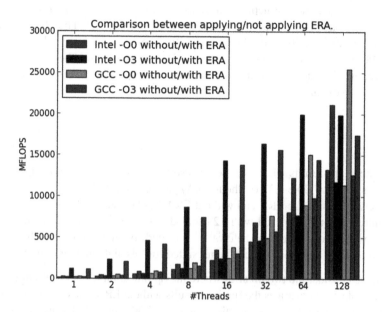

Figure 5.23: This bar chart summarizes the runtime measurements of this test case. The bars are grouped into pairs of the same color. Considering a pair of bars with the same color, the first bar is the runtime without using ERA, and the second bar is the results for the adjoint code obtained with ERA. The first four bars (red and blue) show results from the Intel compiler whereby the last four bars (yellow and green) display the corresponding results obtained by compiling with the g++. Both compilers were called once without optimization (-O0) and once with optimization (-O3).

and 1.59 for all the parallel executions and the best improvement is achieved with 128 threads.

The results where the code optimization (O3) is used are shown in Table 5.26 and Table 5.25. The improvements are completely different as the one with optimization level 0. The best improvement where the execution was over seven times faster (7.74) was achieved during the execution with only one thread. Then the value of the improvement falls until it reaches a factor of 1.69 at the execution with 128 threads.

The reason why the speedup factor declines with an increasing number of threads is likely the case because the assembler instruction that implements the atomic statement needs more CPU cycles than the regular instruction. For example, let us assume that the atomic instruction consumes two CPU cycles and the regular instruction uses one cycle. In case that these instructions are executed hundred times the overall runtime is 200 and 100, respectively. Therefore, we have a difference of 50 CPU cycles between the two executions. With increasing number of threads, this difference is reduced and therefore also the corresponding improvement between the two executions.

g++ 4.8.0: The results where we did not use any code optimization of the g++ (O0) are shown in Table 5.27 (without ERA) and in Table 5.28 (with ERA). The ration between using the ERA and without ERA lies between 1.50 and 2.23. The best improvement is achieved with 128 threads. Interesting is the fact that the runtime of 54 seconds is clearly below the 74 seconds of the Intel binary. The g++ seems to produce code that is much better suited for this high number of threads.

Next, we consider the two binaries where the optimization level 3 (-O3) of g++ is used. Table 5.29 illustrates the runtime results without ERA, the corresponding results where we use ERA are contained in Table 5.30. As with the Intel binary, we achieve the peak improvement while using one thread. Here the improvement factor is 5.54 against the 7.74 of the Intel binary. The improvement values go down until they reach 1.38 when the execution uses 128 threads.

In summary, the average improvement factor of 3.11 is a strong argument for using the ERA before the source transformation in order to create a more efficient adjoint code.

#Threads	Speedup	MFLOPS	Time (sec)
1	1.00	144	8179.88
2	1.93	279	4192.75
4	3.86	559	2121.86
8	7.81	1130	1020.59
16	15.75	2281	521.7068
32	31.15	4512	270.2727
64	55.69	8068	156.1339
128	91.43	13250	106.4187

Table 5.23: SPLc *without* ERA, *Intel* with *-O0*.

#Threads	Speedup	MFLOPS	Time (sec)	Improvement
1	1.00	224	5540.15	1.55
2	1.96	440	2795.94	1.57
4	3.95	887	1323.95	1.58
8	7.80	1751	697.83	1.54
16	15.62	3508	343.6312	1.53
32	30.44	6837	180.6267	1.51
64	54.61	12268	100.4416	1.52
128	93.92	21100	74.2759	1.59

Table 5.24: SPLc *with* ERA, *Intel* with *-O0*. The improvement value refers to the corresponding MFLOPS value in Table 5.23.

#Threads	Speedup	MFLOPS	Time (sec)
1	1.00	157	5162.30
2	1.95	308	2679.30
4	3.90	616	1298.59
8	7.93	1252	660.82
16	15.47	2441	328.7670
32	29.28	4622	182.1834
64	48.55	7667	113.4152
128	74.23	11726	100.3644

Table 5.25: SPLc *without* ERA, *Intel* with *-O3*.

#Threads	Speedup	MFLOPS	Time (sec)	Improvement
1	1.00	1216	1118.32	7.74
2	1.89	2293	522.90	7.44
4	3.78	4594	211.89	7.45
8	7.13	8660	134.46	6.91
16	11.80	14337	73.9313	5.87
32	13.48	16378	58.3417	3.54
64	16.43	19963	45.1974	2.60
128	16.33	19857	52.8118	1.69

Table 5.26: SPLc *with* ERA, *Intel* with *-O3*. The improvement value refers to the corresponding MFLOPS value in Table 5.25.

#Threads	Speedup	MFLOPS	Time (sec)
1	1.00	160	7233.33
2	1.96	313	3775.79
4	3.90	625	1860.15
8	7.81	1251	952.11
16	15.76	2525	470.5537
32	30.78	4934	250.1492
64	55.83	8949	139.8871
128	70.86	11359	117.6648

Table 5.27: SPLc *without* ERA, *g++* with *-O0*.

#Threads	Speedup	MFLOPS	Time (sec)	Improvement
1	1.00	245	4897.72	1.53
2	1.99	487	2410.48	1.55
4	3.88	953	1240.39	1.52
8	7.90	1940	624.93	1.55
16	15.53	3811	307.2994	1.50
32	30.96	7600	153.3394	1.54
64	61.35	15061	79.6251	1.68
128	103.61	25439	53.5939	2.23

Table 5.28: SPLc *with* ERA, *g++* with *-O0*. The improvement value refers to the corresponding MFLOPS value in Table 5.27.

#Threads	Speedup	MFLOPS	Time (sec)
1	1.00	200	4960.79
2	1.99	399	2438.66
4	3.96	792	1272.69
8	7.95	1590	629.06
16	15.32	3065	331.83
32	28.65	5732	178.87
64	49.20	9840	108.68
128	63.07	12600	100.17

Table 5.29: SPLc *without* ERA, *g++* with *-O3*.

#Threads	Speedup	MFLOPS	Time (sec)	Improvement
1	1.00	1109	1080.18	5.54
2	1.89	2093	659.93	5.24
4	3.75	4164	297.97	5.25
8	6.73	7462	172.61	4.69
16	12.47	13838	88.26	4.51
32	14.08	15637	70.02	2.72
64	12.99	14443	64.71	1.46
128	15.68	17427	69.37	1.38

Table 5.30: SPLc *with* ERA, g++ with *-O3*. The improvement value refers to the corresponding MFLOPS value in Table 5.29.

5.4 Least-Squares Problem

In Section 1.1.1 we introduced a least-squares problem where the implementation of the objective function contains an OpenMP construct. This OpenMP parallel for construct distributes the computations connected to independent equations among a team of threads. Algorithm 1 from page 9 has been implemented and compiled with g++ and the Intel compiler. The runtime results in Table 5.31 belong to the binary compiled with g++ while Table 5.32 presents the results from the binary that was compiled with the Intel compiler. In contrast to the previous test cases, we only measured the runtime results and not the MFLOPS values. The bar chart in Figure 5.24 helps to put the results into a context.

From a first look at Figure 5.24 one recognizes that the speedup value for the Intel binary does not grow much once the number of threads reaches 24 threads. The g++ version scales much better than the one from Intel. However, when we consider the overall runtime values in the tables, it turns out that the g++ binary needs over 8000 seconds for its sequential run whereby the binary obtained from the Intel compiler needs only 2594 seconds. Thus, the sequential runtime almost differs by a factor of three. The runtime with 128 threads is 506 seconds for the file compiled with g++ and the Intel compiler produces a binary that consumes 371 seconds for its runtime.

5.5 Nonlinear Constrained Optimization Problem

In this test case we examine the runtime results of the implementation of the constrained optimization problem that we introduced in Section 1.2.3. We assume to

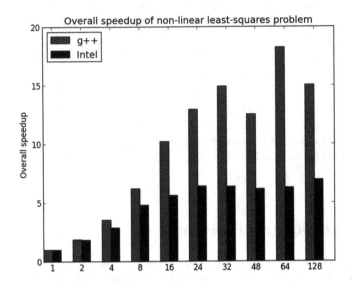

Figure 5.24: Runtime results of the least squares problem. The blue bars represent the speedup factors of the binaries compiled with the g++ compiler, the red bars represent the results for the binary compiled with the Intel compiler. The detailed results are shown in Table 5.31 (g++) and in Table 5.32 (Intel).

have 1024 equations that are all independent of one another and therefore the values for these equations can be computed concurrently. A possible implementation of the objective function is shown in Listing 1.5.

The *Ipopt* software[4] has been used to solve this problem. The first- and second derivative codes of Listing 1.5 were obtained with the help of SPLc where the forward-over-reverse mode was used to obtain the second-order adjoint model. Therefore, the Hessian of the objective function is computed line by line through the second-order adjoint model. The computation of the Hessian lines does not have any data dependencies among each other and can be computed independently. This fact shows, besides the OpenMP parallel region inside the derivative code, another possibility to parallelize the computation. Thus, we use two nested OpenMP parallel regions in the implementation of this test case. One parallel region starts in front of the loop that is responsible for computing the Hessian line by line. We de-

---

[4]https://projects.coin-or.org/Ipopt

#Threads	Overall runtime	Overall speedup
1	8042.61	1.00
2	4277.58	1.88
4	2269.60	3.54
8	1299.08	6.19
16	786.47	10.23
24	621.06	12.95
32	553.64	14.53
48	640.50	12.56
64	441.08	18.23
128	506.17	15.89

Table 5.31: Runtime results from the least squares problem. The compiler for this run was the g++. For each number of threads between 1 and 128, we display the runtime in seconds that the optimization took and its corresponding speedup ratio.

#Threads	Overall runtime	Overall speedup
1	2594.50	1.00
2	1419.80	1.83
4	907.83	2.86
8	540.72	4.80
16	460.23	5.64
24	403.47	6.43
32	407.94	6.36
48	421.55	6.15
64	412.22	6.29
128	371.82	6.98

Table 5.32: Runtime results from the least squares problem. The compiler for this run was the Intel. For each number of threads between 1 and 128, we display the runtime in seconds that the optimization took and its corresponding speedup ratio.

scribe this parallel region as the outer region of this example. The second parallel region is the one introduced by our source transformation inside the derivative code. This parallel region is meant when we speak about the inner region.

The OpenMP standard determines that the nested parallelism is disabled by default. This means that the inner region has no effect without enabling nested

OpenMP. Similar to the previous test cases, we use the g++ and the Intel compiler to obtain the executable files. In order to compare the effect of nested OpenMP, we used two different test setups. The first test run does not use nested OpenMP which means that the inner level is executed sequentially. Afterwards, the second test run tries to achieve better performance by enabling nested OpenMP. The results where nested OpenMP was disabled are shown in Table 5.33.

#Threads	g++ compiler		Intel compiler	
	Runtime (secs)	Speedup	Runtime (secs)	Speedup
1	7511.93	1.00	6899.10	1.00
2	3793.19	1.98	3497.54	1.97
4	1892.27	3.97	1800.42	3.83
8	950.90	7.90	915.31	7.54
16	475.25	15.81	459.03	15.03
32	268.37	27.99	236.58	29.16
64	216.12	34.76	1110.19	6.21
128	222.49	33.76	2158.10	3.20

Table 5.33: Concurrent computation of the lines of the Hessian. Nested OpenMP was disabled in this example.

The results are quite good until the maximal number of threads for one physical node, namely 32, is reached. Up to this number of threads the results for the g++ and the Intel binary do not diverge much. The g++ binary needs 268 seconds runtime whereby the Intel binary consumes 237 seconds of runtime. However, the runtime results for 64 and 128 threads are completely different for both binaries. The g++ binary is able to further reduce the runtime to 222 seconds with 128 threads whereby the Intel binary let rise the runtime up to 2158 seconds with 128 threads.

One may assume that the runtime gets better when enabling nested OpenMP. The test setup that we used determines that the inner OpenMP level uses two threads and the outer level uses the number of threads that is presented in the first column of Table 5.34. The results for the g++ binary are sobering. Even the execution with one thread on the outer and two threads on the inner level consumes 15.000 seconds and therefore needs more than twice as much time as the sequential execution of the g++ binary in Table 5.33. If one increases the number of threads this gets even worse where one achieves a slowdown instead of an performance improvement. The Intel compiler, on the other hand, produces a binary file that has a runtime of 5154 seconds with one thread on the outer level and two threads on the inner level. This runtime is 34 percent faster than the 6899 seconds of the sequential execution from Table 5.34. The execution with four threads in total -

#Threads	g++ compiler		Intel compiler	
	Runtime (secs)	Speedup	Runtime (secs)	Speedup
1	15492.49	1.00	5154.22	1.00
2	57136.79	0.27	2783.40	1.85
4	47680.71	0.32	1675.95	3.08
8	86791.24	0.18	1138.07	4.53
16	65804.87	0.24	2055.03	2.51
32	99851.46	0.16	3892.76	1.32
64	161757.15	0.10	9149.60	0.56

Table 5.34: Concurrent computation on two levels. The outer level comprises of the simultaneous computation of the lines of the Hessian. In addition, the parallel region inside second derivative code is executed by two threads which we denote with the inner OpenMP level.

two threads on the outer and two on the inner level - reveals a speedup value of 1.85. The runtime is 26 percent faster (2783 seconds) than the 3497.54 seconds from Table 5.33. With four threads on the outer level and two threads on the inner level (eight threads overall), the runtime is 1676 seconds. Eight threads is the maximum of one CPU socket of our testing machine. This is probably the reason why a further increase of the number of threads does not improve the performance much.

The comparison above shows that we can achieve a better performance by using the inherent parallelism in the derivative codes. However, when we consider the total number of threads and the runtime results that we achieved with nested OpenMP then the results from Table 5.33 clearly indicate that it is better to use only one level of OpenMP. Figure 5.25 summarizes the speedup results from the current test case. It would be interesting to see how the runtime results are for heterogeneous parallel setups. For example, when MPI is used on the outer level and OpenMP is used on the inner level. Another possibility is to use OpenMP on the outer level and a GPU on the inner level. This should be examined in further studies.

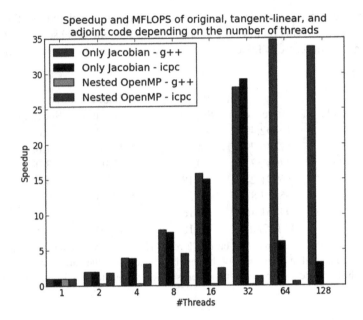

Figure 5.25: Runtime results of four different binaries. Two binaries were compiled by the g++ compiler and two by the Intel compiler. The first and second bar according to a certain number of threads, show the speedup values for the code that uses only the parallelization on the outer level. The third and the fourth bar present the speedup values for the nested OpenMP approach. The details are shown in Table 5.33 and Table 5.34.

5.6 Summary

This chapter contains several experimental results for testing the scaling properties of the derivative codes that we achieved from our source transformations. The source code package of our implementation tool SPLc (see Appendix A) contains a test suite comprising of different scenarios for an occurring OpenMP parallel region. Besides the case of a pure parallel region, it was important to examine the influence of synchronization constructs on the derivative code.

All in all, the test suite reinforces that the first derivative code of an OpenMP parallel region can carry over the parallel performance from the original code to its derivative code model. However, there are partly big differences in the results concerning the binaries from the g++ and from the Intel compiler. One example for this is the adjoint code of the critical region. The Intel binary achieves more than 3500 MFLOPS whereby the g++ binary supplies less than 1300 MFLOPS.

To investigate the scaling properties of the different second derivative codes, the forward-over-forward, the forward-over-reverse, and the reverse-over-forward model of the 'plain-parallel' example from the test suite were generated by SPLc. The results showed that the forward-over-reverse provides the best speedup of 70 while using 128 threads. The forward-over-forward mode achieves a speedup of 63 and the reverse-over-forward a speedup of 53. In addition, the reverse-over-forward model seems to suffer under the big memory usage of the adjoint code because we had to reduce the input size in order to run the test case.

To measure the impact of the exclusive read analysis, we compared two different executions. On the one hand, we performed a test where the contained adjoint code was generated by conservatively augmenting each adjoint assignment with an atomic construct. On the other hand, the adjoint code was generated by using the information obtained by the ERA. The MFLOPS values of the execution without using the ERA could be improved by an average factor of 3.11. The lowest improvement was a factor 1.50 and the best improvement factor was 7.74. This strongly emphasizes the use of static analysis to prevent the synchronization overhead in the adjoint computation.

The two motivating examples from the introduction chapter were also part of these experimental results. The implementation of the least-squares problem exhibits that we achieve a speedup but the fact that these speedup values were all below 20 with a maximum number of 128 threads indicates that further investigations should be made to find the bottleneck. This also holds for the constrained optimization problem where we achieved a relatively good speedup when we stay below 32 threads. Another fact that we saw during the last test case was that nested OpenMP does not seem to be well suited for this kind of problems. The g++ was not able to show a speedup at all if nested OpenMP was used. Further investigations should examine these observations.

6 Conclusions

In this dissertation we developed rules which define a source code transformation for OpenMP parallel regions. This source transformation implements the methods that are known as algorithmic differentiation (AD). Our focus was mainly on OpenMP parallel regions but the basic approach can be applied to further parallel programming APIs in case that they base on an approach where code regions are declared as parallelizable by a compiler directive.

The main contribution of this thesis is the evidence that the defined source transformations are closed. This means that the transformation fulfills two properties. On the one hand, the source transformation has to supply code that is contained in the same language as the one that serves as input language. On the other hand, the source transformation must maintain the property of the original code of being correct in the terms of a concurrent execution.

In the next section, we summarize the main results of this work. We conclude with an outlook that discusses possible research directions to broaden the results of this dissertation.

6.1 Results

The main results of this dissertation can be summarized as follows.

A Methodology for abstracting the parallel execution of parallel regions. The formalism of this methodology was used to show the closure property of the source transformations. In addition, the formalism provides mechanisms to show the correctness in case that synchronization plays a role during runtime. This shows that this approach is widely applicable in the domain of parallel programming.

Source transformation rules for the *SPL* language as well as for constructs occurring in the OpenMP standard. A simple parallel language (*SPL*) was defined where two facts were important. On the one hand, *SPL* is simplified in order to avoid the combinatorial explosion of possible statement combinations during the correctness proof. On the other hand, *SPL* is sophisticated enough to cover most of the occurring numerical kernels.

This work presented source transformation rules starting from a point where the input code was assumed to be a pure parallel region that does not contain any further pragmas. Later we extended the *SPL* language and the corresponding source transformation rules step by step to cover most of the possible OpenMP constructs. To cover most of the occurring OpenMP codes we covered synchronization constructs such as barrier and critical but also worksharing constructs such as the loop construct and the sections construct. The worksharing constructs combined with a parallel construct were also covered.

The simple structure and the formal definitions of the source transformation allowed us to prove certain properties of the derivative codes. In addition, a software engineer who wants to implement AD by source transformation may adapt these formal rules easily to another programming language or to a different parallel programming model.

Our tangent-linear and the adjoint source transformation fulfill the closure property. As long as the software engineer assures to implement the source transformation rules of this work correctly, it is ensured that the resulting derivative code can be executed concurrently and that the source transformation can be applied again to its own output. This is guaranteed because we proved the closure property of the source transformations. We showed how important the closure property is as we implemented the code for solving a nonlinear constrained optimization problem.

A static program analysis called exclusive read analysis. In case that we do not want to use any synchronization in the adjoint code, the input code for the adjoint source transformation must fulfill the exclusive read property as we showed during the proof of the closure property. Otherwise, the adjoint source transformation emits code that has a race condition at runtime. A parallel region fulfills the exclusive read property if it is ensured that a thread never uses a right-hand side reference that is simultaneously used by another thread.

There is often the possibility to adjust the original code such that it fulfills the exclusive read property, for example, by memory expansion. However, we presented two solutions for this problem without changing the original code. The first solution is that the source transformation augments each adjoint assignment by an atomic construct of OpenMP which prevents the race condition connected with the exclusive read property. This solution implies that we require an OpenMP parallel region as output. In case that the derivative code should run on a GPU, this solution is not applicable because there is no atomic floating-point assignment defined in the GPU assembly at the moment. The second solution is to use the exclusive read analysis that has been developed in this work. With this static analysis, we obtain information that indicates whether or not a certain adjoint assignment has to be synchronized or not. The experimental results showed that we achieved runtime improvements of a factor that lies between 1.5 and 7.74. The average improvement factor was 3.11.

An Implementation called SPLc and a test suite to show the scalability. The SPLc tool implements the source transformation rules from Chapter 2 and from Section 4.2. The exclusive read analysis from Chapter 3 has also been implemented. The documentation in Appendix A describes the steps for building the SPLc tool from source.

Extensive experimental results to show the scalability. In Section 5.1, we presented a test suite for obtaining speedup and MFLOPS results for given codes which contain several different setups of OpenMP codes. Appendix B contains a description of the test suite that we used. There are several code listings presented. For each test case we present the original code and the corresponding first-order tangent-linear code and the first-order adjoint code. For one test case we also present the second-order tangent-linear and the second-order adjoint code. All these codes have been obtained by using SPLc.

6.2 Future Work

In this section we give a brief discussion about possibilities to extend the work from this dissertation.

Automatic generation of first- and second-order derivative code frameworks: It should be possible to include these frameworks into a numerical optimization software without further adjustments. During the implementation of the constrained optimization problem we had to perform some coding in order to include the derivative code obtained by SPLc into the code structure provided by Ipopt. For a software engineer who wants to implement a solution for a numerical optimization problem it would be the relieve of a burden if the AD source transformation tool would provide code that can easily be included in a framework such as Ipopt[1].

---

[1] https://projects.coin-or.org/Ipopt

Derivative code execution on accelerator devices: In recent years the approach where computationally intensive problems are calculated on accelerator devices such as GPUs, FPGAs, or the Intel Xeon Phi is rising. Nowadays supercomputer architectures tends to be based on heterogeneous systems. The experimental results showed that one can achieve further performance improvements if one introduces two levels of parallelism, one on the level of the Hessian computation and one on the level of the derivative code. Further investigations should try the approach where the parallelism on the level of the derivative code is performed by an accelerator device.

An efficient second-order source transformation: The main computation effort during most of the numerical optimization algorithms such as the Newton, trusted region, or the sequential quadratic programming method lies in the iterative computation of the Hessian matrix. Therefore, the Hessian computation should be as efficient as possible. A possible direction for further research is to examine how far a static analysis can provide information about the Hessian structure. Since each individual Hessian entry can be computed concurrently, this improvement could be significant for many iterative optimization algorithms.

Extension of the exclusive read analysis (ERA):   During the development of the ERA we saw that things are important that typically lie in the domain of computer algebra systems. We adjusted, for example, integer expressions such that distributivity is exploited. Or we compared inequalities of symbolic expressions where each symbol has a certain value range. Further research may examine if these steps can be performed by a computer algebra system such as Axiom[2] or Maxima[3]. In this case the static program analysis would provide the symbolic inequalities together with the value ranges of each symbol to the computer algebra system. The algebra system computes a possible solution which is returned to the static program analysis.

The downside of the static analysis is often the conservatism that ensures, on the one hand, correct code but, on the other hand, reduces the possibilities to emit efficient code. This conservatism can be soften by giving the compiler more information about the code, for example, in form of an assert statement. For example, if a function has an input variable x and the known interval for this variable is between zero and hundred, the assert statement

```
assert( 0<=x && x<=100 );
```

---

[2]http://www.axiom-developer.org/
[3]http://maxima.sourceforge.net/

ensures that this variable has the correct value range. In case the static analysis knows the semantic of this assert statement, the possible value range of x can be recognized as $[0, 100]$ during compile time analysis. The approach where the software engineer writes code where certain assumptions are made in form of assert statements is known as *programming by contract*.

Another interesting approach is probably a mixture of static analysis and dynamic information that is obtained during runtime. For example, the intervals that the ERA obtains as a result during compile time can lead to different versions of adjoint code. At runtime, the execution can decide between different the versions depending on certain dynamic information obtained during the execution. This increases the code size but could allow a customized adjoint code execution where synchronization is only used when necessary.

A SPLc - **A** *SPL* **compiler**

The implementation of this dissertation is a compiler tool called *simple parallel language compiler* (SPLc). The SPLc source package can be obtained from the author[1].

A.1 Building SPLc

For a successful build of SPLc, one has to make sure that the following utilities are installed on the target system.

1. A C++ compiler.

2. cmake

3. make

4. flex

5. bison

To build the binary SPLc please perform the following steps where we assume that one is currently in the folder which contains the files of the CD.

1. `cd SPLc`

2. `cmake CMakeLists.txt`
   This should lead to an output that ends with the following message.

   ```
   ...
   -- Generating done
   -- Build files have been written to: ...
   ```

3. `make`
   In case that the build is successful the output ends with

   ```
   ...
   Linking CXX executable SPLc
   [100%] Built target SPLc
   ```

A.2 User Guide

You can provide SPLc a parallel region interactively or as file. To use SPLc interactively just type:

---

[1]The email address is Michael.Foerster@com-science.de

```
./SPLc
```

Afterwards, one can type in a parallel region and one finishes the input with 'END'.

```
Type in a SPL code and finish your input with END:
int tid;
#pragma omp parallel
{
   tid = omp_get_thread_num();
}
END
Parsing was successful.
```

SPLc will generate the tangent-linear code in `t1_stdin.spl` and the adjoint code can be found in `a1_stdin.spl`. In addition, several files are generated such as `symbol_table.txt` which contains an overview about the used symbols in the original code. The files with extension '.dot' are input files for the `dot` tool that is part of the software package *graphviz*[2]. In case that `dot` is installed, the files with the '.dot' extension are converted into PDF files automatically. The following list explains the generated files and their content.

- ast.dot (Abstract syntax tree of the original code)
- cfg.dot (Control flow graph of the original code)
- ast-tl.dot (Abstract syntax tree of the tangent-linear code)
- ast-adj.dot (Abstract syntax tree of the adjoint code)
- ast-adj-forward-section.dot (Just the AST of the forward section)
- ast-adj-reverse-section.dot (Just the AST of the reverse section)

The common usage is likely the case where a parallel region is contained in a file. To provide this file provided to SPLc, the syntax is as follows:

```
./SPLc   <input file>   <options>
```

The name of the file that contains the parallel region is provided to SPLc as first argument. As second argument the following options are possible:

- `-O0` : Do not use the exclusive read analysis before the transformation.
- `-O1` : Use the exclusive read analysis to provide better adjoint code.
- `--suppress-atomic` : Do not use an atomic construct before the adjoint statements. One should do this only if it is ensured that the original code fulfills the exclusive read property.

The option `-O1` enables the exclusive read analysis. The results of this static analysis can be examined in the files

---

[2]http://www.graphviz.org/

- `exclusive-read-results-001.txt`
- `exclusive-read-results-002.txt`
- ⋮
- `exclusive-read-results-endresult.txt`

Each file in the list represents a result of the fixed point iteration including widening and narrowing. The last file represents the approximation of the fixed point. This file also contains the result whether or not a certain floating-point memory reference fulfills the exclusive read property.

The reader wants probably to see some examples for getting a feel for the source transformation procedure. In this case the test suite provides a perfect starting point. The first steps and the contained examples are explained in Appendix B.

A.3 Developer Guide

The following list briefly describe the source files of SPLc and their purpose.

- `scanner.lpp`: flex input file.
- `parser.ypp`: bison input file.
- `ast_vertex.{h,cpp}`: Implementation of an abstract syntax tree node.
- `AST.{h,cpp}`: Implementation of the abstract syntax tree.
- `symbol_table.{h,cpp}`: Class that represents the symbol table.
- `unparser.{h,cpp}`: Class that provides the output of the SPLc.
- `sigma.cpp`: Implementation of the forward section transformation.
- `rho.cpp`: Implementation of the reverse section transformation.
- `tau.cpp`: Implementation of the tangent-linear transformation.
- `main.cpp`: main routine of the SPLc.
- `prg-analysis/exclusive-read/exclusive-read.{h,cpp}`: This is the implementation of the exclusive read analysis.

B Test Suite

This chapter contains the source codes of the examples that we used for the test suite in Section 5.1 on page 271.

There are no further requirements to the ones that we listed in Appendix A. The installation of the PAPI 5.3.0 software is optional. In case that one is interested in runtime results such as CPU cycles or million floating-point operations per second (MFLOPS) then the PAPI installation is the right choice.

The test suite is contained in a subfolder below the folder where the source code from SPLc is located. Suppose that the current working directory is the main folder of the CD containing the SPLc software, you can type

```
cd SPLc/test-suite
```

to enter the folder of the test suite.

The Makefile contained in the test suite requires that the shell environment variable SPLDIR points to the folder of the SPLc tool. One can define the shell variable by:

```
export SPLDIR=$PWD/..
```

If one wants to use the PAPI software please execute the following install script.

```
./install-papi.sh
```

This script creates a dummy file to indicate whether or not the PAPI software is installed. If one does not want to use the PAPI software then the dummy can be created by hand inside the main folder of the test suite.

```
touch PAPI_NOT_SUCCESSFUL
```

After these setup steps the test suite is ready to use. For example, enter the folder `atomic` and start the scaling test with optimization level 0 and 3 by typing:

```
cd atomic;
make
```

If one wants to run through the whole test suite then just type `make` inside the test suite folder.

```
make
```

Inside a certain test case, one can use further options to run certain tests.

```
cd atomic
make scale_test_00  # Run scale test with optimization level 0.
make scale_test_03  # Run scale test with optimization level 3.
make scale_test     # Run both of the above scale tests.
make finite_differences_test
                    # Compares the AD results with the one from FD.
```

Examples for the exclusive read analysis are contained in the folder

```
exclusive-read-examples.
```

For example, the reader may try the example that shows the impact of the exclusive read analysis.

```
cd exclusive-read-examples
SPLc second-example.in -O0
grep atomic a1_second-example.in.spl
```

Compare the number of atomic assignments in the adjoint code with the following output.

```
SPLc second-example.in -O1
tail -n 18 exclusive-read-results-endresult.txt
grep atomic a1_second-example.in.spl
```

The results for the individual memory references can be considered by:

```
tail -n 18 exclusive-read-results-endresult.txt
```

The individual listings for the examples are presented on the following pages.

B.1 Test Case Plain Parallel Region

This test example consists of a pure parallel region which means that the parallel region does not contain any further OpenMP pragmas inside the parallel region.

The dummy function serves only as a function that takes the output value of the parallel region as input. Since this function is only linked and not compiled, the code optimization of the backend compiler cannot delete code from the parallel region that influences the output value of the parallel region. This ensures the correct measurement of the performance of the parallel region.

The runtime details of the first derivative code of Listing B.1 can be found in Section 5.1.1 on page 272. For this example we also performed a runtime test of the second derivative codes. These results are shown in Section 5.2 on page 315.

B.1.1 Original Code 'plain-parallel'

```
#pragma omp parallel
{
    int tid←0;
    int p←0;
    int c←0;
    int lb←0;
    int ub←0;
    int i←0;
    int j←0;
    double y;
    double z;

    tid←omp_get_thread_num();
    p←omp_get_num_threads();
    c←n/p;
    lb←tid*c;
    ub←(tid+1)*c−1;
    i←lb;
    thread_result_tid←0.;
    while( i ≤ ub) {
        y←0.;
        z←0.;
        j←0;
        while(j<n) {
            y+←sin(A_{i*n+j}*A_{i*n+j});
            z+←cos(A_{i*n+j}*A_{i*n+j});
            j←j+1;
        }
        thread_result_tid+←y*z;
        i←i+1;
    }
    dummy("", (void*)thread_result);
}
```

Listing B.1: This code is the original code for the test case 'plain-parallel'.

B.1.2 First-Order Tangent-Linear Code 'plain-parallel'

```
#pragma omp parallel
{
    int tid←0;
    int p←0;
    int c←0;
    int lb←0;
    int ub←0;
    int i←0;
    int j←0;
    double y;
    double y^{(1)}←0.;
    double z;
    double z^{(1)}←0.;
    double v_0^0;
    double v_0^{(1)0};
    double v_1^0;
```

```
double  v_1^{(1)0} ;
double  v_2^0 ;
double  v_2^{(1)0} ;
double  v_3^0 ;
double  v_3^{(1)0} ;

tid←omp_get_thread_num();
p←omp_get_num_threads();
c←n/p;
lb←tid*c;
ub←(tid+1)*c-1;
i←lb;
v_0^{(1)0}←0.;
v_0^0←0.;
thread_result_{tid}^{(1)}←v_0^{(1)0};
thread_result_{tid}←v_0^0;
while( i ≤ ub) {
    v_0^{(1)0}←0.;
    v_0^0←0.;
    y^{(1)}←v_0^{(1)0};
    y←v_0^0;
    v_0^{(1)0}←0.;
    v_0^0←0.;
    z^{(1)}←v_0^{(1)0};
    z←v_0^0;
    j←0;
    while(j<n) {
        v_0^{(1)0}←A_{i*n+j}^{(1)};
        v_0^0←A_{i*n+j};
        v_1^{(1)0}←A_{i*n+j}^{(1)};
        v_1^0←A_{i*n+j};
        v_2^{(1)0}←v_0^{(1)0}*v_1^0+v_0^0*v_1^{(1)0};
        v_2^0←v_0^0*v_1^0;
        v_3^{(1)0}←v_2^{(1)0}*cos(v_2^0);
        v_3^0←sin(v_2^0);
        y^{(1)}+←v_3^{(1)0};
        y+←v_3^0;
        v_0^{(1)0}←A_{i*n+j}^{(1)};
        v_0^0←A_{i*n+j};
        v_1^{(1)0}←A_{i*n+j}^{(1)};
        v_1^0←A_{i*n+j};
        v_2^{(1)0}←v_0^{(1)0}*v_1^0+v_0^0*v_1^{(1)0};
        v_2^0←v_0^0*v_1^0;
        v_3^{(1)0}←v_2^{(1)0}*(0.-sin(v_2^0));
        v_3^0←cos(v_2^0);
        z^{(1)}+←v_3^{(1)0};
        z+←v_3^0;
        j←j+1;
    }
```

```
v₀^(1)0←y^(1) ;
v₀⁰←y ;
v₁^(1)0←z^(1) ;
v₁⁰←z ;
v₂^(1)0←v₀^(1)0*v₁⁰+v₀⁰*v₁^(1)0 ;
v₂⁰←v₀⁰*v₁⁰ ;
thread_result^(1)_tid+←v₂^(1)0 ;
thread_result_tid+←v₂⁰ ;
i←i+1;
}
dummy("", (void*)thread_result, (void*)thread_result^(1));
}
```

Listing B.2: This listing shows the tangent-linear code of the 'plain-parallel' test case (Listing B.1).

B.1.3 First-Order Adjoint Code 'plain-parallel'

```
#pragma omp parallel
{
  int tid←0;
  int p←0;
  int c←0;
  int lb←0;
  int ub←0;
  int i←0;
  int j←0;
  double y;
  double y₍₁₎←0.;
  double z;
  double z₍₁₎←0.;
  double v₀⁰;
  double v₍₁₎₀⁰;
  double v₁⁰;
  double v₍₁₎₁⁰;
  double v₂⁰;
  double v₍₁₎₂⁰;
  double v₃⁰;
  double v₍₁₎₃⁰;

  STACK₍₁₎c.push(48);
  STACK₍₁₎i.push(tid);
  tid←omp_get_thread_num();
  STACK₍₁₎i.push(p);
  p←omp_get_num_threads();
  STACK₍₁₎i.push(c);
  c←n/p;
  STACK₍₁₎i.push(lb);
  lb←tid*c;
  STACK₍₁₎i.push(ub);
  ub←(tid+1)*c−1;
  STACK₍₁₎i.push(i);
```

```
i←lb ;
STACK(1)f . push ( thread _ result_tid ) ;
thread _ result_tid←0 . ;
while ( i ≤ ub ) {
  STACK(1)c . push ( 86 ) ;
  STACK(1)f . push ( y ) ;
  y←0 . ;
  STACK(1)f . push ( z ) ;
  z←0 . ;
  STACK(1)i . push ( j ) ;
  j←0 ;
  while ( j<n ) {
    STACK(1)c . push ( 118 ) ;
    STACK(1)f . push ( y ) ;
    y+←sin ( A_{i*n+j}*A_{i*n+j} ) ;
    STACK(1)f . push ( z ) ;
    z+←cos ( A_{i*n+j}*A_{i*n+j} ) ;
    STACK(1)i . push ( j ) ;
    j←j +1;
  }
  STACK(1)c . push ( 154 ) ;
  STACK(1)f . push ( thread _ result_tid ) ;
  thread _ result_tid+←y*z ;
  STACK(1)i . push ( i ) ;
  i←i +1;
}
dummy ( "" ,  ( void* ) thread _ result ) ;
while (  not STACK(1)c . empty ( ) ) {
  if ( STACK(1)c . top ( ) = 48 ) {
    STACK(1)c . pop ( ) ;
    thread _ result_tid←STACK(1)f . top ( ) ;
    STACK(1)f . pop ( ) ;
    v_0^0←0 . ;
    v_{(1)0}^0←thread _ result_{(1)tid} ;
    thread _ result_{(1)tid}←0 . ;
    i←STACK(1)i . top ( ) ;
    STACK(1)i . pop ( ) ;
    ub←STACK(1)i . top ( ) ;
    STACK(1)i . pop ( ) ;
    lb←STACK(1)i . top ( ) ;
    STACK(1)i . pop ( ) ;
    c←STACK(1)i . top ( ) ;
    STACK(1)i . pop ( ) ;
    p←STACK(1)i . top ( ) ;
    STACK(1)i . pop ( ) ;
    tid←STACK(1)i . top ( ) ;
    STACK(1)i . pop ( ) ;
  }
  if ( STACK(1)c . top ( ) = 86 ) {
    STACK(1)c . pop ( ) ;
    j←STACK(1)i . top ( ) ;
    STACK(1)i . pop ( ) ;
    z←STACK(1)f . top ( ) ;
```

```
        STACK(1)f . pop ( ) ;
        v_0^0 ← 0 . ;
        v_(1)0^0 ← z_(1) ;
        z_(1) ← 0 . ;
        y ← STACK(1)f . top ( ) ;
        STACK(1)f . pop ( ) ;
        v_0^0 ← 0 . ;
        v_(1)0^0 ← y_(1) ;
        y_(1) ← 0 . ;
}
if ( STACK(1)c . top ( ) = 118 ) {
        STACK(1)c . pop ( ) ;
        j ← STACK(1)i . top ( ) ;
        STACK(1)i . pop ( ) ;
        z ← STACK(1)f . top ( ) ;
        STACK(1)f . pop ( ) ;
        v_0^0 ← A_{i*n+j} ;
        v_1^0 ← A_{i*n+j} ;
        v_2^0 ← v_0^0 * v_1^0 ;
        v_3^0 ← cos ( v_2^0 ) ;
        v_(1)3^0 ← z_(1) ;
        v_(1)2^0 ← v_(1)3^0 * ( 0 . − sin ( v_2^0 ) ) ;
        v_(1)0^0 ← v_(1)2^0 * v_1^0 ;
        v_(1)1^0 ← v_(1)2^0 * v_0^0 ;
        A_(1)i*n+j ← + ← v_(1)0^0 ;
        A_(1)i*n+j ← + ← v_(1)1^0 ;
        y ← STACK(1)f . top ( ) ;
        STACK(1)f . pop ( ) ;
        v_0^0 ← A_{i*n+j} ;
        v_1^0 ← A_{i*n+j} ;
        v_2^0 ← v_0^0 * v_1^0 ;
        v_3^0 ← sin ( v_2^0 ) ;
        v_(1)3^0 ← y_(1) ;
        v_(1)2^0 ← v_(1)3^0 * cos ( v_2^0 ) ;
        v_(1)0^0 ← v_(1)2^0 * v_1^0 ;
        v_(1)1^0 ← v_(1)2^0 * v_0^0 ;
        A_(1)i*n+j ← + ← v_(1)0^0 ;
        A_(1)i*n+j ← + ← v_(1)1^0 ;
}
if ( STACK(1)c . top ( ) = 154 ) {
        STACK(1)c . pop ( ) ;
        i ← STACK(1)i . top ( ) ;
        STACK(1)i . pop ( ) ;
        thread_result_{tid} ← STACK(1)f . top ( ) ;
        STACK(1)f . pop ( ) ;
        v_0^0 ← y ;
        v_1^0 ← z ;
        v_2^0 ← v_0^0 * v_1^0 ;
        v_(1)2^0 ← thread_result_{(1)tid} ;
```

$$v^0_{(1)0} \twoheadleftarrow v^0_{(1)2} * v^0_1 \,;$$
$$v^0_{(1)1} \twoheadleftarrow v^0_{(1)2} * v^0_0 \,;$$
$$y_{(1)} + \!\!\leftarrow v^0_{(1)0} \,;$$
$$z_{(1)} + \!\!\leftarrow v^0_{(1)1} \,;$$

```
    }
    dummy("", (void*)thread_result, (void*)thread_result(1));
  }
}
```

Listing B.3: The adjoint code of the 'plain-parallel' test case (Listing B.1) is presented in this listing where the exclusive read analysis has not been used to obtain static program information. With help of the exclusive read analysis the four atomic constructs in this code can be prevented.

B.1.4 Second-Order Tangent-Linear Code 'plain-parallel'

```
#pragma omp parallel
{
    int tid ← 0;
    int p ← 0;
    int c ← 0;
    int lb ← 0;
    int ub ← 0;
    int i ← 0;
    int j ← 0;
    double y;
    double y⁽²⁾ ← 0.;
    double y⁽¹⁾ ← 0.;
    double y⁽¹,²⁾ ← 0.;
    double z;
    double z⁽²⁾ ← 0.;
    double z⁽¹⁾ ← 0.;
    double z⁽¹,²⁾ ← 0.;
    double v₀⁰;
    double v₀⁽²⁾⁰ ← 0.;
    double v₀⁽¹⁾⁰;
    double v₀⁽¹,²⁾⁰ ← 0.;
    double v₁⁰;
    double v₁⁽²⁾⁰ ← 0.;
    double v₁⁽¹⁾⁰;
    double v₁⁽¹,²⁾⁰ ← 0.;
    double v₂⁰;
    double v₂⁽²⁾⁰ ← 0.;
    double v₂⁽¹⁾⁰;
    double v₂⁽¹,²⁾⁰ ← 0.;
    double v₃⁰;
    double v₃⁽²⁾⁰ ← 0.;
    double v₃⁽¹⁾⁰;
    double v₃⁽¹,²⁾⁰ ← 0.;
```

```
double  v₀¹;
double  v₀⁽²⁾¹;
double  v₁¹;
double  v₁⁽²⁾¹;
double  v₂¹;
double  v₂⁽²⁾¹;
double  v₃¹;
double  v₃⁽²⁾¹;
double  v₄¹;
double  v₄⁽²⁾¹;
double  v₅¹;
double  v₅⁽²⁾¹;
double  v₆¹;
double  v₆⁽²⁾¹;

tid←omp_get_thread_num();
p←omp_get_num_threads();
c←n/p;
lb←tid*c;
ub←(tid+1)*c-1;
i←lb;
```

$v_0^{(2)1} \leftarrow 0.;$
$v_0^1 \leftarrow 0.;$
$v_0^{(1,2)0} \leftarrow v_0^{(2)1};$
$v_0^{(1)0} \leftarrow v_0^1;$
$v_0^{(2)1} \leftarrow 0.;$
$v_0^1 \leftarrow 0.;$
$v_0^{(2)0} \leftarrow v_0^{(2)1};$
$v_0^0 \leftarrow v_0^1;$
$v_0^{(2)1} \leftarrow v_0^{(1,2)0};$
$v_0^1 \leftarrow v_0^{(1)0};$
$\text{thread\_result}_{tid}^{(1,2)} \leftarrow v_0^{(2)1};$
$\text{thread\_result}_{tid}^{(1)} \leftarrow v_0^1;$
$v_0^{(2)1} \leftarrow v_0^{(2)0};$
$v_0^1 \leftarrow v_0^0;$
$\text{thread\_result}_{tid}^{(2)} \leftarrow v_0^{(2)1};$
$\text{thread\_result}_{tid} \leftarrow v_0^1;$

```
while( i ≤ ub ) {
```
$\quad v_0^{(2)1} \leftarrow 0.;$
$\quad v_0^1 \leftarrow 0.;$
$\quad v_0^{(1,2)0} \leftarrow v_0^{(2)1};$
$\quad v_0^{(1)0} \leftarrow v_0^1;$
$\quad v_0^{(2)1} \leftarrow 0.;$
$\quad v_0^1 \leftarrow 0.;$
$\quad v_0^{(2)0} \leftarrow v_0^{(2)1};$
$\quad v_0^0 \leftarrow v_0^1;$
$\quad v_0^{(2)1} \leftarrow v_0^{(1,2)0};$
$\quad v_0^1 \leftarrow v_0^{(1)0};$

$$y^{(1,2)} \leftarrow v_0^{(2)1} ;$$
$$y^{(1)} \leftarrow v_0^1 ;$$
$$v_0^{(2)1} \leftarrow v_0^{(2)0} ;$$
$$v_0^1 \leftarrow v_0^0 ;$$
$$y^{(2)} \leftarrow v_0^{(2)1} ;$$
$$y \leftarrow v_0^1 ;$$
$$v_0^{(2)1} \leftarrow 0 . ;$$
$$v_0^1 \leftarrow 0 . ;$$
$$v_0^{(1,2)0} \leftarrow v_0^{(2)1} ;$$
$$v_0^{(1)0} \leftarrow v_0^1 ;$$
$$v_0^{(2)1} \leftarrow 0 . ;$$
$$v_0^1 \leftarrow 0 . ;$$
$$v_0^{(2)0} \leftarrow v_0^{(2)1} ;$$
$$v_0^0 \leftarrow v_0^1 ;$$
$$v_0^{(2)1} \leftarrow v_0^{(1,2)0} ;$$
$$v_0^1 \leftarrow v_0^{(1)0} ;$$
$$z^{(1,2)} \leftarrow v_0^{(2)1} ;$$
$$z^{(1)} \leftarrow v_0^1 ;$$
$$v_0^{(2)1} \leftarrow v_0^{(2)0} ;$$
$$v_0^1 \leftarrow v_0^0 ;$$
$$z^{(2)} \leftarrow v_0^{(2)1} ;$$
$$z \leftarrow v_0^1 ;$$
$$j \leftarrow 0 ;$$
while (j < n) {
$$v_0^{(2)1} \leftarrow A_{i*n+j}^{(1,2)} ;$$
$$v_0^1 \leftarrow A_{i*n+j}^{(1)} ;$$
$$v_0^{(1,2)0} \leftarrow v_0^{(2)1} ;$$
$$v_0^{(1)0} \leftarrow v_0^1 ;$$
$$v_0^{(2)1} \leftarrow A_{i*n+j}^{(2)} ;$$
$$v_0^1 \leftarrow A_{i*n+j} ;$$
$$v_0^{(2)0} \leftarrow v_0^{(2)1} ;$$
$$v_0^0 \leftarrow v_0^1 ;$$
$$v_0^{(2)1} \leftarrow A_{i*n+j}^{(1,2)} ;$$
$$v_0^1 \leftarrow A_{i*n+j}^{(1)} ;$$
$$v_1^{(1,2)0} \leftarrow v_0^{(2)1} ;$$
$$v_1^{(1)0} \leftarrow v_0^1 ;$$
$$v_0^{(2)1} \leftarrow A_{i*n+j}^{(2)} ;$$
$$v_0^1 \leftarrow A_{i*n+j} ;$$
$$v_1^{(2)0} \leftarrow v_0^{(2)1} ;$$
$$v_1^0 \leftarrow v_0^1 ;$$
$$v_0^{(2)1} \leftarrow v_0^{(1,2)0} ;$$
$$v_0^1 \leftarrow v_0^{(1)0} ;$$
$$v_1^{(2)1} \leftarrow v_1^{(2)0} ;$$
$$v_1^1 \leftarrow v_1^0 ;$$
$$v_2^{(2)1} \leftarrow v_0^{(2)1} * v_1^1 + v_0^1 * v_1^{(2)1} ;$$
$$v_2^1 \leftarrow v_0^1 * v_1^1 ;$$

$$v_3^{(2)1} \leftarrow v_0^{(2)0} \, ;$$
$$v_3^1 \leftarrow v_0^0 \, ;$$
$$v_4^{(2)1} \leftarrow v_1^{(1,2)0} \, ;$$
$$v_4^1 \leftarrow v_1^{(1)0} \, ;$$
$$v_5^{(2)1} \leftarrow v_3^{(2)1} * v_4^1 + v_3^1 * v_4^{(2)1} \, ;$$
$$v_5^1 \leftarrow v_3^1 * v_4^1 \, ;$$
$$v_6^{(2)1} \leftarrow v_2^{(2)1} + v_5^{(2)1} \, ;$$
$$v_6^1 \leftarrow v_2^1 + v_5^1 \, ;$$
$$v_2^{(1,2)0} \leftarrow v_6^{(2)1} \, ;$$
$$v_2^{(1)0} \leftarrow v_6^1 \, ;$$
$$v_0^{(2)1} \leftarrow v_0^{(2)0} \, ;$$
$$v_0^1 \leftarrow v_0^0 \, ;$$
$$v_1^{(2)1} \leftarrow v_1^{(2)0} \, ;$$
$$v_1^1 \leftarrow v_1^0 \, ;$$
$$v_2^{(2)1} \leftarrow v_0^{(2)1} * v_1^1 + v_0^1 * v_1^{(2)1} \, ;$$
$$v_2^1 \leftarrow v_0^1 * v_1^1 \, ;$$
$$v_2^{(2)0} \leftarrow v_2^{(2)1} \, ;$$
$$v_2^0 \leftarrow v_2^1 \, ;$$
$$v_0^{(2)1} \leftarrow v_2^{(1,2)0} \, ;$$
$$v_0^1 \leftarrow v_2^{(1)0} \, ;$$
$$v_1^{(2)1} \leftarrow v_2^{(2)0} \, ;$$
$$v_1^1 \leftarrow v_2^0 \, ;$$
$$v_2^{(2)1} \leftarrow v_1^{(2)1} * (0. - \sin(v_1^1)) \, ;$$
$$v_2^1 \leftarrow \cos(v_1^1) \, ;$$
$$v_3^{(2)1} \leftarrow v_0^{(2)1} * v_2^1 + v_0^1 * v_2^{(2)1} \, ;$$
$$v_3^1 \leftarrow v_0^1 * v_2^1 \, ;$$
$$v_3^{(1,2)0} \leftarrow v_3^{(2)1} \, ;$$
$$v_3^{(1)0} \leftarrow v_3^1 \, ;$$
$$v_0^{(2)1} \leftarrow v_2^{(2)0} \, ;$$
$$v_0^1 \leftarrow v_2^0 \, ;$$
$$v_1^{(2)1} \leftarrow v_0^{(2)1} * \cos(v_0^1) \, ;$$
$$v_1^1 \leftarrow \sin(v_0^1) \, ;$$
$$v_3^{(2)0} \leftarrow v_1^{(2)1} \, ;$$
$$v_3^0 \leftarrow v_1^1 \, ;$$
$$v_0^{(2)1} \leftarrow v_3^{(1,2)0} \, ;$$
$$v_0^1 \leftarrow v_3^{(1)0} \, ;$$
$$y^{(1,2)} + \!\! \leftarrow v_0^{(2)1} \, ;$$
$$y^{(1)} + \!\! \leftarrow v_0^1 \, ;$$
$$v_0^{(2)1} \leftarrow v_3^{(2)0} \, ;$$
$$v_0^1 \leftarrow v_3^0 \, ;$$
$$y^{(2)} + \!\! \leftarrow v_0^{(2)1} \, ;$$
$$y + \!\! \leftarrow v_0^1 \, ;$$
$$v_0^{(2)1} \leftarrow A_{i*n+j}^{(1,2)} \, ;$$
$$v_0^1 \leftarrow A_{i*n+j}^{(1)} \, ;$$
$$v_0^{(1,2)0} \leftarrow v_0^{(2)1} \, ;$$
$$v_0^{(1)0} \leftarrow v_0^1 \, ;$$

$$v_0^{(2)1} \leftarrow A_{i*n+j}^{(2)} \,;$$
$$v_0^1 \leftarrow A_{i*n+j} \,;$$
$$v_0^{(2)0} \leftarrow v_0^{(2)1} \,;$$
$$v_0^0 \leftarrow v_0^1 \,;$$
$$v_0^{(2)1} \leftarrow A_{i*n+j}^{(1,2)} \,;$$
$$v_0^1 \leftarrow A_{i*n+j}^{(1)} \,;$$
$$v_1^{(1,2)0} \leftarrow v_0^{(2)1} \,;$$
$$v_1^{(1)0} \leftarrow v_0^1 \,;$$
$$v_0^{(2)1} \leftarrow A_{i*n+j}^{(2)} \,;$$
$$v_0^1 \leftarrow A_{i*n+j} \,;$$
$$v_1^{(2)0} \leftarrow v_0^{(2)1} \,;$$
$$v_1^0 \leftarrow v_0^1 \,;$$
$$v_0^{(2)1} \leftarrow v_0^{(1,2)0} \,;$$
$$v_0^1 \leftarrow v_0^{(1)0} \,;$$
$$v_1^{(2)1} \leftarrow v_1^{(2)0} \,;$$
$$v_1^1 \leftarrow v_1^0 \,;$$
$$v_2^{(2)1} \leftarrow v_0^{(2)1} * v_1^1 + v_0^1 * v_1^{(2)1} \,;$$
$$v_2^1 \leftarrow v_0^1 * v_1^1 \,;$$
$$v_3^{(2)1} \leftarrow v_0^{(2)0} \,;$$
$$v_3^1 \leftarrow v_0^0 \,;$$
$$v_4^{(2)1} \leftarrow v_1^{(1,2)0} \,;$$
$$v_4^1 \leftarrow v_1^{(1)0} \,;$$
$$v_5^{(2)1} \leftarrow v_3^{(2)1} * v_4^1 + v_3^1 * v_4^{(2)1} \,;$$
$$v_5^1 \leftarrow v_3^1 * v_4^1 \,;$$
$$v_6^{(2)1} \leftarrow v_2^{(2)1} + v_5^{(2)1} \,;$$
$$v_6^1 \leftarrow v_2^1 + v_5^1 \,;$$
$$v_2^{(1,2)0} \leftarrow v_6^{(2)1} \,;$$
$$v_2^{(1)0} \leftarrow v_6^1 \,;$$
$$v_0^{(2)1} \leftarrow v_0^{(2)0} \,;$$
$$v_0^1 \leftarrow v_0^0 \,;$$
$$v_1^{(2)1} \leftarrow v_1^{(2)0} \,;$$
$$v_1^1 \leftarrow v_1^0 \,;$$
$$v_2^{(2)1} \leftarrow v_0^{(2)1} * v_1^1 + v_0^1 * v_1^{(2)1} \,;$$
$$v_2^1 \leftarrow v_0^1 * v_1^1 \,;$$
$$v_2^{(2)0} \leftarrow v_2^{(2)1} \,;$$
$$v_2^0 \leftarrow v_2^1 \,;$$
$$v_0^{(2)1} \leftarrow v_2^{(1,2)0} \,;$$
$$v_0^1 \leftarrow v_2^{(1)0} \,;$$
$$v_1^{(2)1} \leftarrow 0 \,.\,;$$
$$v_1^1 \leftarrow 0 \,.\,;$$
$$v_2^{(2)1} \leftarrow v_2^{(2)0} \,;$$
$$v_2^1 \leftarrow v_2^0 \,;$$
$$v_3^{(2)1} \leftarrow v_2^{(2)1} * \cos\left(v_2^1\right) \,;$$
$$v_3^1 \leftarrow \sin\left(v_2^1\right) \,;$$
$$v_4^{(2)1} \leftarrow v_1^{(2)1} - v_3^{(2)1} \,;$$
$$v_4^1 \leftarrow v_1^1 - v_3^1 \,;$$

$$v_5^{(2)1} \leftarrow v_0^{(2)1} * v_4^1 + v_0^1 * v_4^{(2)1} \ ;$$
$$v_5^1 \leftarrow v_0^1 * v_4^1 \ ;$$
$$v_3^{(1,2)0} \leftarrow v_5^{(2)1} \ ;$$
$$v_3^{(1)0} \leftarrow v_5^1 \ ;$$
$$v_0^{(2)1} \leftarrow v_2^{(2)0} \ ;$$
$$v_0^1 \leftarrow v_2^0 \ ;$$
$$v_1^{(2)1} \leftarrow v_0^{(2)1} * (0.-\sin(v_0^1)) \ ;$$
$$v_1^1 \leftarrow \cos(v_0^1) \ ;$$
$$v_3^{(2)0} \leftarrow v_1^{(2)1} \ ;$$
$$v_3^0 \leftarrow v_1^1 \ ;$$
$$v_0^{(2)1} \leftarrow v_3^{(1,2)0} \ ;$$
$$v_0^1 \leftarrow v_3^{(1)0} \ ;$$
$$z^{(1,2)} + \leftarrow v_0^{(2)1} \ ;$$
$$z^{(1)} + \leftarrow v_0^1 \ ;$$
$$v_0^{(2)1} \leftarrow v_3^{(2)0} \ ;$$
$$v_0^1 \leftarrow v_3^0 \ ;$$
$$z^{(2)} + \leftarrow v_0^{(2)1} \ ;$$
$$z + \leftarrow v_0^1 \ ;$$
$$j \leftarrow j+1 ;$$
$$\}$$
$$v_0^{(2)1} \leftarrow y^{(1,2)} \ ;$$
$$v_0^1 \leftarrow y^{(1)} \ ;$$
$$v_0^{(1,2)0} \leftarrow v_0^{(2)1} \ ;$$
$$v_0^{(1)0} \leftarrow v_0^1 \ ;$$
$$v_0^{(2)1} \leftarrow y^{(2)} \ ;$$
$$v_0^1 \leftarrow y \ ;$$
$$v_0^{(2)0} \leftarrow v_0^{(2)1} \ ;$$
$$v_0^0 \leftarrow v_0^1 \ ;$$
$$v_0^{(2)1} \leftarrow z^{(1,2)} \ ;$$
$$v_0^1 \leftarrow z^{(1)} \ ;$$
$$v_1^{(1,2)0} \leftarrow v_0^{(2)1} \ ;$$
$$v_1^{(1)0} \leftarrow v_0^1 \ ;$$
$$v_0^{(2)1} \leftarrow z^{(2)} \ ;$$
$$v_0^1 \leftarrow z \ ;$$
$$v_1^{(2)0} \leftarrow v_0^{(2)1} \ ;$$
$$v_1^0 \leftarrow v_0^1 \ ;$$
$$v_0^{(2)1} \leftarrow v_0^{(1,2)0} \ ;$$
$$v_0^1 \leftarrow v_0^{(1)0} \ ;$$
$$v_1^{(2)1} \leftarrow v_1^{(2)0} \ ;$$
$$v_1^1 \leftarrow v_1^0 \ ;$$
$$v_2^{(2)1} \leftarrow v_0^{(2)1} * v_1^1 + v_0^1 * v_1^{(2)1} \ ;$$
$$v_2^1 \leftarrow v_0^1 * v_1^1 \ ;$$
$$v_3^{(2)1} \leftarrow v_0^{(2)0} \ ;$$
$$v_3^1 \leftarrow v_0^0 \ ;$$
$$v_4^{(2)1} \leftarrow v_1^{(1,2)0} \ ;$$
$$v_4^1 \leftarrow v_1^{(1)0} \ ;$$

$$v_5^{(2)1} \leftarrow v_3^{(2)1} * v_4^1 + v_3^1 * v_4^{(2)1};$$
$$v_5^1 \leftarrow v_3^1 * v_4^1;$$
$$v_6^{(2)1} \leftarrow v_2^{(2)1} + v_5^{(2)1};$$
$$v_6^1 \leftarrow v_2^1 + v_5^1;$$
$$v_2^{(1,2)0} \leftarrow v_6^{(2)1};$$
$$v_2^{(1)0} \leftarrow v_6^1;$$
$$v_0^{(2)1} \leftarrow v_0^{(2)0};$$
$$v_0^1 \leftarrow v_0^0;$$
$$v_1^{(2)1} \leftarrow v_1^{(2)0};$$
$$v_1^1 \leftarrow v_1^0;$$
$$v_2^{(2)1} \leftarrow v_0^{(2)1} * v_1^1 + v_0^1 * v_1^{(2)1};$$
$$v_2^1 \leftarrow v_0^1 * v_1^1;$$
$$v_2^{(2)0} \leftarrow v_2^{(2)1};$$
$$v_2^0 \leftarrow v_2^1;$$
$$v_0^{(2)1} \leftarrow v_2^{(1,2)0};$$
$$v_0^1 \leftarrow v_2^{(1)0};$$

```
thread_result_{tid}^{(1,2)} ++← v_0^{(2)1};
thread_result_{tid}^{(1)} ++← v_0^1;
```
$$v_0^{(2)1} \leftarrow v_2^{(2)0};$$
$$v_0^1 \leftarrow v_2^0;$$
```
thread_result_{tid}^{(2)} ++← v_0^{(2)1};
thread_result_{tid} ++← v_0^1;
i ← i+1;
    }
    dummy("", (void*)thread_result, (void*)thread_result^{(1)}, (void*)
        thread_result^{(2)}, (void*)thread_result^{(1,2)});
}
```

Listing B.4: This listing shows the second-order tangent-linear code of the 'plain-parallel' test case (Listing B.1). This code is obtained by applying the tangent-linear source transformation twice, namely $\tau(\tau(P))$.

B.1.5 Second-Order Adjoint Code 'plain-parallel'

Forward-over-Reverse

```
#pragma omp parallel
{
    int tid←0;
    int p←0;
    int c←0;
    int lb←0;
    int ub←0;
    int i←0;
    int j←0;
    double y;
    double y^{(2)}←0.;
    double y_{(1)}←0.;
    double y_{(1)}^{(2)}←0.;
    double z;
```

```
double z(2) ← 0.;
double z(1) ← 0.;
double z(2)(1) ← 0.;
double v00;
double v(2)00 ← 0.;
double v0(1)0;
double v(2)0(1)0 ← 0.;
double v01;
double v(2)01 ← 0.;
double v0(1)1;
double v(2)0(1)1 ← 0.;
double v02;
double v(2)02 ← 0.;
double v0(1)2;
double v(2)0(1)2 ← 0.;
double v03;
double v(2)03 ← 0.;
double v0(1)3;
double v(2)0(1)3 ← 0.;
double v10;
double v(2)10;
double v11;
double v(2)11;
double v12;
double v(2)12;
double v13;
double v(2)13;
double v14;
double v(2)14;
double v15;
double v(2)15;
```

```
STACK(1)c . push (48);
STACK(1)i . push (tid);
tid ← omp_get_thread_num();
STACK(1)i . push (p);
p ← omp_get_num_threads();
STACK(1)i . push (c);
c ← n/p;
STACK(1)i . push (lb);
lb ← tid*c;
STACK(1)i . push (ub);
ub ← (tid+1)*c−1;
STACK(1)i . push (i);
i ← lb;
STACK(2)f . push (thread_result(2)tid);
STACK(1)f . push (thread_resulttid);
v(2)10 ← 0.;
```

```
v_0^1 ← 0.;
thread_result_{tid}^{(2)} ← v_0^{(2)1};
thread_result_{tid} ← v_0^1;
while( i ≤ ub ) {
  STACK_{(1)c} . push ( 86 );
  STACK_{(2)f} . push ( y^{(2)} );
  STACK_{(1)f} . push ( y );
  v_0^{(2)1} ← 0.;
  v_0^1 ← 0.;
  y^{(2)} ← v_0^{(2)1};
  y ← v_0^1;
  STACK_{(2)f} . push ( z^{(2)} );
  STACK_{(1)f} . push ( z );
  v_0^{(2)1} ← 0.;
  v_0^1 ← 0.;
  z^{(2)} ← v_0^{(2)1};
  z ← v_0^1;
  STACK_{(1)i} . push ( j );
  j ← 0;
  while( j<n ) {
    STACK_{(1)c} . push ( 118 );
    STACK_{(2)f} . push ( y^{(2)} );
    STACK_{(1)f} . push ( y );
    v_0^{(2)1} ← A_{i*n+j}^{(2)};
    v_0^1 ← A_{i*n+j};
    v_1^{(2)1} ← A_{i*n+j}^{(2)};
    v_1^1 ← A_{i*n+j};
    v_2^{(2)1} ← v_0^{(2)1} * v_1^1 + v_0^1 * v_1^{(2)1};
    v_2^1 ← v_0^1 * v_1^1;
    v_3^{(2)1} ← v_2^{(2)1} * cos ( v_2^1 );
    v_3^1 ← sin ( v_2^1 );
    y^{(2)} +← v_3^{(2)1};
    y +← v_3^1;
    STACK_{(2)f} . push ( z^{(2)} );
    STACK_{(1)f} . push ( z );
    v_0^{(2)1} ← A_{i*n+j}^{(2)};
    v_0^1 ← A_{i*n+j};
    v_1^{(2)1} ← A_{i*n+j}^{(2)};
    v_1^1 ← A_{i*n+j};
    v_2^{(2)1} ← v_0^{(2)1} * v_1^1 + v_0^1 * v_1^{(2)1};
    v_2^1 ← v_0^1 * v_1^1;
    v_3^{(2)1} ← v_2^{(2)1} * ( 0. - sin ( v_2^1 ) );
    v_3^1 ← cos ( v_2^1 );
    z^{(2)} +← v_3^{(2)1};
    z +← v_3^1;
    STACK_{(1)i} . push ( j );
    j ← j +1;
  }
  STACK_{(1)c} . push ( 154 );
```

```
    STACK(2)f . push ( thread _ result(2)tid ) ;
    STACK(1)f . push ( thread _ resulttid ) ;
    v(2)1 0 ← y(2) ;
    v1 0 ← y ;
    v(2)1 1 ← z(2) ;
    v1 1 ← z ;
    v(2)1 2 ← v(2)1 0 * v1 1 + v1 0 * v(2)1 1 ;
    v1 2 ← v1 0 * v1 1 ;
    thread _ result(2)tid + ← v(2)1 2 ;
    thread _ resulttid + ← v1 2 ;
    STACK(1)i . push ( i ) ;
    i ← i +1;
}
dummy ( " " , ( void *) thread _ result , ( void *) thread _ result(2) ) ;
while ( not STACK(1)c . empty ( ) ) {
    if ( STACK(1)c . top ( ) = 48 ) {
        STACK(1)c . pop ( ) ;
        thread _ result(2)tid ← STACK(2)f . top ( ) ;
        thread _ resulttid ← STACK(1)f . top ( ) ;
        STACK(2)f . pop ( ) ;
        STACK(1)f . pop ( ) ;
        v(2)1 0 ← 0. ;
        v1 0 ← 0. ;
        v(2)0 0 ← v(2)1 0 ;
        v0 0 ← v1 0 ;
        v(2)1 0 ← thread _ result(2)(1)tid ;
        v1 0 ← thread _ result(1)tid ;
        v(2)0 (1)0 ← v(2)1 0 ;
        v0 (1)0 ← v1 0 ;
        v(2)1 0 ← 0. ;
        v1 0 ← 0. ;
        thread _ result(2)(1)tid ← v(2)1 0 ;
        thread _ result(1)tid ← v1 0 ;
        i ← STACK(1)i . top ( ) ;
        STACK(1)i . pop ( ) ;
        ub ← STACK(1)i . top ( ) ;
        STACK(1)i . pop ( ) ;
        lb ← STACK(1)i . top ( ) ;
        STACK(1)i . pop ( ) ;
        c ← STACK(1)i . top ( ) ;
        STACK(1)i . pop ( ) ;
        p ← STACK(1)i . top ( ) ;
        STACK(1)i . pop ( ) ;
        tid ← STACK(1)i . top ( ) ;
        STACK(1)i . pop ( ) ;
    }
    if ( STACK(1)c . top ( ) = 86 ) {
        STACK(1)c . pop ( ) ;
        j ← STACK(1)i . top ( ) ;
```

```
    STACK_(1)i . pop ( ) ;
    z^(2)←STACK_(2)f . top ( ) ;
    z←STACK_(1)f . top ( ) ;
    STACK_(2)f . pop ( ) ;
    STACK_(1)f . pop ( ) ;
    v_0^(2)1 ← 0 . ;
    v_0^1 ← 0 . ;
    v_0^(2)0 ← v_0^(2)1 ;
    v_0^0 ← v_0^1 ;
    v_0^(2)1 ← z_(1)^(2) ;
    v_0^1 ← z_(1) ;
    v_(1)0^(2)0 ← v_0^(2)1 ;
    v_(1)0^0 ← v_0^1 ;
    v_0^(2)1 ← 0 . ;
    v_0^1 ← 0 . ;
    z_(1)^(2) ← v_0^(2)1 ;
    z_(1) ← v_0^1 ;
    y^(2) ← STACK_(2)f . top ( ) ;
    y ← STACK_(1)f . top ( ) ;
    STACK_(2)f . pop ( ) ;
    STACK_(1)f . pop ( ) ;
    v_0^(2)1 ← 0 . ;
    v_0^1 ← 0 . ;
    v_0^(2)0 ← v_0^(2)1 ;
    v_0^0 ← v_0^1 ;
    v_0^(2)1 ← y_(1)^(2) ;
    v_0^1 ← y_(1) ;
    v_(1)0^(2)0 ← v_0^(2)1 ;
    v_(1)0^0 ← v_0^1 ;
    v_0^(2)1 ← 0 . ;
    v_0^1 ← 0 . ;
    y_(1)^(2) ← v_0^(2)1 ;
    y_(1) ← v_0^1 ;
}
if ( STACK_(1)c . top ( ) = 118 ) {
    STACK_(1)c . pop ( ) ;
    j ← STACK_(1)i . top ( ) ;
    STACK_(1)i . pop ( ) ;
    z^(2) ← STACK_(2)f . top ( ) ;
    z ← STACK_(1)f . top ( ) ;
    STACK_(2)f . pop ( ) ;
    STACK_(1)f . pop ( ) ;
    v_0^(2)1 ← A_{i*n+j}^(2) ;
    v_0^1 ← A_{i*n+j} ;
    v_0^(2)0 ← v_0^(2)1 ;
    v_0^0 ← v_0^1 ;
    v_0^(2)1 ← A_{i*n+j}^(2) ;
```

$$v_0^1 \leftarrow A_{i*n+j} \,;$$
$$v_1^{(2)0} \leftarrow v_0^{(2)1} \,;$$
$$v_1^0 \leftarrow v_0^1 \,;$$
$$v_0^{(2)1} \leftarrow v_0^{(2)0} \,;$$
$$v_0^1 \leftarrow v_0^0 \,;$$
$$v_1^{(2)1} \leftarrow v_1^{(2)0} \,;$$
$$v_1^1 \leftarrow v_1^0 \,;$$
$$v_2^{(2)1} \leftarrow v_0^{(2)1} * v_1^1 + v_0^1 * v_1^{(2)1} \,;$$
$$v_2^1 \leftarrow v_0^1 * v_1^1 \,;$$
$$v_2^{(2)0} \leftarrow v_2^{(2)1} \,;$$
$$v_2^0 \leftarrow v_2^1 \,;$$
$$v_0^{(2)1} \leftarrow v_2^{(2)0} \,;$$
$$v_0^1 \leftarrow v_2^0 \,;$$
$$v_1^{(2)1} \leftarrow v_0^{(2)1} * (0. - \sin(v_0^1)) \,;$$
$$v_1^1 \leftarrow \cos(v_0^1) \,;$$
$$v_3^{(2)0} \leftarrow v_1^{(2)1} \,;$$
$$v_3^0 \leftarrow v_1^1 \,;$$
$$v_0^{(2)1} \leftarrow z_{(1)}^{(2)} \,;$$
$$v_0^1 \leftarrow z_{(1)} \,;$$
$$v_{(1)3}^{(2)0} \leftarrow v_0^{(2)1} \,;$$
$$v_{(1)3}^0 \leftarrow v_0^1 \,;$$
$$v_0^{(2)1} \leftarrow v_{(1)3}^{(2)0} \,;$$
$$v_0^1 \leftarrow v_{(1)3}^0 \,;$$
$$v_1^{(2)1} \leftarrow 0. \,;$$
$$v_1^1 \leftarrow 0. \,;$$
$$v_2^{(2)1} \leftarrow v_2^{(2)0} \,;$$
$$v_2^1 \leftarrow v_2^0 \,;$$
$$v_3^{(2)1} \leftarrow v_2^{(2)1} * \cos(v_2^1) \,;$$
$$v_3^1 \leftarrow \sin(v_2^1) \,;$$
$$v_4^{(2)1} \leftarrow v_1^{(2)1} - v_3^{(2)1} \,;$$
$$v_4^1 \leftarrow v_1^1 - v_3^1 \,;$$
$$v_5^{(2)1} \leftarrow v_0^{(2)1} * v_4^1 + v_0^1 * v_4^{(2)1} \,;$$
$$v_5^1 \leftarrow v_0^1 * v_4^1 \,;$$
$$v_{(1)2}^{(2)0} \leftarrow v_5^{(2)1} \,;$$
$$v_{(1)2}^0 \leftarrow v_5^1 \,;$$
$$v_0^{(2)1} \leftarrow v_{(1)2}^{(2)0} \,;$$
$$v_0^1 \leftarrow v_{(1)2}^0 \,;$$
$$v_1^{(2)1} \leftarrow v_1^{(2)0} \,;$$
$$v_1^1 \leftarrow v_1^0 \,;$$
$$v_2^{(2)1} \leftarrow v_0^{(2)1} * v_1^1 + v_0^1 * v_1^{(2)1} \,;$$
$$v_2^1 \leftarrow v_0^1 * v_1^1 \,;$$
$$v_{(1)0}^{(2)0} \leftarrow v_2^{(2)1} \,;$$
$$v_{(1)0}^0 \leftarrow v_2^1 \,;$$
$$v_0^{(2)1} \leftarrow v_{(1)2}^{(2)0} \,;$$
$$v_0^1 \leftarrow v_{(1)2}^0 \,;$$

$$v_1^{(2)1} \leftarrow v_0^{(2)0} \,;$$
$$v_1^1 \leftarrow v_0^0 \,;$$
$$v_2^{(2)1} \leftarrow v_0^{(2)1} * v_1^1 + v_0^1 * v_1^{(2)1} \,;$$
$$v_2^1 \leftarrow v_0^1 * v_1^1 \,;$$
$$v_{(1)1}^{(2)0} \leftarrow v_2^{(2)1} \,;$$
$$v_{(1)1}^0 \leftarrow v_2^1 \,;$$
$$v_0^{(2)1} \leftarrow v_{(1)0}^{(2)0} \,;$$
$$v_0^1 \leftarrow v_{(1)0}^0 \,;$$

#pragma omp atomic
$$A_{(1)i*n+j}^{(2)} + \leftarrow v_0^{(2)1} \,;$$
#pragma omp atomic
$$A_{(1)i*n+j} + \leftarrow v_0^1 \,;$$
$$v_0^{(2)1} \leftarrow v_{(1)1}^{(2)0} \,;$$
$$v_0^1 \leftarrow v_{(1)1}^0 \,;$$

#pragma omp atomic
$$A_{(1)i*n+j}^{(2)} + \leftarrow v_0^{(2)1} \,;$$
#pragma omp atomic
$$A_{(1)i*n+j} + \leftarrow v_0^1 \,;$$
$$y^{(2)} \leftarrow \mathrm{STACK}_{(2)f}.\mathrm{top}()\,;$$
$$y \leftarrow \mathrm{STACK}_{(1)f}.\mathrm{top}()\,;$$
$$\mathrm{STACK}_{(2)f}.\mathrm{pop}()\,;$$
$$\mathrm{STACK}_{(1)f}.\mathrm{pop}()\,;$$
$$v_0^{(2)1} \leftarrow A_{i*n+j}^{(2)} \,;$$
$$v_0^1 \leftarrow A_{i*n+j} \,;$$
$$v_0^{(2)0} \leftarrow v_0^{(2)1} \,;$$
$$v_0^0 \leftarrow v_0^1 \,;$$
$$v_0^{(2)1} \leftarrow A_{i*n+j}^{(2)} \,;$$
$$v_0^1 \leftarrow A_{i*n+j} \,;$$
$$v_1^{(2)0} \leftarrow v_0^{(2)1} \,;$$
$$v_1^0 \leftarrow v_0^1 \,;$$
$$v_0^{(2)1} \leftarrow v_0^{(2)0} \,;$$
$$v_0^1 \leftarrow v_0^0 \,;$$
$$v_1^{(2)1} \leftarrow v_1^{(2)0} \,;$$
$$v_1^1 \leftarrow v_1^0 \,;$$
$$v_2^{(2)1} \leftarrow v_0^{(2)1} * v_1^1 + v_0^1 * v_1^{(2)1} \,;$$
$$v_2^1 \leftarrow v_0^1 * v_1^1 \,;$$
$$v_2^{(2)0} \leftarrow v_2^{(2)1} \,;$$
$$v_2^0 \leftarrow v_2^1 \,;$$
$$v_0^{(2)1} \leftarrow v_2^{(2)0} \,;$$
$$v_0^1 \leftarrow v_2^0 \,;$$
$$v_1^{(2)1} \leftarrow v_0^{(2)1} * \cos(v_0^1)\,;$$
$$v_1^1 \leftarrow \sin(v_0^1)\,;$$
$$v_3^{(2)0} \leftarrow v_1^{(2)1} \,;$$
$$v_3^0 \leftarrow v_1^1 \,;$$
$$v_0^{(2)1} \leftarrow y_{(1)}^{(2)} \,;$$
$$v_0^1 \leftarrow y_{(1)} \,;$$

$$v_{(1)3}^{(2)0} \leftarrow v_0^{(2)1} \, ;$$

$$v_{(1)3}^0 \leftarrow v_0^1 \, ;$$

$$v_0^{(2)1} \leftarrow v_{(1)3}^{(2)0} \, ;$$

$$v_0^1 \leftarrow v_{(1)3}^0 \, ;$$

$$v_1^{(2)1} \leftarrow v_2^{(2)0} \, ;$$

$$v_1^1 \leftarrow v_2^0 \, ;$$

$$v_2^{(2)1} \leftarrow v_1^{(2)1} * (0. - \sin (v_1^1)) \, ;$$

$$v_2^1 \leftarrow \cos (v_1^1) \, ;$$

$$v_3^{(2)1} \leftarrow v_0^{(2)1} * v_2^1 + v_0^1 * v_2^{(2)1} \, ;$$

$$v_3^1 \leftarrow v_0^1 * v_2^1 \, ;$$

$$v_{(1)2}^{(2)0} \leftarrow v_3^{(2)1} \, ;$$

$$v_{(1)2}^0 \leftarrow v_3^1 \, ;$$

$$v_0^{(2)1} \leftarrow v_{(1)2}^{(2)0} \, ;$$

$$v_0^1 \leftarrow v_{(1)2}^0 \, ;$$

$$v_1^{(2)1} \leftarrow v_1^{(2)0} \, ;$$

$$v_1^1 \leftarrow v_1^0 \, ;$$

$$v_2^{(2)1} \leftarrow v_0^{(2)1} * v_1^1 + v_0^1 * v_1^{(2)1} \, ;$$

$$v_2^1 \leftarrow v_0^1 * v_1^1 \, ;$$

$$v_{(1)0}^{(2)0} \leftarrow v_2^{(2)1} \, ;$$

$$v_{(1)0}^0 \leftarrow v_2^1 \, ;$$

$$v_0^{(2)1} \leftarrow v_{(1)2}^{(2)0} \, ;$$

$$v_0^1 \leftarrow v_{(1)2}^0 \, ;$$

$$v_1^{(2)1} \leftarrow v_0^{(2)0} \, ;$$

$$v_1^1 \leftarrow v_0^0 \, ;$$

$$v_2^{(2)1} \leftarrow v_0^{(2)1} * v_1^1 + v_0^1 * v_1^{(2)1} \, ;$$

$$v_2^1 \leftarrow v_0^1 * v_1^1 \, ;$$

$$v_{(1)1}^{(2)0} \leftarrow v_2^{(2)1} \, ;$$

$$v_{(1)1}^0 \leftarrow v_2^1 \, ;$$

$$v_0^{(2)1} \leftarrow v_{(1)0}^{(2)0} \, ;$$

$$v_0^1 \leftarrow v_{(1)0}^0 \, ;$$

```
#pragma omp atomic
```

$$A_{(1)i*n+j}^{(2)} + \leftarrow v_0^{(2)1} \, ;$$

```
#pragma omp atomic
```

$$A_{(1)i*n+j} + \leftarrow v_0^1 \, ;$$

$$v_0^{(2)1} \leftarrow v_{(1)1}^{(2)0} \, ;$$

$$v_0^1 \leftarrow v_{(1)1}^0 \, ;$$

```
#pragma omp atomic
```

$$A_{(1)i*n+j}^{(2)} + \leftarrow v_0^{(2)1} \, ;$$

```
#pragma omp atomic
```

$$A_{(1)i*n+j} + \leftarrow v_0^1 \, ;$$

```
}
if (STACK(1)c . top () = 154) {
  STACK(1)c . pop () ;
  i ←STACK(1)i . top () ;
  STACK(1)i . pop () ;
```

```
thread_result(2)tid ← STACK(2)f . top ( ) ;
thread_resulttid ← STACK(1)f . top ( ) ;
STACK(2)f . pop ( ) ;
STACK(1)f . pop ( ) ;
```

$v_0^{(2)1} \leftarrow y^{(2)}$;

$v_0^1 \leftarrow y$;

$v_0^{(2)0} \leftarrow v_0^{(2)1}$;

$v_0^0 \leftarrow v_0^1$;

$v_0^{(2)1} \leftarrow z^{(2)}$;

$v_0^1 \leftarrow z$;

$v_1^{(2)0} \leftarrow v_0^{(2)1}$;

$v_1^0 \leftarrow v_0^1$;

$v_0^{(2)1} \leftarrow v_0^{(2)0}$;

$v_0^1 \leftarrow v_0^0$;

$v_1^{(2)1} \leftarrow v_1^{(2)0}$;

$v_1^1 \leftarrow v_1^0$;

$v_2^{(2)1} \leftarrow v_0^{(2)1} * v_1^1 + v_0^1 * v_1^{(2)1}$;

$v_2^1 \leftarrow v_0^1 * v_1^1$;

$v_2^{(2)0} \leftarrow v_2^{(2)1}$;

$v_2^0 \leftarrow v_2^1$;

$v_0^{(2)1} \leftarrow$ `thread_result`$_{(1)tid}^{(2)}$;

$v_0^1 \leftarrow$ `thread_result`$_{(1)tid}$;

$v_{(1)2}^{(2)0} \leftarrow v_0^{(2)1}$;

$v_{(1)2}^0 \leftarrow v_0^1$;

$v_0^{(2)1} \leftarrow v_{(1)2}^{(2)0}$;

$v_0^1 \leftarrow v_{(1)2}^0$;

$v_1^{(2)1} \leftarrow v_1^{(2)0}$;

$v_1^1 \leftarrow v_1^0$;

$v_2^{(2)1} \leftarrow v_0^{(2)1} * v_1^1 + v_0^1 * v_1^{(2)1}$;

$v_2^1 \leftarrow v_0^1 * v_1^1$;

$v_{(1)0}^{(2)0} \leftarrow v_2^{(2)1}$;

$v_{(1)0}^0 \leftarrow v_2^1$;

$v_0^{(2)1} \leftarrow v_{(1)2}^{(2)0}$;

$v_0^1 \leftarrow v_{(1)2}^0$;

$v_1^{(2)1} \leftarrow v_0^{(2)0}$;

$v_1^1 \leftarrow v_0^0$;

$v_2^{(2)1} \leftarrow v_0^{(2)1} * v_1^1 + v_0^1 * v_1^{(2)1}$;

$v_2^1 \leftarrow v_0^1 * v_1^1$;

$v_{(1)1}^{(2)0} \leftarrow v_2^{(2)1}$;

$v_{(1)1}^0 \leftarrow v_2^1$;

$v_0^{(2)1} \leftarrow v_{(1)0}^{(2)0}$;

$v_0^1 \leftarrow v_{(1)0}^0$;

$y_{(1)}^{(2)} + \leftarrow v_0^{(2)1}$;

$y_{(1)} + \leftarrow v_0^1$;

$v_0^{(2)1} \leftarrow v_{(1)1}^{(2)0}$;

$$v_0^1 \leftarrow v_{(1)1}^0 \,;$$
$$z_{(1)}^{(2)} + \leftarrow v_0^{(2)1} \,;$$
$$z_{(1)} + \leftarrow v_0^1 \,;$$

```
    }
    dummy ( " " , ( void *) thread _ result , ( void *) thread _ result (1) , ( void *)
        thread _ result (2) , ( void *) thread _ result (2)(1) ) ;
    }
}
```

Listing B.5: This code represents the second-order adjoint code of the 'plain-parallel' test case (Listing B.1) in forward-over-reverse mode. This code is obtained by applying first the reverse mode and then the forward mode, in other words one applies the source transformation $\tau(\sigma(P))$.

Reverse-over-Forward

```
#pragma omp parallel
{
    int  tid←0;
    int  p←0;
    int  c←0;
    int  lb←0;
    int  ub←0;
    int  i←0;
    int  j←0;
    double  y;
    double  y(2)←0.;
    double  y(1)←0.;
    double  y(1)(2)←0.;
    double  z;
    double  z(2)←0.;
    double  z(1)←0.;
    double  z(1)(2)←0.;
    double  v0;
    double  v(2)0←0.;
    double  v(1)0;
    double  v(1)0(2)0←0.;
    double  v0;
    double  v(2)1←0.;
    double  v(1)0;
    double  v(1)0(2)1←0.;
    double  v0;
    double  v(2)2←0.;
    double  v(1)0;
    double  v(1)0(2)2←0.;
    double  v0;
    double  v(2)3←0.;
    double  v(1)0;
    double  v(1)0(2)3←0.;
```

```
double  v_0^1 ;
double  v_{(2)0}^1 ;
double  v_1^1 ;
double  v_{(2)1}^1 ;
double  v_2^1 ;
double  v_{(2)2}^1 ;
double  v_3^1 ;
double  v_{(2)3}^1 ;
double  v_4^1 ;
double  v_{(2)4}^1 ;
double  v_5^1 ;
double  v_{(2)5}^1 ;
double  v_6^1 ;
double  v_{(2)6}^1 ;

STACK_{(2)c} . push ( 78 ) ;
STACK_{(2)i} . push ( tid ) ;
tid←omp_get_thread_num ( ) ;
STACK_{(2)i} . push ( p ) ;
p←omp_get_num_threads ( ) ;
STACK_{(2)i} . push ( c ) ;
c←n/p ;
STACK_{(2)i} . push ( lb ) ;
lb←tid*c ;
STACK_{(2)i} . push ( ub ) ;
ub←( tid +1 )*c −1 ;
STACK_{(2)i} . push ( i ) ;
i←lb ;
STACK_{(2)f} . push ( v_0^{(1)0} ) ;
v_0^{(1)0}←0. ;
STACK_{(2)f} . push ( v_0^0 ) ;
v_0^0←0. ;
STACK_{(2)f} . push ( thread_result_{tid}^{(1)} ) ;
thread_result_{tid}^{(1)}←v_0^{(1)0} ;
STACK_{(2)f} . push ( thread_result_{tid} ) ;
thread_result_{tid}←v_0^0 ;
while ( i ≤ ub ) {
  STACK_{(2)c} . push ( 127 ) ;
  STACK_{(2)f} . push ( v_0^{(1)0} ) ;
  v_0^{(1)0}←0. ;
  STACK_{(2)f} . push ( v_0^0 ) ;
  v_0^0←0. ;
  STACK_{(2)f} . push ( y^{(1)} ) ;
  y^{(1)}←v_0^{(1)0} ;
  STACK_{(2)f} . push ( y ) ;
  y←v_0^0 ;
  STACK_{(2)f} . push ( v_0^{(1)0} ) ;
  v_0^{(1)0}←0. ;
  STACK_{(2)f} . push ( v_0^0 ) ;
```

```
v_0^0 ← 0.;
STACK_(2)f . push ( z^(1) ) ;
z^(1) ← v_0^(1)0 ;
STACK_(2)f . push ( z ) ;
z ← v_0^0 ;
STACK_(2)i . push ( j ) ;
j ← 0 ;
while ( j < n )  {
  STACK_(2)c . push ( 166 ) ;
  STACK_(2)f . push ( v_0^(1)0 ) ;
  v_0^(1)0 ← A_(i*n+j)^(1) ;
  STACK_(2)f . push ( v_0^0 ) ;
  v_0^0 ← A_(i*n+j) ;
  STACK_(2)f . push ( v_1^(1)0 ) ;
  v_1^(1)0 ← A_(i*n+j)^(1) ;
  STACK_(2)f . push ( v_1^0 ) ;
  v_1^0 ← A_(i*n+j) ;
  STACK_(2)f . push ( v_2^(1)0 ) ;
  v_2^(1)0 ← v_0^(1)0 * v_1^0 + v_0^0 * v_1^(1)0 ;
  STACK_(2)f . push ( v_2^0 ) ;
  v_2^0 ← v_0^0 * v_1^0 ;
  STACK_(2)f . push ( v_3^(1)0 ) ;
  v_3^(1)0 ← v_2^(1)0 * cos ( v_2^0 ) ;
  STACK_(2)f . push ( v_3^0 ) ;
  v_3^0 ← sin ( v_2^0 ) ;
  STACK_(2)f . push ( y^(1) ) ;
  y^(1) + ← v_3^(1)0 ;
  STACK_(2)f . push ( y ) ;
  y + ← v_3^0 ;
  STACK_(2)f . push ( v_0^(1)0 ) ;
  v_0^(1)0 ← A_(i*n+j)^(1) ;
  STACK_(2)f . push ( v_0^0 ) ;
  v_0^0 ← A_(i*n+j) ;
  STACK_(2)f . push ( v_1^(1)0 ) ;
  v_1^(1)0 ← A_(i*n+j)^(1) ;
  STACK_(2)f . push ( v_1^0 ) ;
  v_1^0 ← A_(i*n+j) ;
  STACK_(2)f . push ( v_2^(1)0 ) ;
  v_2^(1)0 ← v_0^(1)0 * v_1^0 + v_0^0 * v_1^(1)0 ;
  STACK_(2)f . push ( v_2^0 ) ;
  v_2^0 ← v_0^0 * v_1^0 ;
  STACK_(2)f . push ( v_3^(1)0 ) ;
  v_3^(1)0 ← v_2^(1)0 * ( 0. - sin ( v_2^0 ) ) ;
  STACK_(2)f . push ( v_3^0 ) ;
  v_3^0 ← cos ( v_2^0 ) ;
  STACK_(2)f . push ( z^(1) ) ;
  z^(1) + ← v_3^(1)0 ;
  STACK_(2)f . push ( z ) ;
```

```
        z+←v₃⁰ ;
        STACK₍₂₎ᵢ . push ( j ) ;
        j←j +1;
    }
    STACK₍₂₎꜀ . push ( 316 ) ;
    STACK₍₂₎f . push ( v₀⁽¹⁾⁰ ) ;
    v₀⁽¹⁾⁰←y⁽¹⁾ ;
    STACK₍₂₎f . push ( v₀⁰ ) ;
    v₀⁰←y ;
    STACK₍₂₎f . push ( v₁⁽¹⁾⁰ ) ;
    v₁⁽¹⁾⁰←z⁽¹⁾ ;
    STACK₍₂₎f . push ( v₁⁰ ) ;
    v₁⁰←z ;
    STACK₍₂₎f . push ( v₂⁽¹⁾⁰ ) ;
    v₂⁽¹⁾⁰←v₀⁽¹⁾⁰*v₁⁰+v₀⁰*v₁⁽¹⁾⁰ ;
    STACK₍₂₎f . push ( v₂⁰ ) ;
    v₂⁰←v₀⁰*v₁⁰ ;
    STACK₍₂₎f . push ( thread _ result_tid⁽¹⁾ ) ;
    thread _ result_tid⁽¹⁾+←v₂⁽¹⁾⁰ ;
    STACK₍₂₎f . push ( thread _ result_tid ) ;
    thread _ result_tid+←v₂⁰ ;
    STACK₍₂₎ᵢ . push ( i ) ;
    i←i +1;
}
dummy ( "" , ( void *) thread _ result , ( void *) thread _ result⁽¹⁾ ) ;
while ( not STACK₍₂₎꜀ . empty ( )) {
    if ( STACK₍₂₎꜀ . top ( ) = 78 ) {
        STACK₍₂₎꜀ . pop ( ) ;
        thread _ result_tid←STACK₍₂₎f . top ( ) ;
        STACK₍₂₎f . pop ( ) ;
        v₀¹←v₀⁰ ;
        v₍₂₎₀¹←thread _ result₍₂₎tid ;
        thread _ result₍₂₎tid←0 . ;
        v₍₂₎₀⁰+←v₍₂₎₀¹ ;
        thread _ result_tid⁽¹⁾←STACK₍₂₎f . top ( ) ;
        STACK₍₂₎f . pop ( ) ;
        v₀¹←v₀⁽¹⁾⁰ ;
        v₍₂₎₀¹←thread _ result₍₂₎tid⁽¹⁾ ;
        thread _ result₍₂₎tid⁽¹⁾←0 . ;
        v₍₂₎₀⁽¹⁾⁰+←v₍₂₎₀¹ ;
        v₀⁰←STACK₍₂₎f . top ( ) ;
        STACK₍₂₎f . pop ( ) ;
        v₀¹←0 . ;
        v₍₂₎₀¹←v₍₂₎₀⁰ ;
        v₍₂₎₀⁰←0 . ;
        v₀⁽¹⁾⁰←STACK₍₂₎f . top ( ) ;
        STACK₍₂₎f . pop ( ) ;
        v₀¹←0 . ;
```

```
        v^1_{(2)0} ← v^{(1)0}_{(2)0} ;
        v^{(1)0}_{(2)0} ← 0. ;
        i ← STACK_{(2)i} . top ( ) ;
        STACK_{(2)i} . pop ( ) ;
        ub ← STACK_{(2)i} . top ( ) ;
        STACK_{(2)i} . pop ( ) ;
        lb ← STACK_{(2)i} . top ( ) ;
        STACK_{(2)i} . pop ( ) ;
        c ← STACK_{(2)i} . top ( ) ;
        STACK_{(2)i} . pop ( ) ;
        p ← STACK_{(2)i} . top ( ) ;
        STACK_{(2)i} . pop ( ) ;
        tid ← STACK_{(2)i} . top ( ) ;
        STACK_{(2)i} . pop ( ) ;
    }
    if ( STACK_{(2)c} . top ( ) = 127) {
        STACK_{(2)c} . pop ( ) ;
        j ← STACK_{(2)i} . top ( ) ;
        STACK_{(2)i} . pop ( ) ;
        z ← STACK_{(2)f} . top ( ) ;
        STACK_{(2)f} . pop ( ) ;
        v^1_0 ← v^0_0 ;
        v^1_{(2)0} ← z_{(2)} ;
        z_{(2)} ← 0. ;
        v^0_{(2)0} +← v^1_{(2)0} ;
        z^{(1)} ← STACK_{(2)f} . top ( ) ;
        STACK_{(2)f} . pop ( ) ;
        v^1_0 ← v^{(1)0}_0 ;
        v^1_{(2)0} ← z^{(1)}_{(2)} ;
        z^{(1)}_{(2)} ← 0. ;
        v^{(1)0}_{(2)0} +← v^1_{(2)0} ;
        v^0_0 ← STACK_{(2)f} . top ( ) ;
        STACK_{(2)f} . pop ( ) ;
        v^1_0 ← 0. ;
        v^1_{(2)0} ← v^0_{(2)0} ;
        v^0_{(2)0} ← 0. ;
        v^{(1)0}_0 ← STACK_{(2)f} . top ( ) ;
        STACK_{(2)f} . pop ( ) ;
        v^1_0 ← 0. ;
        v^1_{(2)0} ← v^{(1)0}_{(2)0} ;
        v^{(1)0}_{(2)0} ← 0. ;
        y ← STACK_{(2)f} . top ( ) ;
        STACK_{(2)f} . pop ( ) ;
        v^1_0 ← v^0_0 ;
        v^1_{(2)0} ← y_{(2)} ;
        y_{(2)} ← 0. ;
        v^0_{(2)0} +← v^1_{(2)0} ;
        y^{(1)} ← STACK_{(2)f} . top ( ) ;
```

```
    STACK_{(2)f} . pop ( ) ;
```

$v_0^1 \leftarrow v_0^{(1)0}$;

$v_{(2)0}^1 \leftarrow y_{(2)}^{(1)}$;

$y_{(2)}^{(1)} \leftarrow 0 .$;

$v_{(2)0}^{(1)0} + \leftarrow v_{(2)0}^1$;

$v_0^0 \leftarrow$ STACK$_{(2)f}$. top ( ) ;

```
    STACK_{(2)f} . pop ( ) ;
```

$v_0^1 \leftarrow 0 .$;

$v_{(2)0}^1 \leftarrow v_{(2)0}^0$;

$v_{(2)0}^0 \leftarrow 0 .$;

$v_0^{(1)0} \leftarrow$ STACK$_{(2)f}$. top ( ) ;

```
    STACK_{(2)f} . pop ( ) ;
```

$v_0^1 \leftarrow 0 .$;

$v_{(2)0}^1 \leftarrow v_{(2)0}^{(1)0}$;

$v_{(2)0}^{(1)0} \leftarrow 0 .$;

```
}
if ( STACK_{(2)c} . top ( ) = 166 ) {
    STACK_{(2)c} . pop ( ) ;
    j ← STACK_{(2)i} . top ( ) ;
    STACK_{(2)i} . pop ( ) ;
    z ← STACK_{(2)f} . top ( ) ;
    STACK_{(2)f} . pop ( ) ;
```

$v_0^1 \leftarrow v_3^0$;

$v_{(2)0}^1 \leftarrow z_{(2)}$;

$v_{(2)3}^0 + \leftarrow v_{(2)0}^1$;

$z^{(1)} \leftarrow$ STACK$_{(2)f}$. top ( ) ;

```
    STACK_{(2)f} . pop ( ) ;
```

$v_0^1 \leftarrow v_3^{(1)0}$;

$v_{(2)0}^1 \leftarrow z_{(2)}^{(1)}$;

$v_{(2)3}^{(1)0} + \leftarrow v_{(2)0}^1$;

$v_3^0 \leftarrow$ STACK$_{(2)f}$. top ( ) ;

```
    STACK_{(2)f} . pop ( ) ;
```

$v_0^1 \leftarrow v_2^0$;

$v_1^1 \leftarrow \cos (v_0^1)$;

$v_{(2)1}^1 \leftarrow v_{(2)3}^0$;

$v_{(2)3}^0 \leftarrow 0 .$;

$v_{(2)0}^1 \leftarrow v_{(2)1}^1 * (0 . - \sin (v_0^1))$;

$v_{(2)2}^0 + \leftarrow v_{(2)0}^1$;

$v_3^{(1)0} \leftarrow$ STACK$_{(2)f}$. top ( ) ;

```
    STACK_{(2)f} . pop ( ) ;
```

$v_0^1 \leftarrow v_2^{(1)0}$;

$v_1^1 \leftarrow 0 .$;

$v_2^1 \leftarrow v_2^0$;

$v_3^1 \leftarrow \sin (v_2^1)$;

$v_4^1 \leftarrow v_1^1 - v_3^1$;

$v_5^1 \leftarrow v_0^1 * v_4^1$;

$v_{(2)5}^1 \leftarrow v_{(2)3}^{(1)0}$;

$$v_{(2)3}^{(1)0} \leftarrow 0.\,;$$
$$v_{(2)0}^{1} \leftarrow v_{(2)5}^{1} * v_4^1\,;$$
$$v_{(2)4}^{1} \leftarrow v_{(2)5}^{1} * v_0^1\,;$$
$$v_{(2)2}^{(1)0} + \leftarrow v_{(2)0}^1\,;$$
$$v_{(2)1}^{1} \leftarrow v_{(2)4}^1\,;$$
$$v_{(2)3}^{1} \leftarrow 0. - v_{(2)4}^1\,;$$
$$v_{(2)2}^{1} \leftarrow v_{(2)3}^1 * \cos(v_2^1)\,;$$
$$v_{(2)2}^{0} + \leftarrow v_{(2)2}^1\,;$$
$$v_2^0 \leftarrow STACK_{(2)f}.top()\,;$$
$$STACK_{(2)f}.pop()\,;$$
$$v_0^1 \leftarrow v_0^0\,;$$
$$v_1^1 \leftarrow v_1^0\,;$$
$$v_2^1 \leftarrow v_0^1 * v_1^1\,;$$
$$v_{(2)2}^1 \leftarrow v_{(2)2}^0\,;$$
$$v_{(2)2}^0 \leftarrow 0.\,;$$
$$v_{(2)0}^1 \leftarrow v_{(2)2}^1 * v_1^1\,;$$
$$v_{(2)1}^1 \leftarrow v_{(2)2}^1 * v_0^1\,;$$
$$v_{(2)0}^0 + \leftarrow v_{(2)0}^1\,;$$
$$v_{(2)1}^0 + \leftarrow v_{(2)1}^1\,;$$
$$v_2^{(1)0} \leftarrow STACK_{(2)f}.top()\,;$$
$$STACK_{(2)f}.pop()\,;$$
$$v_0^1 \leftarrow v_0^{(1)0}\,;$$
$$v_1^1 \leftarrow v_1^0\,;$$
$$v_2^1 \leftarrow v_0^1 * v_1^1\,;$$
$$v_3^1 \leftarrow v_0^1\,;$$
$$v_4^1 \leftarrow v_1^{(1)0}\,;$$
$$v_5^1 \leftarrow v_3^1 * v_4^1\,;$$
$$v_6^1 \leftarrow v_2^1 + v_5^1\,;$$
$$v_{(2)6}^1 \leftarrow v_{(2)2}^{(1)0}\,;$$
$$v_{(2)2}^{(1)0} \leftarrow 0.\,;$$
$$v_{(2)2}^1 \leftarrow v_{(2)6}^1\,;$$
$$v_{(2)5}^1 \leftarrow v_{(2)6}^1\,;$$
$$v_{(2)0}^1 \leftarrow v_{(2)2}^1 * v_1^1\,;$$
$$v_{(2)1}^1 \leftarrow v_{(2)2}^1 * v_0^1\,;$$
$$v_{(2)0}^{(1)0} + \leftarrow v_{(2)0}^1\,;$$
$$v_{(2)1}^0 + \leftarrow v_{(2)1}^1\,;$$
$$v_{(2)3}^1 \leftarrow v_{(2)5}^1 * v_4^1\,;$$
$$v_{(2)4}^1 \leftarrow v_{(2)5}^1 * v_3^1\,;$$
$$v_{(2)0}^0 + \leftarrow v_{(2)3}^1\,;$$
$$v_{(2)1}^{(1)0} + \leftarrow v_{(2)4}^1\,;$$
$$v_1^0 \leftarrow STACK_{(2)f}.top()\,;$$
$$STACK_{(2)f}.pop()\,;$$
$$v_0^1 \leftarrow A_{l*n+j}\,;$$
$$v_{(2)0}^1 \leftarrow v_{(2)1}^0\,;$$
$$v_{(2)1}^0 \leftarrow 0.\,;$$

```
#pragma omp atomic
```

$A_{(2)i*n+j} \mathrel{+}\!\!\leftarrow v^1_{(2)0}$;
$v^{(1)0}_1 \leftarrow \text{STACK}_{(2)f}.\text{top}()$;
$\text{STACK}_{(2)f}.\text{pop}()$;
$v^1_0 \leftarrow A^{(1)}_{i*n+j}$;
$v^1_{(2)0} \leftarrow v^{(1)0}_{(2)1}$;
$v^{(1)0}_{(2)1} \leftarrow 0.$;
#**pragma omp atomic**
$A^{(1)}_{(2)i*n+j} \mathrel{+}\!\!\leftarrow v^1_{(2)0}$;
$v^0_0 \leftarrow \text{STACK}_{(2)f}.\text{top}()$;
$\text{STACK}_{(2)f}.\text{pop}()$;
$v^1_0 \leftarrow A_{i*n+j}$;
$v^1_{(2)0} \leftarrow v^0_{(2)0}$;
$v^0_{(2)0} \leftarrow 0.$;
#**pragma omp atomic**
$A_{(2)i*n+j} \mathrel{+}\!\!\leftarrow v^1_{(2)0}$;
$v^{(1)0}_0 \leftarrow \text{STACK}_{(2)f}.\text{top}()$;
$\text{STACK}_{(2)f}.\text{pop}()$;
$v^1_0 \leftarrow A^{(1)}_{i*n+j}$;
$v^1_{(2)0} \leftarrow v^{(1)0}_{(2)0}$;
$v^{(1)0}_{(2)0} \leftarrow 0.$;
#**pragma omp atomic**
$A^{(1)}_{(2)i*n+j} \mathrel{+}\!\!\leftarrow v^1_{(2)0}$;
$y \leftarrow \text{STACK}_{(2)f}.\text{top}()$;
$\text{STACK}_{(2)f}.\text{pop}()$;
$v^1_0 \leftarrow v^0_3$;
$v^1_{(2)0} \leftarrow y_{(2)}$;
$v^0_{(2)3} \mathrel{+}\!\!\leftarrow v^1_{(2)0}$;
$y^{(1)} \leftarrow \text{STACK}_{(2)f}.\text{top}()$;
$\text{STACK}_{(2)f}.\text{pop}()$;
$v^1_0 \leftarrow v^{(1)0}_3$;
$v^1_{(2)0} \leftarrow y^{(1)}_{(2)}$;
$v^{(1)0}_{(2)3} \mathrel{+}\!\!\leftarrow v^1_{(2)0}$;
$v^0_3 \leftarrow \text{STACK}_{(2)f}.\text{top}()$;
$\text{STACK}_{(2)f}.\text{pop}()$;
$v^1_0 \leftarrow v^0_2$;
$v^1_1 \leftarrow \sin(v^1_0)$;
$v^1_{(2)1} \leftarrow v^0_{(2)3}$;
$v^0_{(2)3} \leftarrow 0.$;
$v^1_{(2)0} \leftarrow v^1_{(2)1} * \cos(v^1_0)$;
$v^0_{(2)2} \mathrel{+}\!\!\leftarrow v^1_{(2)0}$;
$v^{(1)0}_3 \leftarrow \text{STACK}_{(2)f}.\text{top}()$;
$\text{STACK}_{(2)f}.\text{pop}()$;
$v^1_0 \leftarrow v^{(1)0}_2$;
$v^1_1 \leftarrow v^0_2$;
$v^1_2 \leftarrow \cos(v^1_1)$;
$v^1_3 \leftarrow v^1_0 * v^1_2$;

$$v^1_{(2)3} \leftarrow v^{(1)0}_{(2)3} \; ;$$
$$v^{(1)0}_{(2)3} \leftarrow 0 . \; ;$$
$$v^1_{(2)0} \leftarrow v^1_{(2)3} * v^1_2 \; ;$$
$$v^1_{(2)2} \leftarrow v^1_{(2)3} * v^1_0 \; ;$$
$$v^{(1)0}_{(2)2} + \leftarrow v^1_{(2)0} \; ;$$
$$v^1_{(2)1} \leftarrow v^1_{(2)2} * (0 . - \sin(v^1_1)) \; ;$$
$$v^0_{(2)2} + \leftarrow v^1_{(2)1} \; ;$$
$$v^0_2 \leftarrow \text{STACK}_{(2)f} . \text{top}() \; ;$$
$$\text{STACK}_{(2)f} . \text{pop}() \; ;$$
$$v^1_0 \leftarrow v^0_0 \; ;$$
$$v^1_1 \leftarrow v^0_1 \; ;$$
$$v^1_2 \leftarrow v^1_0 * v^1_1 \; ;$$
$$v^1_{(2)2} \leftarrow v^0_{(2)2} \; ;$$
$$v^0_{(2)2} \leftarrow 0 . \; ;$$
$$v^1_{(2)0} \leftarrow v^1_{(2)2} * v^1_1 \; ;$$
$$v^1_{(2)1} \leftarrow v^1_{(2)2} * v^1_0 \; ;$$
$$v^0_{(2)0} + \leftarrow v^1_{(2)0} \; ;$$
$$v^0_{(2)1} + \leftarrow v^1_{(2)1} \; ;$$
$$v^{(1)0}_2 \leftarrow \text{STACK}_{(2)f} . \text{top}() \; ;$$
$$\text{STACK}_{(2)f} . \text{pop}() \; ;$$
$$v^1_0 \leftarrow v^{(1)0}_0 \; ;$$
$$v^1_1 \leftarrow v^0_1 \; ;$$
$$v^1_2 \leftarrow v^1_0 * v^1_1 \; ;$$
$$v^1_3 \leftarrow v^0_0 \; ;$$
$$v^1_4 \leftarrow v^{(1)0}_1 \; ;$$
$$v^1_5 \leftarrow v^1_3 * v^1_4 \; ;$$
$$v^1_6 \leftarrow v^1_2 + v^1_5 \; ;$$
$$v^1_{(2)6} \leftarrow v^{(1)0}_{(2)2} \; ;$$
$$v^{(1)0}_{(2)2} \leftarrow 0 . \; ;$$
$$v^1_{(2)2} \leftarrow v^1_{(2)6} \; ;$$
$$v^1_{(2)5} \leftarrow v^1_{(2)6} \; ;$$
$$v^1_{(2)0} \leftarrow v^1_{(2)2} * v^1_1 \; ;$$
$$v^1_{(2)1} \leftarrow v^1_{(2)2} * v^1_0 \; ;$$
$$v^{(1)0}_{(2)0} + \leftarrow v^1_{(2)0} \; ;$$
$$v^0_{(2)1} + \leftarrow v^1_{(2)1} \; ;$$
$$v^1_{(2)3} \leftarrow v^1_{(2)5} * v^1_4 \; ;$$
$$v^1_{(2)4} \leftarrow v^1_{(2)5} * v^1_3 \; ;$$
$$v^0_{(2)0} + \leftarrow v^1_{(2)3} \; ;$$
$$v^{(1)0}_{(2)1} + \leftarrow v^1_{(2)4} \; ;$$
$$v^0_1 \leftarrow \text{STACK}_{(2)f} . \text{top}() \; ;$$
$$\text{STACK}_{(2)f} . \text{pop}() \; ;$$
$$v^1_0 \leftarrow A_{l*n+j} \; ;$$
$$v^1_{(2)0} \leftarrow v^0_{(2)1} \; ;$$
$$v^0_{(2)1} \leftarrow 0 . \; ;$$

```
#pragma omp atomic
```
$$A_{(2)i*n+j} + \leftarrow v^1_{(2)0} \; ;$$

```
v_1^{(1)0} ← STACK_{(2)f} . top ( ) ;
STACK_{(2)f} . pop ( ) ;
v_0^1 ← A_{i*n+j}^{(1)} ;
v_{(2)0}^1 ← v_{(2)1}^{(1)0} ;
v_{(2)1}^{(1)0} ← 0 . ;
```

#pragma omp atomic
```
A_{(2)i*n+j}^{(1)} +← v_{(2)0}^1 ;
v_0^0 ← STACK_{(2)f} . top ( ) ;
STACK_{(2)f} . pop ( ) ;
v_0^1 ← A_{i*n+j} ;
v_{(2)0}^1 ← v_{(2)0}^0 ;
v_{(2)0}^0 ← 0 . ;
```

#pragma omp atomic
```
A_{(2)i*n+j} +← v_{(2)0}^1 ;
v_0^{(1)0} ← STACK_{(2)f} . top ( ) ;
STACK_{(2)f} . pop ( ) ;
v_0^1 ← A_{i*n+j}^{(1)} ;
v_{(2)0}^1 ← v_{(2)0}^{(1)0} ;
v_{(2)0}^{(1)0} ← 0 . ;
```

#pragma omp atomic
```
A_{(2)i*n+j}^{(1)} +← v_{(2)0}^1 ;
}
if ( STACK_{(2)c} . top ( ) = 316 ) {
    STACK_{(2)c} . pop ( ) ;
    i ← STACK_{(2)i} . top ( ) ;
    STACK_{(2)i} . pop ( ) ;
    thread_result_{tid} ← STACK_{(2)f} . top ( ) ;
    STACK_{(2)f} . pop ( ) ;
    v_0^1 ← v_2^0 ;
    v_{(2)0}^1 ← thread_result_{(2)tid} ;
    v_{(2)2}^0 +← v_{(2)0}^1 ;
    thread_result_{tid}^{(1)} ← STACK_{(2)f} . top ( ) ;
    STACK_{(2)f} . pop ( ) ;
    v_0^1 ← v_2^{(1)0} ;
    v_{(2)0}^1 ← thread_result_{(2)tid}^{(1)} ;
    v_{(2)2}^{(1)0} +← v_{(2)0}^1 ;
    v_2^0 ← STACK_{(2)f} . top ( ) ;
    STACK_{(2)f} . pop ( ) ;
    v_0^1 ← v_0^0 ;
    v_1^1 ← v_1^0 ;
    v_2^1 ← v_0^1 * v_1^1 ;
    v_{(2)2}^1 ← v_{(2)2}^0 ;
    v_{(2)2}^0 ← 0 . ;
    v_{(2)0}^1 ← v_{(2)2}^1 * v_1^1 ;
    v_{(2)1}^1 ← v_{(2)2}^1 * v_0^1 ;
    v_{(2)0}^0 +← v_{(2)0}^1 ;
```

$v_{(2)1}^0 + \leftarrow v_{(2)1}^1 \,;$
$v_2^{(1)0} \leftarrow \text{STACK}_{(2)f} . \text{top}()\,;$
$\text{STACK}_{(2)f} . \text{pop}()\,;$
$v_0^1 \leftarrow v_0^{(1)0}\,;$
$v_1^1 \leftarrow v_1^0\,;$
$v_2^1 \leftarrow v_0^1 * v_1^1\,;$
$v_3^1 \leftarrow v_0^0\,;$
$v_4^1 \leftarrow v_1^{(1)0}\,;$
$v_5^1 \leftarrow v_3^1 * v_4^1\,;$
$v_6^1 \leftarrow v_2^1 + v_5^1\,;$
$v_{(2)6}^1 \leftarrow v_{(2)2}^{(1)0}\,;$
$v_{(2)2}^{(1)0} \leftarrow 0.\,;$
$v_{(2)2}^1 \leftarrow v_{(2)6}^1\,;$
$v_{(2)5}^1 \leftarrow v_{(2)6}^1\,;$
$v_{(2)0}^1 \leftarrow v_{(2)2}^1 * v_1^1\,;$
$v_{(2)1}^1 \leftarrow v_{(2)2}^1 * v_0^1\,;$
$v_{(2)0}^{(1)0} + \leftarrow v_{(2)0}^1\,;$
$v_{(2)1}^0 + \leftarrow v_{(2)1}^1\,;$
$v_{(2)3}^1 \leftarrow v_{(2)5}^1 * v_4^1\,;$
$v_{(2)4}^1 \leftarrow v_{(2)5}^1 * v_3^1\,;$
$v_{(2)0}^0 + \leftarrow v_{(2)3}^1\,;$
$v_{(2)1}^{(1)0} + \leftarrow v_{(2)4}^1\,;$
$v_1^0 \leftarrow \text{STACK}_{(2)f} . \text{top}()\,;$
$\text{STACK}_{(2)f} . \text{pop}()\,;$
$v_0^1 \leftarrow z\,;$
$v_{(2)0}^1 \leftarrow v_{(2)1}^0\,;$
$v_{(2)1}^0 \leftarrow 0.\,;$
$z_{(2)} + \leftarrow v_{(2)0}^1\,;$
$v_1^{(1)0} \leftarrow \text{STACK}_{(2)f} . \text{top}()\,;$
$\text{STACK}_{(2)f} . \text{pop}()\,;$
$v_0^1 \leftarrow z^{(1)}\,;$
$v_{(2)0}^1 \leftarrow v_{(2)1}^{(1)0}\,;$
$v_{(2)1}^{(1)0} \leftarrow 0.\,;$
$z_{(2)}^{(1)} + \leftarrow v_{(2)0}^1\,;$
$v_0^0 \leftarrow \text{STACK}_{(2)f} . \text{top}()\,;$
$\text{STACK}_{(2)f} . \text{pop}()\,;$
$v_0^1 \leftarrow y\,;$
$v_{(2)0}^1 \leftarrow v_{(2)0}^0\,;$
$v_{(2)0}^0 \leftarrow 0.\,;$
$y_{(2)} + \leftarrow v_{(2)0}^1\,;$
$v_0^{(1)0} \leftarrow \text{STACK}_{(2)f} . \text{top}()\,;$
$\text{STACK}_{(2)f} . \text{pop}()\,;$
$v_0^1 \leftarrow y^{(1)}\,;$
$v_{(2)0}^1 \leftarrow v_{(2)0}^{(1)0}\,;$
$v_{(2)0}^{(1)0} \leftarrow 0.\,;$

$$y_{(2)}^{(1)} + \leftarrow v_{(2)0}^1;$$
```
    }
    dummy("", (void*)thread_result, (void*)thread_result^(1), (void*)
        thread_result_(2), (void*)thread_result_{(2)}^{(1)});
  }
}
```

Listing B.6: This code represents the second-order adjoint code of the 'plain-parallel' test case (Listing B.1) in reverse-over-forward mode. This code is obtained by applying first the forward mode and then the reverse mode, more precisely one applies the source transformation $\sigma(\tau(P))$.

B.2 Test Case with a Barrier

The runtime details of this test can be found in Section 5.1.2 on page 281.

B.2.1 Original Code 'barrier'

```
#pragma omp parallel
{
    double y;
    double z;
    unsigned int tid;
    unsigned int p;
    unsigned int c;
    unsigned int i;
    unsigned int j;
    unsigned int lb;
    unsigned int ub;

    tid←omp_get_thread_num();
    p←omp_get_num_threads();
    c←n/p;
    lb←tid*c;
    ub←(tid+1)*c−1;
    i←lb;
    z←0.;
    while(i ≤ ub) {
        j←0;
        y←0.;
        while(j<n) {
            y+←A_{l*n+j}*x_j;
            j←j+1;
        }
        z+←y;
        i←i+1;
    }
    thread_result_{tid}←z;
    #pragma omp barrier
    if(tid = 0) {
        i←0;
        y←0.;
```

```
while ( i <p) {
    y+←( thread _ result_i ) ;
    i ← i +1;
}
thread _ result_0 ← y ;
dummy( "" , (void*) thread _ result ) ;
}
}
```

Listing B.7: This code is the original code for the test case 'barrier'.

B.2.2 First-Order Tangent-Linear Code 'barrier'

```
#pragma omp parallel
{
    double y ;
    double y^{(1)} ← 0 .;
    double z ;
    double z^{(1)} ← 0 .;
    unsigned int tid ;
    unsigned int p ;
    unsigned int c ;
    unsigned int i ;
    unsigned int j ;
    unsigned int lb ;
    unsigned int ub ;
    double v_0^0 ;
    double v_0^{(1)0} ;
    double v_1^0 ;
    double v_1^{(1)0} ;
    double v_2^0 ;
    double v_2^{(1)0} ;

    tid ← omp_get_thread_num ( ) ;
    p ← omp_get_num_threads ( ) ;
    c ← n/p ;
    lb ← tid*c ;
    ub ← ( tid +1)*c −1;
    i ← lb ;
    v_0^{(1)0} ← 0 .;
    v_0^0 ← 0 .;
    z^{(1)} ← v_0^{(1)0} ;
    z ← v_0^0 ;
    while ( i ≤ ub ) {
        j ← 0;
        v_0^{(1)0} ← 0 .;
        v_0^0 ← 0 .;
        y^{(1)} ← v_0^{(1)0} ;
        y ← v_0^0 ;
        while ( j <n) {
            v_0^{(1)0} ← A_{i*n+j}^{(1)} ;
            v_0^0 ← A_{i*n+j} ;
            v_1^{(1)0} ← x_j^{(1)} ;
```

```
        v₁⁰←xⱼ;
        v₂^(1)0←v₀^(1)0*v₁⁰+v₀⁰*v₁^(1)0;
        v₂⁰←v₀⁰*v₁⁰;
        y^(1)+←v₂^(1)0;
        y+←v₂⁰;
        j←j+1;
    }
    v₀^(1)0←y^(1);
    v₀⁰←y;
    z^(1)+←v₀^(1)0;
    z+←v₀⁰;
    i←i+1;
}
v₀^(1)0←z^(1);
v₀⁰←z;
thread_resultₜᵢ𝒹^(1)←v₀^(1)0;
thread_resultₜᵢ𝒹←v₀⁰;
#pragma omp barrier
if(tid = 0) {
    i←0;
    v₀^(1)0←0.;
    v₀⁰←0.;
    y^(1)←v₀^(1)0;
    y←v₀⁰;
    while(i<p) {
        v₀^(1)0←thread_resultᵢ^(1);
        v₀⁰←thread_resultᵢ;
        y^(1)+←v₀^(1)0;
        y+←v₀⁰;
        i←i+1;
    }
    v₀^(1)0←y^(1);
    v₀⁰←y;
    thread_result₀^(1)←v₀^(1)0;
    thread_result₀←v₀⁰;
    dummy("", (void*)thread_result, (void*)thread_result^(1));
    }
}
```

Listing B.8: This listing is the first-order tangent-linear code of the test case 'barrier' (Listing B.7).

B.2.3 First-Order Adjoint Code 'barrier'

```
#pragma omp parallel
{
    double y;
    double y_(1)←0.;
    double z;
    double z_(1)←0.;
    unsigned int tid;
```

```
unsigned int p;
unsigned int c;
unsigned int i;
unsigned int j;
unsigned int lb;
unsigned int ub;
double v_0^0;
double v_{(1)0}^0;
double v_1^0;
double v_{(1)1}^0;
double v_2^0;
double v_{(1)2}^0;

STACK_{(1)c}.push(34);
STACK_{(1)i}.push(tid);
tid←omp_get_thread_num();
STACK_{(1)i}.push(p);
p←omp_get_num_threads();
STACK_{(1)i}.push(c);
c←n/p;
STACK_{(1)i}.push(lb);
lb←tid*c;
STACK_{(1)i}.push(ub);
ub←(tid+1)*c-1;
STACK_{(1)i}.push(i);
i←lb;
STACK_{(1)f}.push(z);
z←0.;
while(i ≤ ub) {
  STACK_{(1)c}.push(70);
  STACK_{(1)i}.push(j);
  j←0;
  STACK_{(1)f}.push(y);
  y←0.;
  while(j<n) {
    STACK_{(1)c}.push(92);
    STACK_{(1)f}.push(y);
    y+←A_{i*n+j}*x_j;
    STACK_{(1)i}.push(j);
    j←j+1;
  }
  STACK_{(1)c}.push(102);
  STACK_{(1)f}.push(z);
  z+←y;
  STACK_{(1)i}.push(i);
  i←i+1;
}
STACK_{(1)c}.push(113);
STACK_{(1)f}.push(thread_result_{tid});
thread_result_{tid}←z;
#pragma omp barrier
if(tid = 0) {
  STACK_{(1)c}.push(120);
  STACK_{(1)i}.push(i);
```

```
      i ← 0;
      STACK(1)f . push ( y ) ;
      y ← 0 . ;
      while ( i < p )  {
        STACK(1)c . push ( 133 ) ;
        STACK(1)f . push ( y ) ;
        y + ← ( thread _ resulti ) ;
        STACK(1)i . push ( i ) ;
        i ← i + 1;
      }
      STACK(1)c . push ( 145 ) ;
      STACK(1)f . push ( thread _ result0 ) ;
      thread _ result0 ← y ;
      dummy ( "π" ,  ( void * ) thread _ result ) ;
}
while (  not  STACK(1)c . empty ( ) )  {
   if ( STACK(1)c . top ( )  =  34 )  {
     STACK(1)c . pop ( ) ;
     z ← STACK(1)f . top ( ) ;
     STACK(1)f . pop ( ) ;
     v0^0 ← 0 . ;
     v(1)0^0 ← z(1) ;
     z(1) ← 0 . ;
     i ← STACK(1)i . top ( ) ;
     STACK(1)i . pop ( ) ;
     ub ← STACK(1)i . top ( ) ;
     STACK(1)i . pop ( ) ;
     lb ← STACK(1)i . top ( ) ;
     STACK(1)i . pop ( ) ;
     c ← STACK(1)i . top ( ) ;
     STACK(1)i . pop ( ) ;
     p ← STACK(1)i . top ( ) ;
     STACK(1)i . pop ( ) ;
     tid ← STACK(1)i . top ( ) ;
     STACK(1)i . pop ( ) ;
   }
   if ( STACK(1)c . top ( )  =  70 )  {
     STACK(1)c . pop ( ) ;
     y ← STACK(1)f . top ( ) ;
     STACK(1)f . pop ( ) ;
     v0^0 ← 0 . ;
     v(1)0^0 ← y(1) ;
     y(1) ← 0 . ;
     j ← STACK(1)i . top ( ) ;
     STACK(1)i . pop ( ) ;
   }
   if ( STACK(1)c . top ( )  =  92 )  {
     STACK(1)c . pop ( ) ;
     j ← STACK(1)i . top ( ) ;
     STACK(1)i . pop ( ) ;
     y ← STACK(1)f . top ( ) ;
     STACK(1)f . pop ( ) ;
```

```
            v_0^0←A_{i*n+j} ;
            v_1^0←x_j ;
            v_2^0←v_0^0*v_1^0 ;
            v_{(1)2}^0←y_{(1)} ;
            v_{(1)0}^0←v_{(1)2}^0*v_1^0 ;
            v_{(1)1}^0←v_{(1)2}^0*v_0^0 ;
            A_{(1)i*n+j}++←v_{(1)0}^0 ;
            #pragma omp atomic
            x_{(1)j}++←v_{(1)1}^0 ;
        }
        if (STACK_{(1)c}.top() = 102) {
            STACK_{(1)c}.pop() ;
            i←STACK_{(1)i}.top() ;
            STACK_{(1)i}.pop() ;
            z←STACK_{(1)f}.top() ;
            STACK_{(1)f}.pop() ;
            v_0^0←y ;
            v_{(1)0}^0←z_{(1)} ;
            y_{(1)}++←v_{(1)0}^0 ;
        }
        if (STACK_{(1)c}.top() = 113) {
            STACK_{(1)c}.pop() ;
            #pragma omp barrier
            thread_result_{tid}←STACK_{(1)f}.top() ;
            STACK_{(1)f}.pop() ;
            v_0^0←z ;
            v_{(1)0}^0←thread_result_{(1)tid} ;
            thread_result_{(1)tid}←0. ;
            z_{(1)}++←v_{(1)0}^0 ;
        }
        if (STACK_{(1)c}.top() = 120) {
            STACK_{(1)c}.pop() ;
            y←STACK_{(1)f}.top() ;
            STACK_{(1)f}.pop() ;
            v_0^0←0. ;
            v_{(1)0}^0←y_{(1)} ;
            y_{(1)}←0. ;
            i←STACK_{(1)i}.top() ;
            STACK_{(1)i}.pop() ;
        }
        if (STACK_{(1)c}.top() = 133) {
            STACK_{(1)c}.pop() ;
            i←STACK_{(1)i}.top() ;
            STACK_{(1)i}.pop() ;
            y←STACK_{(1)f}.top() ;
            STACK_{(1)f}.pop() ;
            v_0^0←thread_result_i ;
            v_{(1)0}^0←y_{(1)} ;
            #pragma omp atomic
            thread_result_{(1)i}++←v_{(1)0}^0 ;
```

```
        }
      if (STACK₍₁₎c . top ( ) = 145) {
        STACK₍₁₎c . pop ( ) ;
        thread_result₀←STACK₍₁₎f . top ( ) ;
        STACK₍₁₎f . pop ( ) ;
        v⁰₀←y ;
        v⁰₍₁₎₀←thread_result₍₁₎₀ ;
        thread_result₍₁₎₀←0. ;
        y₍₁₎++v⁰₍₁₎₀ ;
      }
      dummy("" , (void*)thread_result , (void*)thread_result₍₁₎ ) ;
    }
  }
```

Listing B.9: This code represents the adjoint code of the test case 'barrier' (Listing B.7). The exclusive read analysis recognizes that the atomic construct in line 120 can be neglected but the one in line 122 is necessary. The atomic construct in line 163 is recognized as exclusively since the analysis does not recognize that this code is only performed by the master thread.

B.3 Test Case with a master Construct

The runtime details of this test can be found in Section 5.1.3 on page 290.

B.3.1 Original Code 'master'

```
#pragma omp parallel
{
  int tests←0;
  int tid←0;
  int p←0;
  int i←0;
  int j←0;
  int lb←0;
  int ub←0;
  int c←0;
  double y←0.;

  tid←omp_get_thread_num();
  p←omp_get_num_threads();
  c←n/p;
  lb←tid*c;
  ub←(tid+1)*c−1;
  i←lb;
  y←1.;
  while(i ≤ ub) {
    j←0;
    while(j<n) {
      y←y*sin(A_{i*n+j}*x_j)*cos(A_{i*n+j}*x_j);
      j←j+1;
    }
    i←i+1;
```

```
    }
    thread_result_{tid}←y;
#pragma omp barrier
#pragma omp master
    {
        i←1;
        while(i<p) {
            thread_result_0←thread_result_0*thread_result_i;
            i←i+1;
        }
        dummy("", (void*)thread_result);
    }
}
```

Listing B.10: This code is the original code for the test case 'master'.

B.3.2 First-Order Tangent-Linear Code 'master'

```
#pragma omp parallel
{
    int tests←0;
    int tid←0;
    int p←0;
    int i←0;
    int j←0;
    int lb←0;
    int ub←0;
    int c←0;
    double y←0.;
    double y^{(1)}←0.;
    double v_0^0;
    double v_0^{(1)0};
    double v_1^0;
    double v_1^{(1)0};
    double v_2^0;
    double v_2^{(1)0};
    double v_3^0;
    double v_3^{(1)0};
    double v_4^0;
    double v_4^{(1)0};
    double v_5^0;
    double v_5^{(1)0};
    double v_6^0;
    double v_6^{(1)0};
    double v_7^0;
    double v_7^{(1)0};
    double v_8^0;
    double v_8^{(1)0};
    double v_9^0;
    double v_9^{(1)0};
    double v_{10}^0;
    double v^{(1)0}_{10};
```

```
tid←omp_get_thread_num();
p←omp_get_num_threads();
c←n/p;
lb←tid*c;
ub←(tid+1)*c−1;
i←lb;
v_0^{(1)0}←0.;
v_0^0←1.;
y^{(1)}←v_0^{(1)0};
y←v_0^0;
while(i ≤ ub) {
   j←0;
   while(j<n) {
      v_0^{(1)0}←y^{(1)};
      v_0^0←y;
      v_1^{(1)0}←A_{i*n+j}^{(1)};
      v_1^0←A_{i*n+j};
      v_2^{(1)0}←x_j^{(1)};
      v_2^0←x_j;
      v_3^{(1)0}←v_1^{(1)0}*v_2^0+v_1^0*v_2^{(1)0};
      v_3^0←v_1^0*v_2^0;
      v_4^{(1)0}←v_3^{(1)0}*cos(v_3^0);
      v_4^0←sin(v_3^0);
      v_5^{(1)0}←v_0^{(1)0}*v_4^0+v_0^0*v_4^{(1)0};
      v_5^0←v_0^0*v_4^0;
      v_6^{(1)0}←A_{i*n+j}^{(1)};
      v_6^0←A_{i*n+j};
      v_7^{(1)0}←x_j^{(1)};
      v_7^0←x_j;
      v_8^{(1)0}←v_6^{(1)0}*v_7^0+v_6^0*v_7^{(1)0};
      v_8^0←v_6^0*v_7^0;
      v_9^{(1)0}←v_8^{(1)0}*(0.−sin(v_8^0));
      v_9^0←cos(v_8^0);
      v_{10}^{(1)0}←v_5^{(1)0}*v_9^0+v_5^0*v_9^{(1)0};
      v_{10}^0←v_5^0*v_9^0;
      y^{(1)}←v_{10}^{(1)0};
      y←v_{10}^0;
      j←j+1;
   }
   i←i+1;
}
v_0^{(1)0}←y^{(1)};
v_0^0←y;
thread_result_{tid}^{(1)}←v_0^{(1)0};
thread_result_{tid}←v_0^0;
#pragma omp barrier
#pragma omp master
{
   i←1;
```

```
while ( i <p) {
    v₀^(1)0←thread _ result₀^(1) ;
    v₀^0←thread _ result₀ ;
    v₁^(1)0←thread _ resultᵢ^(1) ;
    v₁^0←thread _ resultᵢ ;
    v₂^(1)0←v₀^(1)0 *v₁^0+v₀^0 *v₁^(1)0 ;
    v₂^0←v₀^0 *v₁^0 ;
    thread _ result₀^(1)←v₂^(1)0 ;
    thread _ result₀←v₂^0 ;
    i←i +1;
}
dummy ( " " , ( void *)thread _ result , ( void *)thread _ result^(1) );
}
}
```

Listing B.11: This listing shows the first-order tangent-linear code for the test case 'master' (Listing B.10).

B.3.3 First-Order Adjoint Code 'master'

```
#pragma omp parallel
{
    int tests←0;
    int tid←0;
    int p←0;
    int i←0;
    int j←0;
    int lb←0;
    int ub←0;
    int c←0;
    double y←0.;
    double y₍₁₎←0.;
    double v₀^0;
    double v₍₁₎₀^0;
    double v₁^0;
    double v₍₁₎₁^0;
    double v₂^0;
    double v₍₁₎₂^0;
    double v₃^0;
    double v₍₁₎₃^0;
    double v₄^0;
    double v₍₁₎₄^0;
    double v₅^0;
    double v₍₁₎₅^0;
    double v₆^0;
    double v₍₁₎₆^0;
    double v₇^0;
    double v₍₁₎₇^0;
    double v₈^0;
    double v₍₁₎₈^0;
    double v₉^0;
```

```
double  v⁰₍₁₎₉ ;
double  v⁰₁₀ ;
double  v₍₁₎₁₀⁰ ;

STACK₍₁₎c . push(54) ;
STACK₍₁₎i . push(tid) ;
tid←omp_get_thread_num() ;
STACK₍₁₎i . push(p) ;
p←omp_get_num_threads() ;
STACK₍₁₎i . push(c) ;
c←n/p ;
STACK₍₁₎i . push(lb) ;
lb←tid*c ;
STACK₍₁₎i . push(ub) ;
ub←(tid+1)*c−1;
STACK₍₁₎i . push(i) ;
i←lb ;
STACK₍₁₎f . push(y) ;
y←1. ;
while( i ≤ ub) {
  STACK₍₁₎c . push(90) ;
  STACK₍₁₎i . push(j) ;
  j←0;
  while(j<n) {
    STACK₍₁₎c . push(127) ;
    STACK₍₁₎f . push(y) ;
    y←y*sin(Aᵢ*ₙ₊ⱼ*xⱼ)*cos(Aᵢ*ₙ₊ⱼ*xⱼ) ;
    STACK₍₁₎i . push(j) ;
    j←j+1;
  }
  STACK₍₁₎c . push(139) ;
  STACK₍₁₎i . push(i) ;
  i←i+1;
}
STACK₍₁₎c . push(145) ;
STACK₍₁₎f . push(thread_resultₜᵢd) ;
thread_resultₜᵢd←y ;
#pragma omp barrier
#pragma omp master
{
  STACK₍₁₎c . push(149) ;
  STACK₍₁₎i . push(i) ;
  i←1;
  while(i<p) {
    STACK₍₁₎c . push(164) ;
    STACK₍₁₎f . push(thread_result₀) ;
    thread_result₀←thread_result₀*thread_resultᵢ ;
    STACK₍₁₎i . push(i) ;
    i←i+1;
  }
  dummy("", (void*)thread_result) ;
}
while( not STACK₍₁₎c . empty()) {
  if(STACK₍₁₎c . top() = 54) {
```

```
        STACK_(1)c . pop ( ) ;
        y←STACK_(1)f . top ( ) ;
        STACK_(1)f . pop ( ) ;
        v_0^0←1 . ;
        v_(1)0^0←y_(1) ;
        y_(1)←0 . ;
        i←STACK_(1)i . top ( ) ;
        STACK_(1)i . pop ( ) ;
        ub←STACK_(1)i . top ( ) ;
        STACK_(1)i . pop ( ) ;
        lb←STACK_(1)i . top ( ) ;
        STACK_(1)i . pop ( ) ;
        c←STACK_(1)i . top ( ) ;
        STACK_(1)i . pop ( ) ;
        p←STACK_(1)i . top ( ) ;
        STACK_(1)i . pop ( ) ;
        tid←STACK_(1)i . top ( ) ;
        STACK_(1)i . pop ( ) ;
}
if (STACK_(1)c . top ( ) = 90) {
        STACK_(1)c . pop ( ) ;
        j←STACK_(1)i . top ( ) ;
        STACK_(1)i . pop ( ) ;
}
if (STACK_(1)c . top ( ) = 127) {
        STACK_(1)c . pop ( ) ;
        j←STACK_(1)i . top ( ) ;
        STACK_(1)i . pop ( ) ;
        y←STACK_(1)f . top ( ) ;
        STACK_(1)f . pop ( ) ;
        v_0^0←y ;
        v_1^0←A_{l*n+j} ;
        v_2^0←x_j ;
        v_3^0←v_1^0*v_2^0 ;
        v_4^0←sin ( v_3^0 ) ;
        v_5^0←v_0^0*v_4^0 ;
        v_6^0←A_{l*n+j} ;
        v_7^0←x_j ;
        v_8^0←v_6^0*v_7^0 ;
        v_9^0←cos ( v_8^0 ) ;
        v_{10}^0←v_5^0*v_9^0 ;
        v_(1)10^0←y_(1) ;
        y_(1)←0 . ;
        v_(1)5^0←v_(1)10^0*v_9^0 ;
        v_(1)9^0←v_(1)10^0*v_5^0 ;
        v_(1)0^0←v_(1)5^0*v_4^0 ;
        v_(1)4^0←v_(1)5^0*v_0^0 ;
        y_(1)+←v_(1)0^0 ;
        v_(1)3^0←v_(1)4^0*cos ( v_3^0 ) ;
        v_(1)1^0←v_(1)3^0*v_2^0 ;
```

```
      v⁰₍₁₎₂←v⁰₍₁₎₃*v⁰₁;
      #pragma omp atomic
      A₍₁₎ᵢ*ₙ₊ⱼ←+←v⁰₍₁₎₁;
      #pragma omp atomic
      x₍₁₎ⱼ←+←v⁰₍₁₎₂;
      v⁰₍₁₎₈←v⁰₍₁₎₉*(0.-sin(v⁰₈));
      v⁰₍₁₎₆←v⁰₍₁₎₈*v⁰₇;
      v⁰₍₁₎₇←v⁰₍₁₎₈*v⁰₆;
      #pragma omp atomic
      A₍₁₎ᵢ*ₙ₊ⱼ←+←v⁰₍₁₎₆;
      #pragma omp atomic
      x₍₁₎ⱼ←+←v⁰₍₁₎₇;
  }
  if(STACK₍₁₎c.top() = 139) {
      STACK₍₁₎c.pop();
      i←STACK₍₁₎ᵢ.top();
      STACK₍₁₎ᵢ.pop();
  }
  if(STACK₍₁₎c.top() = 145) {
      STACK₍₁₎c.pop();
      #pragma omp barrier
      thread_result_tid←STACK₍₁₎f.top();
      STACK₍₁₎f.pop();
      v⁰₀←y;
      v⁰₍₁₎₀←thread_result₍₁₎tid;
      thread_result₍₁₎tid←0.;
      y₍₁₎←+←v⁰₍₁₎₀;
  }
  if(STACK₍₁₎c.top() = 149) {
      STACK₍₁₎c.pop();
      i←STACK₍₁₎ᵢ.top();
      STACK₍₁₎ᵢ.pop();
  }
  if(STACK₍₁₎c.top() = 164) {
      STACK₍₁₎c.pop();
      i←STACK₍₁₎ᵢ.top();
      STACK₍₁₎ᵢ.pop();
      thread_result₀←STACK₍₁₎f.top();
      STACK₍₁₎f.pop();
      v⁰₀←thread_result₀;
      v⁰₁←thread_result_i;
      v⁰₂←v⁰₀*v⁰₁;
      v⁰₍₁₎₂←thread_result₍₁₎₀;
      thread_result₍₁₎₀←0.;
      v⁰₍₁₎₀←v⁰₍₁₎₂*v⁰₁;
      v⁰₍₁₎₁←v⁰₍₁₎₂*v⁰₀;
      #pragma omp atomic
      thread_result₍₁₎₀←+←v⁰₍₁₎₀;
      #pragma omp atomic
      thread_result₍₁₎ᵢ←+←v⁰₍₁₎₁;
```

```
    }
    dummy("", (void*)thread_result, (void*)thread_result_(1));
  }
}
```

Listing B.12: This listing presents the first-order adjoint code for the test case 'master' (Listing B.10). The exclusive read analysis has not been used to achieve static information about the original code. The atomic constructs in the lines 182 and 184 may be neglected because these assignments are only executed by the master thread. This is a possible extension of the exclusive read analysis since it does not recognize this fact at the moment.

B.4 Test Case with a critical Region

The runtime details of this test can be found in Section 5.1.4 on page 291.

B.4.1 Original Code 'critical'

```
#pragma omp parallel
{
    int tid←0;
    int p←0;
    int c←0;
    int lb←0;
    int ub←0;
    int i←0;
    int j←0;
    double y;

    tid←omp_get_thread_num();
    p←omp_get_num_threads();
    c←n/p;
    lb←tid*c;
    ub←(tid+1)*c-1;
    i←lb;
    y←0.;
    while(i ≤ ub) {
        j←0;
        while(j<n) {
            y+←A_{i*n+j}*x_j;
            j←j+1;
        }
        i←i+1;
    }
    #pragma omp critical
    {
        thread_result_0←sin(thread_result_0)*sin(y);
    }
}
```

Listing B.13: This code is the original code for the test case 'critical'.

B.4.2 First-Order Tangent-Linear Code 'critical'

```
#pragma omp parallel
{
    int tid←0;
    int p←0;
    int c←0;
    int lb←0;
    int ub←0;
    int i←0;
    int j←0;
    double y;
    double y^(1)←0.;
    double v_0^0;
    double v_0^(1)0;
    double v_1^0;
    double v_1^(1)0;
    double v_2^0;
    double v_2^(1)0;
    double v_3^0;
    double v_3^(1)0;
    double v_4^0;
    double v_4^(1)0;

    tid←omp_get_thread_num();
    p←omp_get_num_threads();
    c←n/p;
    lb←tid*c;
    ub←(tid+1)*c-1;
    i←lb;
    v_0^(1)0←0.;
    v_0^0←0.;
    y^(1)←v_0^(1)0;
    y←v_0^0;
    while(i ≤ ub) {
        j←0;
        while(j<n) {
            v_0^(1)0←A_{i*n+j}^(1);
            v_0^0←A_{i*n+j};
            v_1^(1)0←x_j^(1);
            v_1^0←x_j;
            v_2^(1)0←v_0^(1)0*v_1^0+v_0^0*v_1^(1)0;
            v_2^0←v_0^0*v_1^0;
            y^(1)+←v_2^(1)0;
            y+←v_2^0;
            j←j+1;
        }
        i←i+1;
    }
    #pragma omp critical
    {
        v_0^(1)0←thread_result_0^(1);
```

```
        v_0^0 ← thread _ result_0 ;
        v_1^{(1)0} ← v_0^{(1)0} * cos ( v_0^0 ) ;
        v_1^0 ← sin ( v_0^0 ) ;
        v_2^{(1)0} ← y^{(1)} ;
        v_2^0 ← y ;
        v_3^{(1)0} ← v_2^{(1)0} * cos ( v_2^0 ) ;
        v_3^0 ← sin ( v_2^0 ) ;
        v_4^{(1)0} ← v_1^{(1)0} * v_3^0 + v_1^0 * v_3^{(1)0} ;
        v_4^0 ← v_1^0 * v_3^0 ;
        thread _ result_0^{(1)} ← v_4^{(1)0} ;
        thread _ result_0 ← v_4^0 ;
    }
}
```

Listing B.14: This listing presents the first-order tangent-linear code for the test case 'critical' (Listing B.13).

B.4.3 First-Order Adjoint Code 'critical'

```
#pragma omp parallel
{
    int tid ← 0;
    int p ← 0;
    int c ← 0;
    int lb ← 0;
    int ub ← 0;
    int i ← 0;
    int j ← 0;
    double y;
    double y_{(1)} ← 0.;
    double v_0^0;
    double v_{(1)0}^0;
    double v_1^0;
    double v_{(1)1}^0;
    double v_2^0;
    double v_{(1)2}^0;
    double v_3^0;
    double v_{(1)3}^0;
    double v_4^0;
    double v_{(1)4}^0;

    STACK_{(1)c} . push ( 46 ) ;
    STACK_{(1)i} . push ( tid ) ;
    tid ← omp _ get _ thread _ num ( ) ;
    STACK_{(1)i} . push ( p ) ;
    p ← omp _ get _ num _ threads ( ) ;
    STACK_{(1)i} . push ( c ) ;
    c ← n / p ;
    STACK_{(1)i} . push ( lb ) ;
    lb ← tid * c ;
    STACK_{(1)i} . push ( ub ) ;
```

```
ub←(tid+1)*c−1;
STACK(1)i.push(i);
i←lb;
STACK(1)f.push(y);
y←0.;
while(i ≤ ub) {
  STACK(1)c.push(82);
  STACK(1)i.push(j);
  j←0;
  while(j<n) {
    STACK(1)c.push(101);
    STACK(1)f.push(y);
    y+←Ai*n+j*xj;
    STACK(1)i.push(j);
    j←j+1;
  }
  STACK(1)c.push(113);
  STACK(1)i.push(i);
  i←i+1;
}
STACK(1)c.push(127);
#pragma omp critical
{
  STACK(1)c.push(125);
  STACK(1)f.push(thread_result0);
  thread_result0←sin(thread_result0)*sin(y);
  STACK(1)c.push(critical_counter_1087);
  critical_counter_1087←critical_counter_1087+1;
}
STACK(1)c.push(127);
#pragma omp barrier
while( not STACK(1)c.empty()) {
  if(STACK(1)c.top() = 46) {
    STACK(1)c.pop();
    y←STACK(1)f.top();
    STACK(1)f.pop();
    v00←0.;
    v0(1)0←y(1);
    y(1)←0.;
    i←STACK(1)i.top();
    STACK(1)i.pop();
    ub←STACK(1)i.top();
    STACK(1)i.pop();
    lb←STACK(1)i.top();
    STACK(1)i.pop();
    c←STACK(1)i.top();
    STACK(1)i.pop();
    p←STACK(1)i.top();
    STACK(1)i.pop();
    tid←STACK(1)i.top();
    STACK(1)i.pop();
  }
  if(STACK(1)c.top() = 82) {
```

```
        STACK_(1)c . pop ( ) ;
        j ← STACK_(1)i . top ( ) ;
        STACK_(1)i . pop ( ) ;
    }
    if (STACK_(1)c . top ( ) = 101) {
        STACK_(1)c . pop ( ) ;
        j ← STACK_(1)i . top ( ) ;
        STACK_(1)i . pop ( ) ;
        y ← STACK_(1)f . top ( ) ;
        STACK_(1)f . pop ( ) ;
        v_0^0 ← A_{i*n+j} ;
        v_1^0 ← x_j ;
        v_2^0 ← v_0^0 * v_1^0 ;
        v_{(1)2}^0 ← y_{(1)} ;
        v_{(1)0}^0 ← v_{(1)2}^0 * v_1^0 ;
        v_{(1)1}^0 ← v_{(1)2}^0 * v_0^0 ;
        A_{(1)i*n+j} + ← v_{(1)0}^0 ;
        #pragma omp atomic
        x_{(1)j} + ← v_{(1)1}^0 ;
    }
    if (STACK_(1)c . top ( ) = 113) {
        STACK_(1)c . pop ( ) ;
        i ← STACK_(1)i . top ( ) ;
        STACK_(1)i . pop ( ) ;
    }
    if (STACK_(1)c . top ( ) = 127) {
        STACK_(1)c . pop ( ) ;
        while (STACK_(1)c . top ( ) ≠ 127) {
            #pragma omp critical
            {
                if (STACK_(1)c . top ( ) = critical_counter_1087 −1) {
                    STACK_(1)c . pop ( ) ;
                    critical_counter_1087 ← critical_counter_1087 −1;
                    while (STACK_(1)c . top ( ) ≠ 127) {
                        if (STACK_(1)c . top ( ) = 125) {
                            STACK_(1)c . pop ( ) ;
                            thread_result_0 ← STACK_(1)f . top ( ) ;
                            STACK_(1)f . pop ( ) ;
                            v_0^0 ← thread_result_0 ;
                            v_1^0 ← sin ( v_0^0 ) ;
                            v_2^0 ← y ;
                            v_3^0 ← sin ( v_2^0 ) ;
                            v_4^0 ← v_1^0 * v_3^0 ;
                            v_{(1)4}^0 ← thread_result_{(1)0} ;
                            thread_result_{(1)0} ← 0. ;
                            v_{(1)1}^0 ← v_{(1)4}^0 * v_3^0 ;
                            v_{(1)3}^0 ← v_{(1)4}^0 * v_1^0 ;
                            v_{(1)0}^0 ← v_{(1)1}^0 * cos ( v_0^0 ) ;
                            thread_result_{(1)0} + ← v_{(1)0}^0 ;
                            v_{(1)2}^0 ← v_{(1)3}^0 * cos ( v_2^0 ) ;
```

$$y_{(1)} + \!\!\leftarrow v^0_{(1)2} \, ;$$
$$\}$$
$$\}$$
$$\}$$
$$\}$$
$$\}$$
$$STACK_{(1)c} \cdot pop \, (\,) \, ;$$
$$\}$$
$$\}$$
$$\}$$

Listing B.15: This listing presents the first-order adjoint code for the test case 'critical' (Listing B.13). The atomic construct in line 102 can be prevented by using the exclusive read analysis.

B.5 Test Case with an atomic Construct

The runtime details of this test can be found in Section 5.1.5 on page 300.

B.5.1 Original Code 'atomic'

```
#pragma omp parallel
{
    int tid←0;
    int p←0;
    int c←0;
    int lb←0;
    int ub←0;
    int i←0;
    int j←0;
    double y;

    tid←omp_get_thread_num();
    p←omp_get_num_threads();
    c←n/p;
    lb←tid*c;
    ub←(tid+1)*c−1;
    i←lb;
    y←0.;
    while(i ≤ ub) {
        j←0;
        while(j<n) {
            y+←Aᵢ*ₙ₊ⱼ*xⱼ;
            j←j+1;
        }
        i←i+1;
    }
    #pragma omp atomic
    thread_result₀+←y;
}
```

Listing B.16: This listings shows the original code for the test case 'atomic'.

B.5.2 First-Order Tangent-Linear Code 'atomic'

```
#pragma omp parallel
{
    int tid←0;
    int p←0;
    int c←0;
    int lb←0;
    int ub←0;
    int i←0;
    int j←0;
    double y;
    double y⁽¹⁾←0.;
    double v₀⁰;
    double v₀⁽¹⁾⁰;
    double v₁⁰;
    double v₁⁽¹⁾⁰;
    double v₂⁰;
    double v₂⁽¹⁾⁰;

    tid←omp_get_thread_num();
    p←omp_get_num_threads();
    c←n/p;
    lb←tid*c;
    ub←(tid+1)*c−1;
    i←lb;
    v₀⁽¹⁾⁰←0.;
    v₀⁰←0.;
    y⁽¹⁾←v₀⁽¹⁾⁰;
    y←v₀⁰;
    while( i ≤ ub) {
        j←0;
        while(j<n) {
            v₀⁽¹⁾⁰←A⁽¹⁾ᵢ*ₙ₊ⱼ;
            v₀⁰←Aᵢ*ₙ₊ⱼ;
            v₁⁽¹⁾⁰←x⁽¹⁾ⱼ;
            v₁⁰←xⱼ;
            v₂⁽¹⁾⁰←v₀⁽¹⁾⁰*v₁⁰+v₀⁰*v₁⁽¹⁾⁰;
            v₂⁰←v₀⁰*v₁⁰;
            y⁽¹⁾+←v₂⁽¹⁾⁰;
            y+←v₂⁰;
            j←j+1;
        }
        i←i+1;
    }
    v₀⁽¹⁾⁰←y⁽¹⁾;
    v₀⁰←y;
    #pragma omp atomic
    thread_result₀⁽¹⁾+←v₀⁽¹⁾⁰;
    #pragma omp atomic
    thread_result₀+←v₀⁰;
```

}

Listing B.17: This listing presents the first-order tangent-linear code for the test case 'atomic' (Listing B.18).

B.5.3 First-Order Adjoint Code 'atomic'

```
#pragma omp parallel
{
   int tid←0;
   int p←0;
   int c←0;
   int lb←0;
   int ub←0;
   int i←0;
   int j←0;
   double y;
   double y(1)←0.;
   double v⁰₀;
   double v⁰(1)0;
   double v⁰₁;
   double v⁰(1)1;
   double v⁰₂;
   double v⁰(1)2;

   #pragma omp master
   {
      atomic_flag_120←0;
   }#pragma omp barrier
   STACK(1)c.push(46);
   STACK(1)i.push(tid);
   tid←omp_get_thread_num();
   STACK(1)i.push(p);
   p←omp_get_num_threads();
   STACK(1)i.push(c);
   c←n/p;
   STACK(1)i.push(lb);
   lb←tid*c;
   STACK(1)i.push(ub);
   ub←(tid+1)*c−1;
   STACK(1)i.push(i);
   i←lb;
   STACK(1)f.push(y);
   y←0.;
   while(i ≤ ub) {
      STACK(1)c.push(82);
      STACK(1)i.push(j);
      j←0;
      while(j<n) {
         STACK(1)c.push(101);
         STACK(1)f.push(y);
         y+←A(l*n+j)*xj;
         STACK(1)i.push(j);
```

```
        j←j+1;
    }
    STACK(1)c.push(113);
    STACK(1)i.push(i);
    i←i+1;
}
STACK(1)c.push(120);
#pragma omp critical
{
    if(atomic_flag_120 = 0) {
        atomic_flag_120←1;
        atomic_storage_120←thread_result0;
    }
    thread_result0+←y;
}
#pragma omp barrier
while( not STACK(1)c.empty()) {
    if(STACK(1)c.top() = 46) {
        STACK(1)c.pop();
        y←STACK(1)f.top();
        STACK(1)f.pop();
        v00←0.;
        v0(1)0←y(1);
        y(1)←0.;
        i←STACK(1)i.top();
        STACK(1)i.pop();
        ub←STACK(1)i.top();
        STACK(1)i.pop();
        lb←STACK(1)i.top();
        STACK(1)i.pop();
        c←STACK(1)i.top();
        STACK(1)i.pop();
        p←STACK(1)i.top();
        STACK(1)i.pop();
        tid←STACK(1)i.top();
        STACK(1)i.pop();
    }
    if(STACK(1)c.top() = 82) {
        STACK(1)c.pop();
        j←STACK(1)i.top();
        STACK(1)i.pop();
    }
    if(STACK(1)c.top() = 101) {
        STACK(1)c.pop();
        j←STACK(1)i.top();
        STACK(1)i.pop();
        y←STACK(1)f.top();
        STACK(1)f.pop();
        v00←Al∗n+j;
        v01←xj;
        v02←v00∗v01;
        v0(1)2←y(1);
```

```
      v^0_{(1)0} ← v^0_{(1)2} * v^0_1 ;
      v^0_{(1)1} ← v^0_{(1)2} * v^0_0 ;
      A_{(1)i*n+f} ++← v^0_{(1)0} ;
      #pragma omp atomic
      x_{(1)f} ++← v^0_{(1)1} ;
   }
   if (STACK_{(1)c} . top () = 113) {
      STACK_{(1)c} . pop () ;
      i ← STACK_{(1)i} . top () ;
      STACK_{(1)i} . pop () ;
   }
   if (STACK_{(1)c} . top () = 120) {
      STACK_{(1)c} . pop () ;
      v^0_0 ← y ;
      v^0_{(1)0} ← thread _ result_{(1)0} ;
      y_{(1)} ++← v^0_{(1)0} ;
      #pragma omp critical
      {
         if (atomic _ flag _ 120 ≠ 2) {
            atomic _ flag _ 120 ← 2 ;
            thread _ result_0 ← atomic_storage_120 ;
         }
      }
    }
  }
 }
}
```

Listing B.18: The first-order adjoint code for the test case 'atomic' is shown in this listing (Listing B.18). The atomic construct in line 102 can be avoided by using the exclusive read analysis.

Bibliography

[1] A. Aho, M. Lam, R. Sethi, and J. Ullman. *Compilers. Principles, Techniques, and Tools (Second Edition)*. Addison-Wesley Educational Publishers, Incorporated, 2007.

[2] R. Altenfeld, M. Apel, D. an Mey, B. Böttger, S. Benke, and C. Bischof. Parallelising Computational Microstructure Simulations for Metallic Materials with OpenMP. 2011.

[3] H. Anton, I.C. Bivens, and S. Davis. *Calculus Early Transcendentals, 10th Edition E-Text*. Wiley, 2011.

[4] A.W. Appel and M. Ginsburg. *Modern Compiler Implementation in C*. Cambridge University Press, 2004.

[5] M. Ben-Ari. *Principles of Concurrent and Distributed Programming*. Prentice-Hall International Series in Computer Science. Addison-Wesley, 2006.

[6] C. Bischof, N. Guertler, A. Kowarz, and A. Walther. *Parallel Reverse Mode Automatic Differentiation for OpenMP Programs with ADOL-C.*, pages 163–173. Berlin: Springer, 2008.

[7] B. Braunschweig and R. Gani. *Software Architectures and Tools for Computer Aided Process Engineering: Computer-Aided Chemical Engineeirng*. Computer Aided Chemical Engineering. Elsevier Science, 2002.

[8] S.C. Brenner and R. Scott. *The Mathematical Theory of Finite Element Methods*. Texts in Applied Mathematics. Springer, 2008.

[9] C. Breshears. *The Art of Concurrency - A Thread Monkey's Guide to Writing Parallel Applications*. O'Reilly, 2009.

[10] M. Bücker, B. Lang, D. an Mey, and C. Bischof. Bringing together automatic differentiation and OpenMP. In *ICS '01: Proceedings of the 15th international conference on Supercomputing*, pages 246–251, New York, 2001. ACM.

[11] M. Bücker, B. Lang, A. Rasch, C. Bischof, and D. an Mey. Explicit Loop Scheduling in OpenMP for Parallel Automatic Differentiation. *High Performance Computing Systems and Applications, Annual International Symposium on*, 0:121, 2002.

[12] M. Bücker, A. Rasch, and A. Wolf. A class of OpenMP applications involving nested parallelism. In *SAC '04: Proceedings of the 2004 ACM symposium on Applied computing*, pages 220–224, New York, 2004. ACM.

[13] D. Butenhof. *Programming with POSIX threads*. Addison-Wesley Longman Publishing Co., Inc., Boston, MA, USA, 1997.

[14] R. Chandra, L. Dagum, D. Kohr, D. Maydan, J. McDonald, and R. Menon. *Parallel Programming in OpenMP*. Morgan Kaufmann Publishers Inc., San Francisco, 2001.

[15] B. Chapman, G. Jost, and R. Pas. *Using OpenMP: Portable Shared Memory Parallel Programming (Scientific and Engineering Computation)*. MIT Press, 2007.

[16] B. Char and Waterloo Maple Software. *Maple 8: learning guide*. Maple 8. Waterloo Maple, 2002.

[17] P.G. Ciarlet. *The Finite Element Method for Elliptic Problems*. Classics in Applied Mathematics. Society for Industrial and Applied Mathematics, 2002.

[18] G. Corliss, A. Griewank, P. Henneberger, G. Kirlinger, F. Potra, and H. Stetter. High-Order Stiff ODE Solvers via Automatic Differentiation and Rational Prediction. In *WNAA*, pages 114–125, 1996.

[19] R. Craig and A. Kurdila. *Fundamentals of Structural Dynamics*. Wiley, 2011.

[20] W. Dahmen and A. Reusken. *Numerik für Ingenieure und Naturwissenschaftler*. Springer-Lehrbuch. Springer, 2008.

[21] P. Droste, K. Nöh, and W. Wiechert. Omix - A Visualization Tool for Metabolic Networks with Highest Usability and Customizability in Focus. *Chemie - Ingenieur - Technik*, 85(6):849–862, 2013.

[22] M. Förster. Verification of Data Dependences in Derivative Codes, 2009.

[23] M. Förster, U. Naumann, and J. Utke. Toward Adjoint OpenMP. Technical Report AIB-2011-13, RWTH Aachen, July 2011.

[24] P. Fritzson. *Principles of Object-Oriented Modeling and Simulation with Modelica 2.1*. Wiley, 2010.

[25] B. Gaster, L. Howes, D.R. Kaeli, P. Mistry, and D. Schaa. *Heterogeneous Computing with OpenCL: Revised OpenCL 1.2 Edition*. Elsevier Science, 2012.

[26] R. Giering, T. Kaminski, B. Eisfeld, and N. Gauger. Automatic Differentiation of FLOWer and MUGRIDO. In *Megadesign and Megaopt-German Initiatives for Aerodynamic Simulation and Optimization in Aircraft Design: Results of the Closing Symposium of the Megadesign and Megaopt Projects, Braunschweig, Germany, May 23 and 24 2007*, volume 107, page 221. Springer, 2009.

[27] R. Giering, T. Kaminski, R. Todling, R. Errico, R. Gelaro, and N. Winslow. Generating tangent linear and adjoint versions of NASA/GMAO's Fortran-90 global weather forecast model. *Automatic Differentiation: Applications, Theory, and Implementations, Lecture Notes in Computational Science and Engineering*, 50:275–284, 2005.

[28] R. Giering, T. Kaminski, and M. Voßbeck. Generating and Maintaining Highly Efficient Adjoint and Hessian Code for Optimisation and Uncertainty Analysis by Automatic Differentiation.

[29] A. Griewank and A. Walter. *Evaluating Derivatives. Principles and Techniques of Algorithmic Differentiation (2nd ed.)*. SIAM, Philadelphia, 2008.

[30] A. Griewank and A. Walther. Algorithm 799: Revolve: An Implementation of Check-point for the Reverse or Adjoint Mode of Computational Differentiation. *ACM Transactions on Mathematical Software*, 26(1):19–45, mar 2000. Also appeared as Technical University of Dresden, Technical Report IOKOMO-04-1997.

[31] D. Grune, C. Jacobs, K. Langendoen, and H. Bal. *Modern Compiler Design*. John Wiley & Sons, Inc., New York, NY, USA, 1st edition, 2000.

[32] R. Hannemann, W. Marquardt, U. Naumann, and B. Gendler. Discrete first-and second-order adjoints and automatic differentiation for the sensitivity analysis of dynamic models. In *Procedia Computer Science*, pages 297–305. Elsevier, 2010.

[33] R. Hannemann, J. Tillack, M. Schmitz, M. Förster, J. Wyes, K. Nöh, E. von Lieres, E. Naumann, W. Wiechert, and W. Marquardt. First- and Second-Order Parameter Sensitivities of a Metabolically and Isotopically Non-Stationary Biochemical Network Model. In M. Otter and D. Zimmer, editors, *Proceedings of the 9th International Modelica Conference*, pages 641–648. Modelica Association, 2012.

[34] P. Heimbach, C. Hill, and R. Giering. Automatic Generation of Efficient Adjoint Code for a Parallel Navier-Stokes Solver. In *Computational Science—ICCS 2002*, pages 1019–1028. Springer, 2002.

[35] P. Heimbach, C. Hill, and R. Giering. An efficient exact adjoint of the parallel {MIT} general circulation model, generated via automatic differentiation. *Future Generation Computer Systems*, 21(8):1356 – 1371, 2005.

[36] W. Hock and K. Schittkowski. Test examples for nonlinear programming codes. *Lecture Notes in Economics and Mathematical Systems, 187*, 1981.

[37] IEEE. IEEE: Threads Extension for Portable Operating Systems (Draft 6). Specification, 1992.

[38] R. Ipanaqué and A. Iglesias. A Mathematica Package for Solving and Displaying Inequalities. In *Computational Science - ICCS 2004*, volume 3039 of *Lecture Notes in Computer Science*, pages 303–310. Springer Berlin Heidelberg, 2004.

[39] T. Kaminski, R. Giering, and C. Othmer. Topological design based on highly efficient adjoints generated by automatic differentiation. In *ERCOFTAC Design Optimization Conf., Las Palmas*, 2006.

[40] P. Kapinos and D. an Mey. Parallel Simulation of Bevel Gear Cutting Processes with OpenMP Tasks. 2009.

[41] D. Kaushik, D. Keyes, S. Balay, and B. Smith. Hybrid Programming Model for Implicit PDE Simulations on Multicore Architectures. 2011.

[42] K. Kennedy and J. Allen. *Optimizing compilers for modern architectures: a dependence-based approach*. Morgan Kaufmann Publishers Inc., San Francisco, CA, USA, 2002.

[43] U. Khedker, A. Sanyal, and B. Sathe. *Data Flow Analysis: Theory and Practice*. Taylor & Francis, 2009.

[44] Khronos OpenCL Working Group. *The OpenCL Specification, version 1.0.29*, 2008.

[45] B. Knaster. Un théorème sur les fonctions d'ensembles. *Annales de la Société Polon-aise de Mathématiques*, 6:133–134, 1928.

[46] D.E. Knuth. *The Art of Computer Programming*. The Art of Computer Programming: Seminumerical Algorithms. Addison-Wesley, 2001.

[47] A. Kowarz and A. Walther. Optimal checkpointing for time-stepping procedures in ADOL-C. In V. Alexandrov, G. van Albada, P. Sloot, and J. Dongarra, editors, *Computational Science – ICCS 2006*, volume 3994 of *Lecture Notes in Computer Science*, pages 541–549, Heidelberg, 2006. Springer.

[48] A. Kowarz and A. Walther. Parallel Derivative Computation using ADOL-C. In W. Nagel, R. Hoffmann, and A. Koch, editors, *9th Workshop on Parallel Systems and Algorithms (PASA) held at the 21st Conference on the Architecture of Computing Systems (ARCS), February 26th, 2008, in Dresden, Germany*, volume 124 of *LNI*, pages 83–92. GI, 2008.

[49] M. Lange, G. Gorman, M. Weiland, L. Mitchell, and J. Southern. Achieving Efficient Strong Scaling with PETSc using Hybrid MPI/OpenMP Optimisation. *CoRR*, abs/1303.5275, 2013.

[50] K. Levenberg. A Method for the Solution of Vertain Problems in Least Squares, 1944.

[51] K. Madsen, B. Nielsen, and O. Tingleff. Methods for Non-Linear Least-Squares Problems, 2004.

[52] Z. Manna and A. Pnueli. *The Temporal Logic of Reactive and Concurrent Systems - Specification*. Springer, 1992.

[53] D. Marquardt. An Algorithm for Least Squares Estimation on Nonlinear Parameters, 1963.

[54] Message Passing Interface Forum. MPI: A Message-Passing Interface Standard, Version 2.1. Specification, 2008.

[55] R. Moona. *"Assembly Language Programming In Gnu/Linux For IA32 Architectures"*. Prentice-Hall, New Delhi, 2007.

[56] S. Muchnick. *Advanced Compiler Design Implementation*. Morgan Kaufmann Publishers, 1997.

[57] U. Naumann. *The Art of Differentiating Computer Programs*. Software, Environments and Tools. Society for Industrial and Applied Mathematics, 2012.

[58] U. Naumann, J. Utke, J. Riehme, P. Hovland, and C. Hill. A Framework for Proving Correctness of Adjoint Message Passing Programs. In *Proceedings of EU-ROPVM/MPI 2008*, pages 316–321, 2008.

[59] F. Nielson, H.R. Nielson, and C. Hankin. *Principles of Program Analysis*. Springer, 1999.

[60] J. Nocedal and S. Wright. *Numerical Optimization*. Springer, New York, 2nd edition, 2006.

[61] NVIDIA. CUDA Technology. Technical report, 2007.

[62] OpenACC Architecture Review Board. OpenACC Application Program Interface. Specification, 2011.

[63] OpenMP Architecture Review Board. OpenMP Application Program Interface. Specification, 2011.

[64] F. Potra and S. Wright. Primal-dual interior-point methods. SIAM, 1997.

[65] W.H. Press. *Numerical Recipes in C++: The Art of Scientific Computing*. Cambridge University Press, 2002.

[66] T. Rauber and G. Rünger. *Parallel Programming for Multicore and Cluster Systems*. Springer Verlag, 2010.

[67] H. Ricardo. *A Modern Introduction to Differential Equations*. Elsevier Science, 2009.

[68] H. Rice. Classes of Recursively Enumerable Sets and Their Decision Problems. *Transactions of the American Mathematical Society*, 74(2):pp. 358–366, 1953.

[69] J. Sanders and E. Kandrot. *CUDA by Example: An Introduction to General-Purpose GPU Programming*. Pearson Education, 2010.

[70] M. Schanen, M. Förster, B. Gendler, and U. Naumann. Compiler-based Differentiation of Higher-Order Numerical Simulation Codes using Interprocedural Checkpointing. *International Journal on Advances in Software*, 5(1&2):27–35, 2012.

[71] M. Schanen, U. Naumann, and M. Förster. Second-order adjoint algorithmic differentiation by source transformation of MPI code. In *Recent Advances in the Message Passing Interface, Lecture Notes in Computer Science*, pages 257–264. Springer, 2010.

[72] M. Schanen, U. Naumann, L. Hascoët, and J. Utke. Interpretative adjoints for numerical simulation codes using MPI. In *Procedia Computer Science*, pages 1819–1827. Elsevier, 2010.

[73] M. Schwartzbach. *Lecture Notes on Static Analysis*, 2014 (accessed February 23, 2014). `http://www.itu.dk/people/brabrand/UFPE/Data-Flow-Analysis/static.pdf`.

[74] A. Silberschatz, P.B. Galvin, and G. Gagne. *Operating System Concepts Essentials*. Wiley, 2010.

[75] A.S. Tanenbaum and A.S. Woodhull. *Operating Systems Design and Implementation*. Pearson Education, 2011.

[76] A. Tarski. A lattice-theoretical fixpoint theorem and its applications. 1955.

[77] J. Utke, L. Hascoët, C. Hill, P. Hovland, and U. Naumann. Toward Adjoinable MPI. In *Proceedings of IPDPS 2009*, 2009.

[78] A. Walther and A. Griewank. Bounding The Number Of Processes And Checkpoints Needed In ... , 2001.

[79] R. Wilhelm and D. Maurer. *Übersetzerbau: Theorie, Konstruktion, Generierung.* Springer-Lehrbuch. Springer, 1997.

[80] M.J. Wolfe. *High Performance Compilers for Parallel Computing.* ADDISON WESLEY Publishing Company Incorporated, 1996.

[81] S. Wolfram. *The Mathematica Book (5. ed.).* Wolfram-Media, 2003.

[82] A. Wächter and L. Biegler. On the implementation of an interior-point filter line-search algorithm for large-scale nonlinear programming. *Mathematical Programming*, 106(1):25–57, 2006.

[83] D. Zill, W. Wright, and M. Cullen. *Differential Equations With Boundary-Value Problems.* Textbooks Available with Cengage Youbook. BROOKS COLE Publishing Company, 2012.

Index

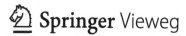